The Shell Bitumen Handbook

Fifth Edition

The authors

Dr John Read

John began his career working for a consultant testing house before moving on to Lafarge Aggregates. After a period of time running asphalt plants on mobile contracts he began studying for his PhD at the University of Nottingham and after graduating he was appointed as a full time academic member of staff.

In 1997 John became the Technical Manager for Croda Bitumen where he was responsible for managing both the QC and R&D laboratories and in 1998 John became the Technical Development Engineer for Shell Bitumen where he was responsible for the development and commercialisation of new innovative products. He was also involved in the day-to-day support of customers.

John is currently the Cluster Technology Manager for Shell Bitumen with responsibility for supplying technical services within the UK and Ireland.

John sits on many asphalt and bitumen related committees and has published over 50 technical papers, publications and articles.

Mr David Whiteoak

David has worked in the road construction industry for over 30 years. He began his career with Lothian Regional Council working in a wide variety of areas, from traffic management to site supervision. In 1977 he left Lothian Region to study Civil Engineering at Heriot-Watt University graduating with a BSc Honours Degree in 1980.

He joined Shell in 1980 working in the Bitumen Group at Thornton Research Centre where he investigated various aspects of the performance of bitumen and asphalt, carrying out technical service activity for customers and the development of new products including Cariphalte DM.

In 1986 he joined the technical department of Shell Bitumen UK and it was during this time that David wrote the 4th edition of the Shell Bitumen Handbook. Following the publication of the handbook in 1990 David had a three-year assignment in the Elastomers group of Shell International Chemical Company before returning to Shell Bitumen as the Technical Manager in 1994.

David is currently the New Technology Manager for Shell Bitumen responsible for a number of activities including the execution of technical service and R & D activity carried out at the Pavement Research Building. This is a purpose-built laboratory established in conjunction with the University of Nottingham and opened September 2001.

The Shell Bitumen Handbook

Fifth Edition

Published for Shell Bitumen by

T Thomas Telford *Publishing*

Thomas Telford Ltd, 1 Heron Quay, London E14 4JD
www.thomastelford.com

Distributors for Thomas Telford books are
USA: ASCE Press, 1801 Alexander Bell Drive, Reston, VA 20191-4400
Japan: Maruzen Co. Ltd, Book Department, 3–10 Nihonbashi 2-chome, Chuo-ku, Tokyo 103
Australia: DA Books and Journals, 648 Whitehorse Road, Mitcham 3132, Victoria

This title has been previously published as
Mexphalte Handbook, First Edition, 1949
Mexphalte Handbook, Second Edition, Jarman A.W. (ed), Shell-Mex and B.P. Ltd, London, 1955
Mexphalte Handbook, Third Edition, 1963
The Shell Bitumen Handbook, Fourth Edition, Whiteoak, D., Shell Bitumen UK, Chertsey, 1990

A catalogue record for this book is available from the British Library

ISBN: 0 7277 3220 X   1003615807

Typeset by Academic + Technical Typesetting
Printed and bound in Great Britain by The University Press, Cambridge

Foreword

In editing the text of this book, I have had considerable assistance from many people. They are listed in the acknowledgements. However, a number of people warrant special mention. David Rockliff of Rock40C and Ian Walsh of Babtie were pestered by me on a number of occasions and always responded with expertise, courtesy and efficiency and I am very grateful to these two giants of the industry. However, there are two other gentleman without whom this enterprise would never have been completed. The first is Dr John Read who never failed to help me through either his own encyclopaedic knowledge or his vast network of contacts on the many occasions when I needed answers or text or whatever. The other is the main reason why this book came into being, Dave Whiteoak. Dave is known in our industry as the font of knowledge on all subjects associated with bitumen. In addition, all who have met him consider him to be the nicest guy you could wish to meet. He produced the 1990 edition and without him this new book and the opportunity which it affords all of us to enhance our knowledge of asphalt technology would simply not exist.

Whilst editing this text, I was constantly reminded of the enormous contribution which has been made by Shell Bitumen to asphalt technology. Indeed, this book demonstrates that continued commitment. This new edition reflects many of the very significant advances which have taken place in the period since the last edition was published. I am confident that you will feel that this is a worthy addition to your asphalt book shelf.

Dr Robert N Hunter
Technical Editor
November 2003

Acknowledgements

John and David would personally like to acknowledge all of the help given to them in writing this book by their colleagues in the Shell European Bitumen Technical Team:

Mr Theo Terlouw
Dr Martin Vodenhof
Mr Pierre-Jean Cerino

Mr Eivind Olav Andersen
Mr Koen Steernberg
Mr Mike Southern

The authors and editor also wish to gratefully acknowledge the contributions made by the following people:

Mr Fredrik Åkesson
Mr John Atkins
Mr John Baxter of the Road Surface
  Dressing Association
Mr Andy Broomfield
Dynapac International High Comp
  Centre, Sweden
Mr Jack Edgar of Hunter & Edgar
Mr Terry Fabb
Mr Jeff Farrington
Mr Derek Fordyce
Dr Mike Gibb
Mr Ray Guthrie
Dr Tony Harrison of the Refined
  Bitumen Association
Mr Bryan Hayton
Ms Delia Harverson
Mr Alistair Jack
Mr Colin Loveday of Tarmac
Mr John Moore of Gencor
  International Ltd
Dr Cliff Nicholls of TRL Ltd

Dr Mike Nunn
Mr Tony Pakenham
Mr Mike Phillips
Mr John Richardson of
  Colas Limited
Mr David Rockliff of Rock40C
Mr Robert Thomas of the Institution
  of Civil Engineers Library
Dr Todd Schole
Mr Martin Schouten
Mr Andrew Scorer of Miles
  Macadam Ltd
Mr Andy Self
Mr Dave Strickland
Mr Nick Toy
Mr Colin Underwood
Mr Willem Vonk
Mr Ian Walsh of Babtie
Mr Maurice White of the Quarry
  Products Association
Professor Alan Woodside
Dr David Woodward

Contents

Shell Bitumen's polymer modified plant at Stanlow, Cheshire, UK

Chapter 1

Introduction

1.1 Preamble

Bitumen is defined in the *Oxford English Reference Dictionary* as 'a tarlike mixture of hydrocarbons derived from petroleum naturally or by distillation, and used for road surfacing and roofing'. However, as will be seen, its use is not restricted to these applications. This chapter considers the history of this material and the various sources from which it can be derived.

It is widely believed that the term 'bitumen' originated in the ancient and sacred language of Hindus in India, Sanskrit, in which '*jatu*' means pitch and '*jatu-krit*' means pitch creating. These terms referred to the pitch produced by some resinous trees. The Latin equivalent is claimed by some to be originally '*gwitu-men*' (pertaining to pitch) and by others to be '*pixtu-men*' (bubbling pitch) which was subsequently shortened to '*bitumen*' then passing via French into English.

There are several references to bitumen in the Bible, although the terminology used is confusing. In Genesis, Noah's ark is 'pitched within and without with pitch', and Moses' juvenile adventure is in 'an ark of bulrushes, daubed with slime and with pitch'. Even more perplexing are the descriptions of the building of the Tower of Babel. The *Authorised Version of the Bible* says 'they had brick for stone, and slime had they for mortar'; the *New International Version* states that 'they used bricks instead of stone and tar instead of mortar'; Moffat's 1935 translation says 'they had bricks for stone and asphalt for mortar'; but the *New English Bible* states that 'they used bricks for stone and bitumen for mortar'. Even today, the meanings of the words bitumen, tar, asphalt and pitch vary between users.

1.2 The earliest uses of bituminous binders

Surface seepage may occur at geological faults in the vicinity of subterranean crude oil deposits. The amount and constitution of this naturally occurring material depends on a number of processes that modify the

1

properties of the substance. This product may be considered to be a 'natural' bitumen, often being accompanied by mineral matter, the amount and nature of which will depend upon the circumstances which caused such an admixture to occur.

There are, of course, extensive crude oil deposits in the Middle East and for thousands of years there has been corresponding surface seepage of 'natural' bitumen. The ancient inhabitants of these parts were quick to appreciate the excellent waterproofing, adhesive and preservative properties of the material which was so readily placed at their disposal. For over 5000 years[1], bitumen, in one form or another, has been used as a waterproofing and/or bonding agent; the earliest recorded use was by the Sumerians whose empire existed from around 3500 BC to approximately 2000 BC. At Mohenjo Daro, in the Indus Valley, there is a particularly well-preserved water tank which dates back to around 3000 BC. In the walls of this tank, not only are the stone blocks bonded with a 'natural' bitumen but there is also a vertical bituminous core in the centre of the wall. This same principle is used in modern dam design. It is believed that Nebuchadnezzar was an able exponent of the use of bitumen because there is evidence that he used the product for waterproofing the masonry of his palace and as a grout for stone roads. The process of mummification used by the ancient Egyptians also testifies to the preservative qualities of bitumen, although it is a matter of dispute as to whether bitumens or resins were actually used.

The ancient uses of 'natural' bitumens undoubtedly continued in those inhabited parts of the world where deposits were readily available. However, there seems to have been little development of usage elsewhere. In the UK, none of the present major uses of bitumen was introduced until the end of the nineteenth century. However, there would appear to have been some knowledge of alternative binders in earlier days as it is recorded that Sir Walter Raleigh, in 1595, proclaimed the lake asphalt he found in Trinidad to be 'most excellent good' for caulking (waterproofing) the seams of ships. In the middle of the nineteenth century, attempts were made to utilise rock asphalt from European deposits for road surfacing and, from this, there was a slow development of the use of natural products for this purpose, followed by the advent of coal tar and later of bitumen manufactured from crude oil.

1.3 The growth of bitumen consumption in Europe

Bitumen was imported into the UK on a small scale early in the twentieth century but it was not until 1920, when the Shell Haven refinery opened,

---

1 ABRAHAM H. *Asphalts and allied substances, their occurrence, modes of production, uses in the arts and methods of testing, 5th ed, vol 1, 1945*. Van Nostrand, New York.

that major bitumen production began in the UK. Subsequent development in the UK and other parts of Europe was rapid and, with the facilities for bulk distribution, the basis was laid for bitumen to be used in a
wide range of modern applications, as illustrated in Table 1.1.

In the early 1920s, bitumen consumption in the UK was around 200 000
tonnes per annum. Demand increased rapidly, reaching 1 million tonnes
per annum by 1960, and with the start of the motorway building programme in the early 1960s, the demand for bitumen soared to 2 million
tonnes per annum by 1970, reaching a peak of 2.4 million tonnes in 1973.
Since 1973, government views on allocating budgets to new construction
and/or maintenance works have resulted in the demand for bitumen
being variable. Bitumen consumption from 1990 is illustrated in Fig. 1.1.
No equivalent consumption levels for Europe are available but there is
no reason to suppose that they have not followed similar trends.

1.4 Sources of binder

The word 'asphalt' has different meanings in Europe and North America.
In Europe, it means a mixture of bitumen and aggregate, e.g. thin
surfacings, high modulus mixtures, mastic asphalt, güssasphalt, etc. In
contrast, in North America, the word 'asphalt' means bitumen. The
following descriptions of binders, which can be obtained from various
sources, may help to clarify the different uses.

1.4.1 Lake asphalt

This is the most extensively used and best known form of 'natural'
asphalt. It is found in well-defined surface deposits, the most important
of which is located in Trinidad. It is generally believed that this deposit
was discovered in 1595 by Sir Walter Raleigh. However, it is now known
that the Portuguese and possibly the Spaniards knew of the existence of
the deposit before this date. The earliest recorded reference to lake
asphalt relates to Sir Walter Raleigh's cousin, Robert Dudley, who
visited Trinidad and 'discovered' the asphalt deposit just three months
before Raleigh's arrival. Since the intervening country was dense and
virtually impenetrable jungle, it is doubtful if either Raleigh or Dudley
saw the 'pitch lake' or 'Trinidad Lake Asphalt' (TLA) during their initial
visits. They probably only found some overflow material on the seashore
approximately one kilometre north of the lake itself.

There are several small deposits of asphalt on the island of Trinidad
but it is the lake in the southern part of the island which constitutes
one of the largest deposits in the world. The lake occupies an area of
approximately 35 hectares and is estimated to be some 90 m deep containing well in excess of 10 million tonnes of material. The surface of the lake
is such that it can support the weight of the crawler tractors and dumper

3

Table 1.1 Some 250 uses of bitumen[2,3]

Agriculture	Miscellaneous	Compositions	Depilatory

Agriculture
Dampproofing and
  waterproofing buildings,
  structures
Disinfectants
Fence post coating
Mulches
Mulching paper
Paved barn floors,
  barnyards, feed
  platforms, etc.
Protecting tanks, vats, etc.
Protection for concrete
  structures
Tree paints
Water and moisture
  barriers (above and
  below ground)
Wind and water erosion
  control
Weather modification
  areas

Buildings
Floors
Dampproofing and
  waterproofing
Floor compositions, tiles,
  coverings
Insulating fabrics, papers
Step treads

Roofing
Building papers
Built-up roof adhesives,
  felts, primers
Caulking compounds
Cement waterproofing
  compounds
Cleats for roofing
Glass wool compositions
Insulating fabrics, felts,
  papers
Joint filler compounds
Laminated roofing shingles
Liquid-roof coatings
Plastic cements
Shingles

Walls, siding, ceilings
Acoustical blocks,
  compositions, felts
Architectural decoration
Bricks
Brick siding
Building blocks, papers
Dampproofing coatings,
  compositions
Insulating board, fabrics,
  felts, paper
Joint filler compounds
Masonry coatings
Plaster boards
Putty
Siding compositions
Soundproofing
Stucco base
Wallboard

Miscellaneous
Air-drying paints, varnishes
Artificial timber
Ebonised timber
Insulating paints
Plumbing, pipes
Treated awnings

**Hydraulics and erosion
control**
Canal linings, sealants
Catchment areas, basins
Dam groutings
Dam linings, protection
Dyke protection
Ditch linings
Drainage gutters, structures
Embankment protection
Groynes
Jetties
Levee protection
Mattresses for levee and
  bank protection
Membrane linings,
  waterproofing
Reservoir linings
Revetments
Sand dune stabilisation
Sewage lagoons, oxidation
  ponds
Swimming pools
Waste ponds
Water barriers

Industrial
*Aluminium foil compositions
  using bitumen*
Backed felts
Conduit insulation,
  lamination
Insulating boards
Paint compositions
Papers
Pipe wrapping
Roofing, shingles

Automotive
Acoustical compositions,
  felts
Brake linings
Clutch facings
Floor sound deadeners
Friction elements
Insulating felts
Panel boards
Shim strips
Tacking strips
Underseal

Electrical
Armature carbons,
  windings
Battery boxes, carbons
Electrical insulating
  compounds, papers,
  tapes, wire coatings
Junction box compounds
Moulded conduits

Compositions
Black grease
Buffing compounds
Cable splicing compound
Coffin linings
Embalming
Etching compositions
Extenders
Explosives
Fire extinguisher
  compounds
Joint fillers
Lap cement
Lubricating grease
Pipe coatings, dips, joint
  seals
Plastic cements
Plasticisers
Preservatives
Printing inks
Well drilling fluid
Wooden cask liners

*Impregnated, treated
  materials*
Armoured bituminised
  fabrics
Burlap impregnation
Canvas treating
Carpeting medium
Deck cloth impregnation
Fabrics, felts
Mildew prevention
Packing papers
Pipes and pipe wrapping
Planks
Rugs, asphalt base
Sawdust, cork, asphalt
  composition
Treated leather
Wrapping papers

Paints, varnishes, etc.
Acid-proof enamels,
  mastics, varnishes
Acid-resistant coatings
Air-drying paints,
  varnishes
Anti-corrosive and
  anti-fouling paints
Antioxidants and solvents
Base for solvent
  compositions
Baking and heat-resistant
  enamels
Boat deck sealing
  compound
Lacquers, japans
Marine enamels

Miscellaneous
Belting
Blasting fuses
Briquette binders
Burial vaults
Casting moulds
Clay articles
Clay pigeons

Depilatory
Expansion joints
Flower pots
Foundry cores
Friction tape
Fuel
Gaskets
Gramophone records
Imitation leather
Mirror backing
Rubber, moulded
  compositions
Shoe fillers, soles
Table tops

Paving
(See also Agriculture,
  Hydraulics, Railways,
  Recreation)

Airport runways, taxiways,
  aprons, etc.
Asphalt blocks
Brick fillers
Bridge deck surfacing
Crack fillers
Floors for buildings,
  warehouses, garages, etc.
Highways, roads, streets,
  shoulders
Kerbs, gutters, drainage
  ditches
Parking lots, driveways
Portland cement concrete
  underseal
Roof-deck parking
Pavements, footpaths
Soil stabilisation

Railways
Ballast treatment
Curve lubricant
Dust laying
Paved ballast, sub-ballast
Paved crossings, freight
  yards, station platforms
Rail fillers
Railway sleepers
Sleeper impregnating,
  stabilisation

Recreation
Paved surfaces for:
Dance pavilions
Drive-in movies
Gymnasiums, sports arenas
Playgrounds, school yards
Race tracks
Running tracks
Skating rinks
Swimming and wading
  pools
Tennis courts, handball
  courts

Bases for:
Synthetic playing field and
  running track surfaces

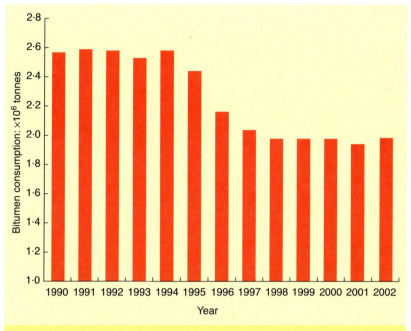

Fig. 1.1 Bitumen consumption in the UK
Courtesy of the Refined Bitumen Association

trucks that transport excavated material from the surface of the lake to railway trucks that run along the edge of the lake.

Several theories have been put forward to explain the origin of the pitch lake. However, it is generally considered to have originated as a surface seepage of a viscous bitumen in the late Miocene epoch (24·6 to 5·1 million years ago). Lowering of the Earth's surface led to an incursion of the sea and, as a result, silt and clay were deposited over the bitumen. Some of the silt and clay penetrated the bitumen, forming a plastic mixture of silt, clay, water and bitumen. Subsequently, the land was elevated above sea level, lateral pressure deforming the material into its present lens-like shape. Erosion removed the overlying silt and clay, exposing the surface of the lake.

The enormous mass of the lake appears to be in constant, very slow, circulatory movement from the centre outwards at the surface. It is assumed that material returns to the centre within the body of the lake. The level of the lake falls significantly less than would be expected

2 SHELL BRIEFING SERVICE. *Shell in the bitumen industry, Jun 1973*. Shell International Petroleum Company Ltd, London.

3 ASPHALT INSTITUTE. *The asphalt handbook, Manual Series No 4, 1989*. AI, Kentucky.

given the quantity of material which is removed and holes dug in the surface slowly fill up and usually disappear within 24 hours.

Excavated material is refined by heating the material to 160°C, vaporising the water. The molten material is passed through fine screens to remove the coarse, foreign and vegetable matter. This residue is usually termed 'Trinidad Epuré' or 'refined TLA' and, typically, has the following composition (in % by weight):

Binder	54%
Mineral matter	36%
Organic matter	10%

Since it has a penetration of about 2 dmm (see Section 7.1.1) and a softening point of about 95°C, the Epuré is too hard to be used in asphalts. Trinidad Epuré and 160/220 pen bitumen were used occasionally in hot rolled asphalt surface courses in a 1:1 blend, the resultant mixture having a penetration of about 50 dmm. The addition of the Epuré improved the 'weatherability' of the bitumen, preventing the possibility of the surface courses becoming unacceptably smooth during trafficking. However, improvements in the techniques used to manufacture bitumens have enhanced their weathering characteristics. This, coupled with the fact that the blending process required several hours heating before use, led to the rapid decline in TLA usage in the mid 1970s.

Trinidad Epuré is now occasionally used in stone mastic asphalt and asphaltic concrete mixtures where it replaces 20 to 30% of the paving grade bitumen. It is also used in mastic asphalt materials for roofing and paving, replacing up to 70% of the binder in some applications. Continuing development has led to TLA being supplied in granulated form.

1.4.2 Rock asphalt

Natural or rock asphalt, i.e. bitumen impregnated rock, has been used in Europe since the early seventeenth century. Its main uses were for waterproofing, caulking ships and the protection of wood against rot and vermin. In the nineteenth century, rock asphalt was used in early road surfacing applications. Through distillation, oils were also extracted from rock asphalt for use as lubricants or in medicines for humans or animals. Today, rock asphalt enjoys only minimal use. Applications include waterproofing and mastic asphalts where ageing characteristics are particularly important. Notwithstanding, its manufacture and characteristics continue to be specified in some national standards (e.g. NFP 13-001[4]

---

4 ASSOCIATION FRANCAISE DE NORMALISATION. *Roches, poudres et fines d'asphalte naturel, NFP 13-001, 1977.* AFNOR, Paris.

in France which defines the aggregate grading and the bitumen content).

Rock asphalt is extracted from mines or quarries depending on the type of deposit. Rock asphalt occurs when bitumen, formed by the same concentration processes as occur during the refining of oil, becomes trapped in impervious rock formations.

The largest deposits in Europe were found at Val de Travers in Switzerland, Seyssel in France and Ragusa in Italy, all of which were mined in underground galleries. These natural asphalts are composed of limestone or sandstone impregnated with bitumen at concentrations up to 12%. In the Ticino region of Italy, bituminous schist was also mined although this was undertaken primarily for the extraction of the mineral oil fraction. In North America, vast bituminous sandstone and schist deposits were mined up to the early twentieth century in Utah and Kentucky. An example is the Sunnyside sandstone deposit in Utah which is estimated to contain 800 million tonnes of rock asphalt, with a bitumen concentration in the range of 8 to 13%. This would be enough to construct a road 22 metres wide three times around the Earth. In both North America and Europe, rock asphalt was extracted from deposits, transported to preparation areas, where it was dried, if necessary, then ground and pulverised in a series of crushers before being packaged and shipped for use.

Although its use is now disappearing, it should be noted that rock asphalt was used in the very first roads and streets to be surfaced with a waterproof asphaltic material. These were in Paris in 1854 using Seyssel rock asphalt and in Union Square in New York City in 1872 using rock asphalt from Switzerland.

At present, only one asphaltic limestone deposit is still being mined at St Jean de Maruejols in France where the rock asphalt has a concentration of 7–10% bitumen.

1.4.3 Gilsonite

The state of Utah in the Mid-West of the United States holds a sizeable deposit of natural bitumen. Discovered in vertical deposits in the 1860s, it was first exploited by Samuel H Gilson in 1880 as a waterproofing agent for timber. The material is very hard, having a penetration of zero with a softening point between 115 and 190°C. The addition of gilsonite to bitumen reduces the penetration and increases the softening point. Due to the labour intensive nature of the mining process, gilsonite is relatively expensive. This makes the material unattractive for widespread use in paving materials. It does, however, enjoy some use in bridge and roof waterproofing materials as a means of altering the softening point and stiffness of mastic asphalt.

1.4.4 Tar

'Tar' is a generic word for the liquid obtained when natural organic materials such as coal or wood are carbonised or destructively distilled in the absence of air[5]. It is customary to prefix the word tar with the name of the material from which it is derived. Thus, the products of this initial carbonisation process are referred to as crude coal tar, crude wood tar, etc. Two types of crude coal tar are produced as a by-product of the carbonisation of coal—coke oven tar and low temperature tar.

Since the 1970s, dramatic changes have occurred in the supply of crude coal tar in the UK. In the mid 1960s, over 2 million tonnes of crude coal tar were produced per annum of which around half was manufactured as a by-product of the operation of carbonisation ovens which were used to produce town gas. However, the introduction of North Sea Gas in the late 1960s resulted in a rapid reduction of tar production from this source and by 1975 it had disappeared completely.

1.4.5 Refined bitumen

Bitumen is manufactured from crude oil. It is generally agreed that crude oil originates from the remains of marine organisms and vegetable matter deposited with mud and fragments of rock on the ocean bed. Over millions of years, organic material and mud accumulated into layers some hundreds of metres thick, the substantial weight of the upper layers compressing the lower layers into sedimentary rock. Conversion of the organisms and vegetable matter into the hydrocarbons of crude oil is thought to be the result of the application of heat from within the Earth's crust and pressure applied by the upper layers of sediments, possibly aided by the effects of bacterial action and radioactive bombardment. As further layers were deposited on the sedimentary rock where the oil had formed, the additional pressure squeezed the oil sideways and upwards through porous rock. Where the porous rock extended to the Earth's surface, oil seeped through to the surface. Fortunately, the majority of the oil and gas was trapped in porous rock which was overlaid by impermeable rock, thus forming gas and oil reservoirs. The oil remained here until its presence was detected by seismic surveys and recovered by drilling through the impermeable rock.

The four main oil producing areas in the world are the United States, the Middle East, Russia and the countries around the Caribbean. Crude oils differ in their physical and chemical properties. Physically, they vary from viscous black liquids to free-flowing straw coloured liquids.

---

5 ABRAHAM H. *Asphalts and allied substances, their occurrence, modes of production, uses in the arts and methods of testing, 5th ed, vol 1, 1945.* Van Nostrand, New York.

Chemically, they may be predominantly paraffinic, naphthenic or aromatic with combinations of the first two being most common. There are nearly 1500 different crudes produced throughout the world. Based on the yield and the quality of the resultant product, only a few of these are considered suitable for the manufacture of bitumen.

Chapter 2

Manufacture, storage, handling and environmental aspects of bitumens

2.1 The manufacture of bitumen

Crude oil is a complex mixture of hydrocarbons differing in molecular weight and, consequently, in boiling range. Before it can be used, crude oil has to be separated, purified, blended and, sometimes, chemically or physically changed.

2.1.1 Fractional distillation of crude oil

The first process in the refining of crude oil is fractional distillation. This is carried out in tall steel towers known as fractionating or distillation columns. The inside of the column is divided at intervals by horizontal steel trays punctured with holes to allow vapour to rise up the column. Over these holes are small domes called bubble caps which deflect the vapour downwards so that it bubbles through liquid condensed on the tray. This improves the efficiency of distillation and also has the advantage of reducing the height of the column.

On entering the distillation plant, the crude oil is heated in a furnace to temperatures between 350 and 380°C before being passed into the lower part of the column operating at a pressure slightly above atmospheric. The material entering the column is a mixture of liquid and vapour; the liquid comprises the higher boiling point fractions of the crude oil and the vapour consists of the lower boiling point fractions. The vapours rise up the column through the holes in the trays losing heat as they rise. When each fraction reaches the tray where the temperature is just below its own boiling point, it condenses and changes back into a liquid. As the fractions condense on the trays, they are continuously drawn off via pipes.

The lightest fractions of the crude oil remain as vapour and are taken from the top of the distillation column; heavier fractions are taken off the column as side-streams; the heaviest fractions remain as liquids which, therefore, leave at the base of the column. The lightest fractions produced by the crude distillation process include propane and butane, both of which are gases under atmospheric conditions. Moving down the column, naphtha – a slightly heavier material – is produced. Naphtha is used as a feedstock for gasoline production and the chemical industry. Further down the column, kerosene is produced. Kerosene is used primarily for aviation fuel and, to a lesser extent, domestic fuel. Heavier again is gas oil which is used as a fuel for diesel engines and central heating. The heaviest fraction taken from the crude oil distillation process is known as the long residue, a complex mixture of high molecular weight hydrocarbons, which requires further processing before it can be used as a feedstock for the manufacture of bitumen.

The long residue is further distilled at reduced pressure in a vacuum distillation column. This is carried out under a vacuum of 10 to 100 mmHg at a temperature between 350 and 425°C to produce gas oil, distillates and short residue. If this second distillation was carried out by simply increasing the temperature, cracking or thermal decomposition of the long residue would occur, hence the need to vary the pressure.

The short residue is the feedstock used in the manufacture of over 20 different grades of bitumen. The viscosity of the short residue is a function of both the origin of the crude oil and the temperature and pressure

Fig. 2.1 Shell's bitumen refinery at Petit Couronne, France

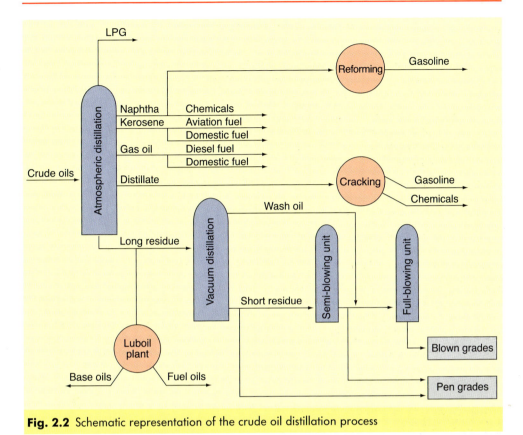

Fig. 2.2 Schematic representation of the crude oil distillation process

in the vacuum column during processing. Depending on the origin of the crude oil, the conditions in the column are adjusted to produce a short residue with a penetration in the range 35 to 300 dmm. A modern bitumen refinery is shown in Fig. 2.1.

Figure 2.2 shows a schematic representation of the distillation process and its relationship with other fundamental refining processes such as reforming and cracking for the manufacture of motor gasoline, and lub oil plants for the manufacture of lubricating oils, waxes, etc.

2.1.2 Air blowing of short residues

The physical properties of the short residue may be modified further by 'air blowing'. This is an oxidation process that involves passing air through the short residue, either on a batch or a continuous basis, with the short residue at a temperature between 240 and 320°C. The main effect of blowing is that it converts some of the relatively low molecular weight 'maltenes' into relatively higher molecular weight 'asphaltenes' (these bitumen components are described in detail in Chapter 3). The

13

result is a reduction in penetration, a comparatively greater increase in softening point and a lower temperature susceptibility.

The continuous blowing process

After preheating, the short residue is introduced into the blowing column just below the normal liquid level. Air is blown through the bitumen by means of an air distributor located at the bottom of the column. The air acts not only as a reactant but also serves to agitate and mix the bitumen thereby increasing both surface area and rate of reaction. The bitumen absorbs oxygen as air ascends through the material. Steam and water are sprayed into the vapour space above the bitumen level, the former to suppress foaming and dilute the oxygen content of waste gases and the latter to cool the vapours in order to prevent after-burning and the resulting formation of coke.

The blown product passes through heat exchangers to achieve the desired 'rundown' temperatures and to provide an economical means of preheating the short residue, before pumping the product to storage. The penetration and softening point of the blown bitumen are affected by the following factors:

- viscosity of the feedstock
- temperature in the blowing column
- residence time in the blowing column
- origin of the crude oil used to manufacture the feedstock
- air-to-feed ratio.

Figure 2.3 shows distillation and blowing 'curves' for a bitumen feedstock. In both the distillation and blowing processes, the softening point increases and the penetration falls. However, in the distillation process, the temperature susceptibility (or penetration index) of the material is largely unchanged. Thus, the distillation line on Fig. 2.3 is a relatively straight line whereas the curves for the blown bitumen flatten substantially as the softening point of the bitumen increases. This demonstrates that the temperature susceptibility of the material is substantially reduced, i.e. the penetration index is increased (the penetration index is described in detail in Section 7.5).

Air-rectified or semi-blown bitumens

Some crude oils produce bitumens that require a limited amount of air blowing in order to generate penetration grade bitumens that are suitable for road construction. This process is termed semi-blowing or air-rectification. Used judiciously, semi-blowing can be applied to reduce the temperature susceptibility of the bitumen, i.e. increase its penetration

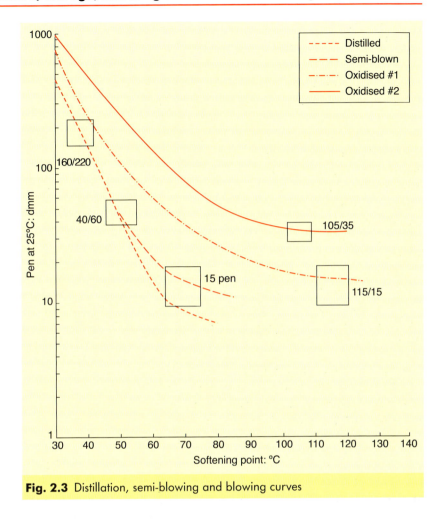

Fig. 2.3 Distillation, semi-blowing and blowing curves

index. This technique has been successfully used in the UK and elsewhere in Europe for the past 30 to 40 years.

Fully blown bitumens

Fully blown or oxidised bitumens are produced by vigorously air-blowing short residue or by blending short residue with a relatively soft flux. The position of the blowing curve on Fig. 2.3 is primarily dependent on the viscosity and chemical nature of the feed. The softer the feed, the higher the curve. The amount of blowing that is required depends on the temperature in the column and, to a lesser extent, on the residence time. Thus, by selection of a suitable bitumen feedstock, control of the viscosity of the feed and the conditions in the column, a range of blown grades of bitumen can be manufactured.

The chemistry of the blowing process

The blowing of bitumen can be described as a conversion process in which oxidation, dehydrogenation and polymerisation take place. The result is an increase in the overall molecular size of the asphaltenes which are already present in the feed and the formation of additional asphaltenes from the maltene phase. The reaction is exothermic which means that the process requires close control of temperatures. This is achieved by regulating the air to short residue ratio in the blowing column.

2.2 Delivery, storage and handling temperatures of bitumens

When handled properly, bitumens can be reheated or maintained at elevated temperatures for a considerable time without adversely affecting their properties. However, mistreatment of bitumens by overheating or by permitting the material to be exposed to conditions that promote oxidation can adversely affect the properties of the bitumen and may influence the long-term performance of mixtures that contain bitumen. The degree of hardening (or, under certain circumstances, softening) that is produced as a result of mishandling is a function of a number of parameters such as temperature, the presence of air, the surface to volume ratio of the bitumen, the method of heating and the duration of exposure to these conditions.

2.2.1 Safe delivery of bitumen

Over 50% of bitumen-related accidents and incidents that result in lost working days occur during the delivery of bitumen. In the UK, the Refined Bitumen Association (RBA) has produced a detailed Code of Practice[6] designed to assist in reducing the frequency of incidents and accidents by raising awareness of their causes. Personnel involved in the delivery and receipt of bitumen are strongly advised to read this document.

2.2.2 Bitumen tanks

All bitumens should be stored in tanks specifically designed for the purpose[7]. In order to minimise the possible hardening of the bitumen during storage, certain aspects of the design of the tank should be considered.

- In order to minimise the risk of overheating the bitumen, the tank should be fitted with accurate temperature sensors and gauges.

---

6 REFINED BITUMEN ASSOCIATION. *Code of practice for the safe delivery of petroleum bitumens, 2001.* RBA, London.

7 INSTITUTE OF PETROLEUM. *Model code of safe practice in the petroleum industry, Bitumen Safety Code, part 11, 3rd ed, 1991.* IP, London.

These should be positioned in the region of the heaters and preferably be removable to facilitate regular cleaning and maintenance.

- Oxidation and the loss of volatile fractions from bitumen are both related to the exposed surface to volume ratio of the storage tank which, for a cylindrical vessel, equals the reciprocal height of the filled part of the tank and is given by:

$$\frac{\text{surface area}}{\text{volume}} = \frac{\pi r^2}{\pi r^2 h} = \frac{1}{h}$$

where h = height of bitumen and r = radius of the tank.

Thus, the dimensions of the bulk storage tank should be such that the surface to volume ratio is minimised. Accordingly, vertical storage tanks with a high height to radius ratio are preferable to horizontal tanks.

It is common practice for bitumen in bulk storage tanks at mixing plants to be recirculated around a ring main in order to heat the pipework that carries the bitumen to the processing point. Return lines in a recirculation system should re-enter the storage tank below the bitumen surface to prevent hot bitumen cascading through the air. Often, the bitumen is returned to the bulk storage tank through a pipe fitted into the upper part of the tank, flush with the side, roof or protruding just into the air space at the top of the tank. If the bitumen enters the tank above the bitumen all the factors that promote oxidation are present:

- high temperature
- access to oxygen
- high exposed surface to volume ratio.

Fortunately, the residence time of bitumen in the tank is usually sufficiently low for any hardening to be insignificant. However, if material is stored for a prolonged period, recirculation should be used only intermittently and the bitumen should be tested before use to ensure its continued suitability for the proposed application. A recommended layout for a bitumen storage tank is shown in Fig. 2.4.

Bitumen storage tanks should be fitted with automatic level indicators together with low- and high-level alarms to avoid having to dip manually. Such an approach avoids exposing the hot heater tubes to a potentially combustible or explosive atmosphere should the bitumen level fall below that of the heater tubes. Automatic level control also ensures that the tank is not overfilled. Regardless of whether a high-level alarm is fitted, a maximum safe filled level for the tank should be predetermined taking into account the effects of thermal expansion of the bitumen in the tank.

17

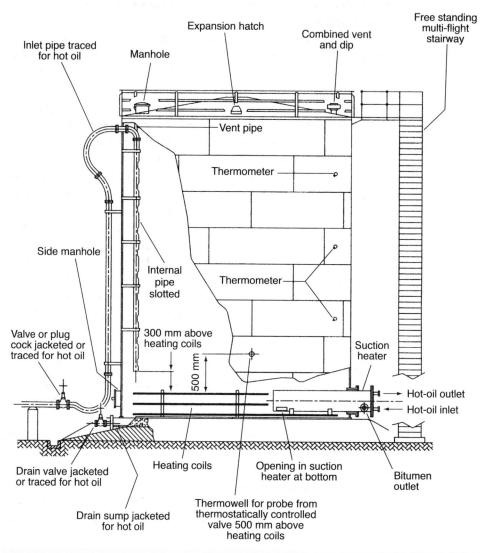

Fig. 2.4 Recommended format for a bitumen storage tank

Before ordering additional bitumen, it is obviously essential to check that the ullage in the tank is capable of taking the delivery without exceeding the maximum safe working level.

Every tank should be clearly labelled with the grade of bitumen it contains. When the grade of bitumen in a tank is changed, it is important to ensure that the tank is empty and relabelled before the new grade is delivered.

2.2.3 *Bitumen storage and pumping temperatures*

Conventional bitumen

Bitumen should always be stored and handled at the lowest temperature possible, consistent with efficient use. As a guide, working temperatures for specific operations are given in the RBA's Technical Bulletin No 7[8]. These temperatures have been calculated on the basis of viscosity measurements and are supported by operational experience. For normal operations, i.e. the blending and transferring of liquid bitumen, temperatures of 10 to 50°C above the minimum pumping temperature are recommended, but the maximum safe handling temperature of 230°C must never be exceeded.

The period during which bitumen resides in a storage tank at elevated temperatures and is recirculated should be minimised to prevent hardening of the bitumen. If bitumen must be stored for an extended period, say for a period exceeding one week without the addition of fresh material, the temperature should be reduced to approximately 20 to 25°C above the softening point of the bitumen and, if possible, recirculation stopped.

When bitumen is being reheated in bulk storage, care must be taken to heat the bitumen intermittently over an extended period to prevent localised overheating of the product around the heating pipes or coils. This is particularly important where direct flame tube heating is used because surface temperatures in excess of 300°C may be reached. In such installations, the amount of heat that is applied should be limited, sufficient only to raise the temperature of the product to just above its softening point. This will allow the material to soften after which further heat can be applied to raise the temperature of the product to the required working value. This technique is beneficial because when the bitumen is a fluid, albeit a viscous fluid, convection currents dissipate the heat throughout the material and localised overheating is thus less of a problem. Circulation of the tank contents should begin as soon as the product is sufficiently fluid, thereby further reducing the likelihood of local overheating. With hot oil, steam or electric heaters that are designed properly, reheating from cold should not cause these problems.

Polymer modified and special bitumens

A wide variety of polymer modified and process modified bitumens are available (see Chapter 5). Different polymers and different polymer/bitumen combinations require different storage regimes and pumping temperatures. Not all polymer modified bitumens are stable in storage and the management of these binders can be radically different to the management of bitumens that *are* stable in storage.

---

8 REFINED BITUMEN ASSOCIATION. *Safe handling of petroleum bitumens, Technical Bulletin No 7, 1998*. RBA, London.

All polymer modified and special bitumens that are currently supplied by Shell Bitumen can be stored in conventional bitumen tanks. Users should always check with their supplier whether this is true of the materials they are purchasing. Essential information on storage should always be ascertained from the relevant product information sheet and such advice must be applied rigidly. Where required, additional information will almost certainly be available from suppliers.

2.3 Health, safety and environmental aspects of bitumens

Bitumen has a long history of being used safely in a wide range of applications. Although it is primarily used in road construction, it can also be employed in roofing felts, reservoir linings and also as an internal lining for potable water pipes. Bitumen presents a low order of potential hazard provided that good handling practices are observed. (These are described in detail in part of the Model Code of Safe Practice[9] which is published by the Institute of Petroleum and are also described in the material safety data sheets. A substantial amount of health, safety and environmental data on bitumen and its derivatives are detailed in the CONCAWE product dossier number 92/104[10].) Notwithstanding, bitumens are generally applied at elevated temperatures and this brings with it a number of hazards that are considered below.

2.3.1 Health aspects of bitumen

Potential hazards associated with the handling and application of hot bitumen

- *Elevated temperatures.* The main hazard associated with bitumen is that the product is held at elevated temperatures during transportation, storage and processing. Thus, it is critical that appropriate personal protective equipment (PPE) is worn and any skin contact with hot bitumen is avoided. Some guidance is given below but detailed advice is available in a number of publications[9–11] and the relevant material safety data sheet for bitumen[12,13].

---

9 INSTITUTE OF PETROLEUM. *Model code of safe practice in the petroleum industry, Bitumen Safety Code, part 11, 3rd ed, 1991.* IP, London.

10 CONCAWE. *Bitumens and bitumen derivatives, Product Dossier No 92/104, Dec 1992.* CONCAWE, The Hague.

11 REFINED BITUMEN ASSOCIATION. *Code of practice for the safe delivery of petroleum bitumens, 2001.* RBA, London.

12 EUROBITUME. *Material safety data sheet template for paving grade bitumen, Jul 2001.* Eurobitume, Brussels.

13 SHELL BITUMEN. *Material safety data sheet for Shell paving grade bitumen, Dec 2001.* Shell International Petroleum Company Ltd.

- *Vapour emissions.* Bitumens are complex mixtures of hydrocarbons that do not have well defined boiling points because their components boil over a wide temperature range. Visible emissions or fumes normally start to develop at approximately 150°C. The amount of fume generated doubles for each 10 to 12°C increase in temperature. Fumes are mainly composed of hydrocarbons[14] and small quantities of hydrogen sulphide (H_2S). The latter is of particular concern as it can accumulate in enclosed spaces such as the tops of storage tanks. Exposure to this gas at concentrations as little as 500 ppm can be fatal so it is essential that any space where H_2S may be present is tested and approved as being gas-free before anyone enters the area. Bitumen fumes also contain small quantities of polycyclic aromatic compounds (PACs); these are discussed below.

- *Combustion.* Very high temperatures are required to make bitumen burn. Certain materials, if hot enough, will ignite when exposed to air, and are sometimes described as being 'pyrophoric'. In the case of bitumens, this auto-ignition temperature is generally around 400°C. However, fires in bitumen tanks have occurred only occasionally. Under conditions of low oxygen content, H_2S from bitumen can react with rust (iron oxide) on the roof and walls of storage tanks to form 'pyrophoric iron oxide'. This material reacts readily with oxygen and can self-ignite if the oxygen content of the tank increases suddenly which, in turn, can ignite coke deposits on the roof and walls of the tank. Coke deposits are the result of condensate from the bitumen which has been deposited on the roof and walls of a tank degrading over a period of time forming carbonaceous material. Under conditions of high temperature and in the presence of oxygen or a sudden increase in available oxygen, an exothermic reaction can occur leading to the risk of fire or explosion. Accordingly, manholes in bitumen tanks should be kept closed and access to tank roofs should be restricted.

- *Contact with water.* It is essential that water does not come into contact with hot bitumen. If this does happen, the water is converted into steam. In the process, its volume increases by a factor of approximately 1400, resulting in spitting, foaming and, depending on the amount of water present, possible boil-over of the hot bitumen.

Composition of bitumen

Bitumens are complex hydrocarbon materials containing components of many chemical forms, the majority of which have relatively high

14 BRANT H C A and P C DE GROOT. *Emission and composition of fumes from current bitumen types, Eurasphalt/Eurobitume Congress, 1996.*

molecular weight. Crude oils normally contain small quantities of polycyclic aromatic compounds (PACs), some of which end up in the bitumen. These chemicals consist of a number of benzene rings that are grouped together. Some of these with three to seven (usually four to six) fused rings are known to cause or suspected of causing cancer in humans. However, the concentrations of these carcinogens in bitumen are extremely low[15].

Potential hazards through skin contact

Other than thermal burns, the hazards associated with skin contact of most bitumens are negligible. Studies by the International Agency for Research on Cancer (IARC)[16], part of the World Health Organization, concluded that there was no direct evidence to associate bitumen with long-term skin disorders in humans despite bitumens having been widely used for many years. Nevertheless, it is prudent to avoid intimate and prolonged skin contact with bitumen.

Cutback bitumens and bitumen emulsions are handled at lower temperatures, which increases the chance of skin contact. If personal hygiene is poor, regular skin contact may occur. However, studies carried out by Shell[17,18] have demonstrated that bitumens are unlikely to be bioavailable (skin penetration and body uptake) and bitumens diluted with solvents are unlikely to present a carcinogenic risk. Nevertheless, bitumen emulsions can cause irritation to the skin and eyes and can produce allergic responses in some individuals.

First aid for skin burns

The following text is taken from the Eurobitume notes for guidance of first aid and medical personnel. (Eurobitume is the European Bitumen Association and comprises a number of bitumen companies and bitumen associations in Europe.) This advice is produced in the form of A5 sized card, copies of which are available from Eurobitume, the RBA and bitumen suppliers. It is intended that the card should accompany a

15 CONCAWE. *Bitumens and bitumen derivatives, Product dossier no 92/104, Dec 1992.* CONCAWE, The Hague.

16 BOFFETTA P T. *Cancer risk in asphalt workers and roofers: Review and meta-analysis of epidemiologic studies, Am J Ind Med, vol 26, pp 721–740, 2001.*

17 BRANDT H C A, E D BOOTH, P C DE GROOT and W P WATSON. *Development of a carcinogenic potency index for dermal exposure to viscous oil products, Arch Toxicol, vol 73, pp 180–188, 1999.*

18 POTTER D, E D BOOTH, H C A BRANDT, R W LOOSE, R A J PRISTON, A S WRIGHT and W P WATSON. *Studies on the dermal and systemic bioavailability of polycyclic aromatic compounds in high viscosity oil products, Arch Toxicol, vol 73, pp 129–140, 1999.*

burns victim to the hospital to provide immediate advice on proper treatment.

First aid

When an accident has occurred, the affected area should be cooled as quickly as possible to prevent the heat causing further damage. The burn should be drenched in cold water for at least ten minutes for skin and at least five minutes for eyes. However, body hypothermia must be avoided. No attempt should be made to remove the bitumen from the area of the burn.

Further treatment, first aid and medical care

The bitumen layer will be firmly attached to the skin and removal should not be attempted unless carried out at a medical facility under the supervision of a doctor. The cold bitumen will form a waterproof, sterile layer over the burn which will prevent the burn from drying out. If the bitumen is removed from the wound, there is the possibility that the skin will be damaged further, bringing with it the possibility of further complications. In addition, by exposing a second degree burn in order to treat it, there is the possibility that infection or drying out will make the wound deeper.

Second degree burns

The bitumen should be left in place and covered with a Tulle dressing containing paraffin or a burn ointment containing paraffin, e.g. Flammazine (silver sulphadiazine). Such treatment will have the effect of softening the bitumen enabling it to be gently removed over a period of days. As a result of the skin reforming naturally (re-epithelialisation), any remaining bitumen will peel off in time.

Third degree burns

Active removal of the bitumen should be avoided unless primary surgical treatment is being considered due to the location and depth of the wound. In such cases, removal of the bitumen is best carried out in the operating theatre between the second and fifth day after the burn occurred. By the second day, the capillary circulation has usually recovered and the bed of the wound is such that a specialist can assess the depth to which the burn has penetrated. There are normally no secondary problems such as infections to contend with before the sixth day. However, it is essential that treatment is started using paraffin-based substances from the day of the accident to facilitate removal during surgery.

Circumferential burns

Where hot bitumen completely encircles a limb or other body part, the cooled and hardened bitumen may cause a tourniquet effect. In the event of this occurring, the adhering bitumen must be softened and/or split to prevent restriction of blood flow.

Eye burns

No attempt should be made to remove the bitumen by unqualified personnel. The patient should be referred urgently for specialist medical assessment and treatment.

Potential hazards via inhalation

Generally, when bitumens are heated in bulk or mixed with hot aggregate, fumes are emitted. The fumes contain particulate bitumen, hydrocarbon vapour and very small amounts of hydrogen sulphide.

The UK occupational exposure standards[19] for bitumen fumes in the working atmosphere are:

long-term exposure limit (8 h time weighted average) $= 5 \, \text{mg/m}^3$
short-term exposure limit (10 min time weighted average) $=$
$10 \, \text{mg/m}^3$.

Under normal operating conditions, exposure is well below these limits.

The UK occupational exposure[19] standards for hydrogen sulphide in the working atmosphere are:

long-term exposure limit (8 h time weighted average) $= 10 \, \text{ppm}$
$(14 \, \text{mg/m}^3)$
short-term exposure limit (10 min time weighted average) $= 15 \, \text{ppm}$
$(21 \, \text{mg/m}^3)$.

Exposure to bitumen fumes can result in irritation to the eyes, nose and respiratory tract, headaches and nausea. The symptoms are usually mild and temporary. Removal of the affected personnel from the source results in rapid recovery. Even though the irritation is usually mild, exposure to bitumen fumes should be minimised and, where there is any doubt, tests, e.g. Drægar tube analysis or personal exposure monitoring, should be undertaken to determine the concentration of bitumen fume or hydrogen sulphide in the working atmosphere.

---

19 HEALTH and SAFETY EXECUTIVE. *Occupational exposure limits 2002, Guidance Note EH 40, 2001*. HSE Books, London.

First aid for inhalation of bitumen fumes

Persons affected by inhalation of bitumen fumes should be removed to fresh air as soon as possible. If the symptoms are severe or if symptoms persist, medical help should be sought without delay.

Studies into health of workers in the asphalt industry

There have been many studies of the health of workers in the asphalt and road construction industry. Despite the known presence of PACs in bitumen and bitumen fume, experience in the asphalt industry has shown that bitumen and bitumen fume do not present a health risk to staff when good working practices are adopted. To provide further support for this view, the asphalt industry is collaborating with independent epidemiological research currently being carried out by the International Agency for Research on Cancer (IARC).

2.3.2 Safe use of bitumen – recommended precautions for personnel

Personal protective equipment

The principal hazard from handling hot bitumen is thermal burns resulting from contact with the product. Thus, it is essential to wear clothing that provides adequate protection (see Fig. 2.5) including:

- long-term exposure limit (8 h time weighted average) $= 10\,\text{ppm}$ $(14\,\text{mg/m}^3)$ – helmet and neck apron to provide head protection

Fig. 2.5 A safe bitumen delivery

- short-term exposure limit (10 min time weighted average) = 15 ppm ($21\,mg/m^3$) – visor to protect the face (goggles only protect eyes)
- heat-resistant gloves (with cuffs worn inside coverall sleeves)
- safety boots
- coveralls (with coverall legs worn over boots).

Personal hygiene

Garments soiled with bitumen should either be replaced or dry cleaned in order to avoid permeation of the product to underclothing. Soiled rags or tools should not be placed in the pockets of overalls as contamination of the lining of the pocket will result.

Personnel handling bitumen and asphalts should be provided with and use barrier creams to protect exposed skin, particularly hands and fingers. Skin should be thoroughly washed after any contamination and always before going to the toilet, eating or drinking.

The application of barrier creams, prior to handling bitumen, assists in subsequent cleaning should accidental contact occur. However, barrier creams are no substitute for gloves or other impermeable clothing. Consequently, they should not be used as the sole form of protection. Solvents such as petrol, diesel oil, white spirit etc., should not be used for removing bitumen from the skin and may spread the contamination. An approved skin cleanser together with warm water should be used.

Fire prevention and fire fighting

The adoption of safe handling procedures will substantially reduce the risk of fire. However, if a fire occurs, it is essential that personnel are properly trained and well equipped to extinguish the fire thereby ensuring that the risk of injury to personnel and damage to plant is minimised. Detailed advice on fire prevention and fire fighting is given in a code of practice published by the Institute of Petroleum[20] and in individual material safety data sheets.

Small bitumen fires can be extinguished using dry chemical powder, foam, vaporising liquid or inert gas extinguishers, fog nozzle spray hoses and steam lances. Direct water jets must not be used because frothing may occur which tends to spread the hot bitumen and, therefore, the fire.

Injection of steam or a 'fog' of water into the vapour space can extinguish internal tank fires where the roof of the tank is largely intact. However, only trained operatives should use this method as the water vaporises instantly on contact with the hot bitumen. This initiates

---

20 INSTITUTE OF PETROLEUM. *Model code of safe practice in the petroleum industry, Bitumen safety code, part 11, 3rd ed, 1991.* IP, London.

foaming which may result in the tank overflowing, creating an additional hazard. Alternatively, foam extinguishers may be used. The foam ensures that the water is well dispersed, thereby reducing the risk of froth-over. The disadvantage of using this type of extinguisher is that the foam breaks down rapidly when applied to hot bitumen.

Portable extinguishers containing either aqueous film-forming foam or dry chemical powder are suitable for dealing with small bitumen fires, at least initially. In bitumen handling areas, these should be placed at strategic, permanent and conspicuous locations. The type and location of equipment to be used if initial attempts fail should be discussed with the local fire brigade before installation.

Sampling

Sampling of hot bitumen is particularly hazardous because of the risk of heat burns from spills and splashes of the material. It is therefore essential that appropriate protective clothing is worn. The area should be well lit and safe access to and egress from the sample point should be provided. Gantry access should be provided where samples are required from the tanks of vehicles.

Dip sampling

In this process a sample of bitumen is obtained by dipping a weighted can or 'thief' on the end of a rope or rod through the access lid into bitumen stored in a bulk tank. The sample is then transferred to a suitable permanent container. The method is simple but is only appropriate for small samples. Dip sampling from cutback tanks should be avoided because of the presence of flammable atmospheres in tank vapour spaces.

Sample valves

Properly designed sample valves are very useful for sampling from pipelines or from tanks. Their design should ensure that they are kept hot by the product in the pipeline or tank in order to avoid blockage when in the closed position.

Sample valves should preferably be the screw-driven plunger type. When closed, the plunger of this type of valve extends into the fresh product. Thus when the valve is opened, a representative sample of product is obtained without 'fore-runnings'. With ball and plug type valves, fore-runnings have to be collected and disposed of before a representative sample can be obtained. Designs for bitumen sample valves are described in detail by the British Standards Institution[21].

---

21 BRITISH STANDARDS INSTITUTION. *Methods for sampling petroleum products, Method for sampling bituminous binders, BS 3195-3: 1987.* BSI, London.

2.3.3 *Environmental aspects of bitumen*
Life cycle assessment of bitumen

Life cycle assessment (LCA) is a tool to investigate the environmental aspects and potential impact of a product, process or activity by identifying and quantifying energy and material flows. LCA covers the entire life cycle including extraction of the raw material, manufacturing, transport and distribution, product use, service and maintenance and disposal (recycling, incineration or landfill). It is a complete cradle-to-grave analysis focusing on the environmental input (based on ecological effects) and resource use.

LCA can be divided into two distinct parts – life cycle inventory (LCI) and life cycle impact. Eurobitume has carried out a partial life cycle inventory of bitumen[22] to generate inventory data on the production of paving grade bitumen for use in future LCI studies where bitumen is used.

Leaching of components from bitumen

Bitumen and asphalt have been used for many years for applications in contact with water, such as reservoir linings, dams and dykes[23]. A number of studies have been carried out both in the USA and Europe to determine if components are leached out of asphalt and bitumen when in prolonged contact with water.

Shell has carried out laboratory studies of leaching on a range of bitumens and asphalt mixtures[24]. These concluded that although prolonged contact with water will result in PACs being leached into water, the levels rapidly reach an equilibrium level that is well below the surface water limits that exist in a number of EU countries and more than an order of magnitude below the EU limits for potable water.

22 EUROBITUME. *Partial life cycle inventory or "Eco-profile" for paving grade bitumen, Eurobitume report 99/007, May 1999.* Eurobitume, Brussels.

23 SCHÖNIAN E. *The Shell Bitumen Hydraulic Engineering Handbook, 1999.* Shell International Petroleum Company Ltd, London.

24 BRANT H C A and P C DE GROOT. *Aqueous leaching of polycyclic aromatic hydrocarbons from bitumen and asphalt, Water Res, vol 35, no 17, pp 4200–4207, 2001.*

Chapter 3

Constitution, structure and rheology of bitumens

Rheology is the science that deals with the flow and deformation of matter. The rheological characteristics of a bitumen at a particular temperature are determined by both the constitution (chemical composition) and the structure (physical arrangement) of the molecules in the material. Changes to the constitution, structure or both will result in a change to the rheology. Thus, to understand changes in bitumen rheology, it is essential to understand how the structure and constitution of a bitumen interact to influence rheology.

3.1 Bitumen constitution

The configuration of the internal structure of a bitumen is largely determined by the chemical constitution of the molecular species present. Bitumen is a complex chemical mixture of molecules that are predominantly hydrocarbons with a small amount of structurally analogous heterocyclic species and functional groups containing sulphur, nitrogen and oxygen atoms[25–27]. Bitumen also contains trace quantities of metals such as vanadium, nickel, iron, magnesium and calcium, which occur in the form of inorganic salts and oxides or in porphyrine structures. Elementary analysis of bitumens manufactured from a variety of crude oils shows that most bitumens contain:

- carbon 82–88%
- hydrogen 8–11%
- sulphur 0–6%

25 TRAXLER R N. *The physical chemistry of asphaltic bitumen, Chem Rev, vol 19, no 2, 1936.*

26 ROMBERG J W, S D NESMITTS and R N TRAXLER. *Some chemical aspects of the components of asphalt, J Chem Eng Data, vol 4, no 2, Apr 1959.*

27 TRAXLER R N and C E COOMBS. *The colloidal nature of asphalt as shown by its flow properties, 13th Colloid Symp, St Louis, Missouri, June 1936.*

- oxygen 0–1·5%
- nitrogen 0–1%.

The precise composition varies according to the source of the crude oil from which the bitumen originates, modification induced by semi-blowing and blowing during manufacture and ageing in service.

The chemical composition of bitumen is extremely complex. Thus, a complete analysis of bitumen (if it were possible) would be extremely laborious and would produce such a large quantity of data that correlation with the rheological properties would be impractical, if not impossible. However, it is possible to separate bitumen into two broad chemical groups called asphaltenes and maltenes. The maltenes can be further sub-divided into saturates, aromatics and resins. The four groups are not well defined and there is some overlap between the groups. However, this does enable bitumen rheology to be set against broad chemical composition.

The methods available for separating bitumens into fractions can be classified as:

- solvent extraction
- adsorption by finely divided solids and removal of unadsorbed solution by filtration
- chromatography
- molecular distillation used in conjunction with one of the above techniques.

Solvent extraction is attractive as it is a relatively rapid technique[28] but the separation obtained is generally poorer than that which results from using chromatography where a solvent effect is combined with selective adsorption. Similarly, simple adsorption methods[29] are not as effective as column chromatography in which the eluting solution is constantly re-exposed to fresh adsorbent and different equilibrium conditions as it progresses down the column. (An eluting solution is one that is used to remove an adsorbed substance by washing.) Molecular distillation is lengthy and has limitations in terms of the extent to which type separation and distillation of high molecular weight components of bitumen can be effected.

Chromatographic techniques[30–32] have, therefore, been most widely used to define bitumen constitution. The basis of the method is to initially

28 TRAXLER R N and H E SCHWEYER. *Oil Gas J, vol 52, p 158, 1953.*

29 MARCUSSON J and Z AGNEW. *Chem, vol 29, p 346, 1916.*

30 SCHWEYER H E, H CHELTON and H H BRENNER. *A chromatographic study of asphalt, Proc Assoc Asph Pav Tech, vol 24, p 3, 1955.* Association Asphalt Paving Technologists, Seattle.

31 MIDDLETON W R. *American Chemistry Society Symp, vol 3, A-45, 1958.*

32 CORBETT L W and R E SWARBRICK. *Clues to asphalt composition, Proc Am Assoc Asph Pav Tech, vol 27, p 107, 1958.* Association Asphalt Paving Technologists, Seattle.

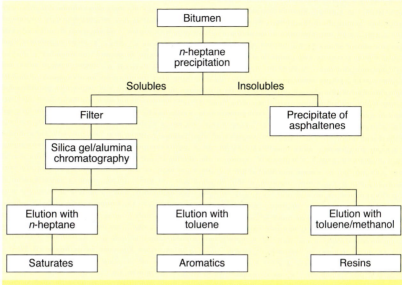

Fig. 3.1 Schematic representation of the analysis for broad chemical composition of bitumen

precipitate asphaltenes using *n*-heptane followed by chromatographic separation of the remaining material. Figure 3.1 shows a schematic representation of the chromatographic method. Using this technique, bitumens can be separated into the four groups: asphaltenes, resins, aromatics and saturates. The main characteristics of these four broad component groups and the metallic constituents are now discussed.

3.1.1 Asphaltenes

These are *n*-heptane insoluble black or brown amorphous solids containing, in addition to carbon and hydrogen, some nitrogen, sulphur and oxygen. Asphaltenes are generally considered to be highly polar and complex aromatic materials of fairly high molecular weight. Different methods of determining molecular weights have led to different values ranging widely from 600 to 300 000 depending on the separation technique employed. However, the majority of test data indicates that the molecular weights of asphaltenes range from 1000 to 100 000; they have a particle size of 5 to 30 nm and a hydrogen/carbon (H/C) atomic ratio of about 1·1. The asphaltene content has a large effect on the rheological characteristics of a bitumen. Increasing the asphaltene content produces a harder, more viscous bitumen with a lower penetration, higher softening point and, consequently, higher viscosity. Asphaltenes constitute 5 to 25% of the bitumen. Figure 3.2 shows a typical chemical structure of an asphaltene.

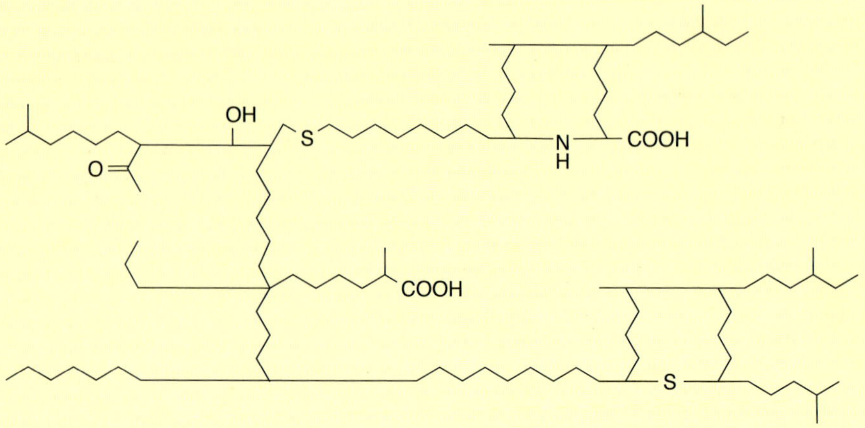

Fig. 3.2 Asphaltene structure

3.1.2 Resins

Resins are soluble in *n*-heptane. Like asphaltenes, they are largely composed of hydrogen and carbon and contain small amounts of oxygen, sulphur and nitrogen. They are dark brown in colour, solid or semi-solid and, being polar in nature, they are strongly adhesive. Resins are dispersing agents or peptisers for the asphaltenes. The proportion of resins to asphaltenes governs, to a degree, the solution (SOL) or gelatinous (GEL) type character of the bitumen. Resins separated from bitumens are found to have molecular weights ranging from 500 to 50 000, a particle size of 1 to 5 nm and a H/C atomic ratio of 1·3 to 1·4.

3.1.3 Aromatics

Aromatics comprise the lowest molecular weight naphthenic aromatic compounds in the bitumen and represent the major proportion of the dispersion medium for the peptised asphaltenes. They constitute 40 to 65% of the total bitumen and are dark brown viscous liquids. The average molecular weight range is in the region of 300 to 2000. They consist of non-polar carbon chains in which the unsaturated ring systems (aromatics) dominate and they have a high dissolving ability for other high molecular weight hydrocarbons (see Fig. 3.3).

3.1.4 Saturates

Saturates consist of straight and branch chain aliphatic hydrocarbons together with alkyl-naphthenes and some alkyl-aromatics. They are non-polar viscous oils which are straw or white in colour. The average molecular weight range is similar to that of aromatics and the components

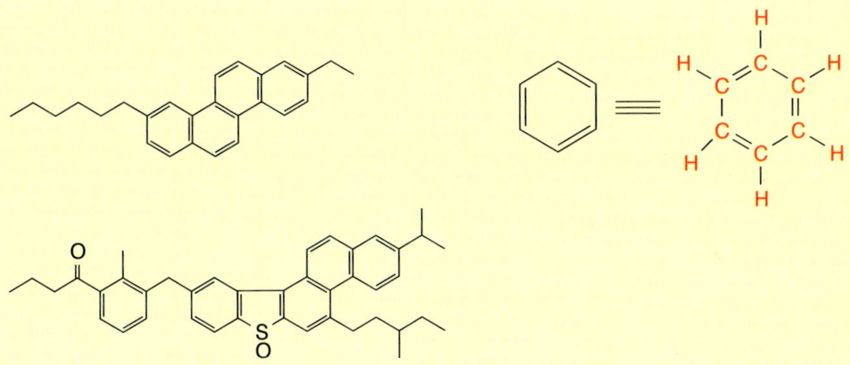

Fig. 3.3 Aromatic structures

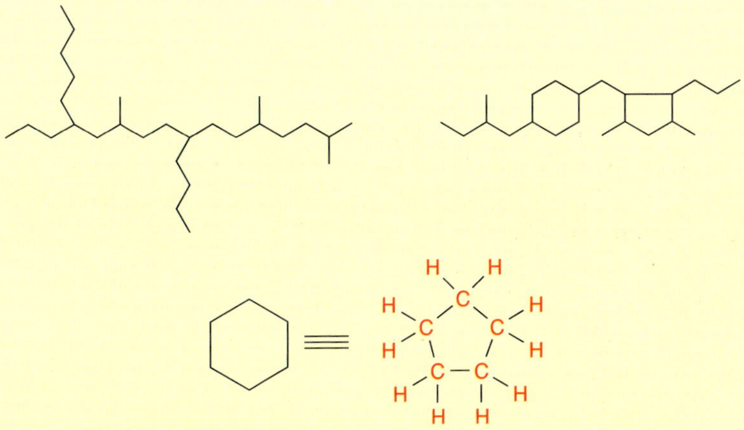

Fig. 3.4 Saturate structures

include both waxy and non-waxy saturates. This fraction forms 5 to 20% of the bitumen. Figure 3.4 shows two different saturate structures.

An elemental analysis of the above four groups from a 100 pen bitumen is detailed in Table 3.1.

3.1.5 Metallic constituents of bitumen

It was known that bitumens contain metalloporphyrins in the 1930s[33] from analytical work that identified iron and vanadium and, consequently, established the link between marine plant chlorophyll and petroleum genesis. Ash from fuel oil on examination by UV emission

---

33 TRIEBS A. *Ann der Chem, vol 42, p 510, 1934.*

Table 3.1 Typical elemental analysis of the four groups of a 100 pen bitumen[34]

	Yield on bitumen: %w	Carbon: %w	Hydrogen: %w	Nitrogen: %w	Sulphur: %w	Oxygen: %w	Atomic ratio: H/C	Molecular weight
Asphaltenes (n-heptane)	5·7	82·0	7·3	1·0	7·8	0·8	1·1	11 300
Resins	19·8	81·6	9·1	1·0	5·2	–	1·4	1270
Aromatics	62·4	83·3	10·4	0·1	5·6	–	1·5	870
Saturates	9·6	85·6	13·2	0·05	0·3	–	1·8	835

Table 3.2 Elemental analysis of bitumens from various sources[35]

	Carbon: %w	Hydrogen: %w	Nitrogen: %w	Sulphur: %w	Oxygen: %w	Nickel: ppm	Vanadium: ppm
Range	80·2–84·3	9·8–10·8	0·2–1·2	0·9–6·6	0·4–1·0	10–139	7–1590
Average	82·8	10·2	0·7	3·8	0·7	83	254

	Iron: ppm	Manganese: ppm	Calcium: ppm	Magnesium: ppm	Sodium: ppm	Atomic ratio: H/C
Range	5–147	0·1–3·7	1–335	1–134	6–159	1·42–1·50
Average	67	1·1	118	26	63	1·47

spectrography has been shown to contain the following metallic elements[36].

aluminium	barium	calcium
chromium	copper	gallium
iron	lanthanum	lead
magnesium	manganese	molybdenum
nickel	potassium	silicon
silver	sodium	strontium
tantalum	tin	uranium
vanadium	zinc	zirconium

These elements occur principally in the heavier or involatile components of the oils, some as inorganic contaminants, possibly in colloidal form,

34 CHIPPERFIELD E H. *Bitumen production properties and uses in relation to occupational exposures, IARC review on bitumen carcinogenicity, Institute of Petroleum Report IP 84-006, 1984.* IP, London.

35 GOODRICH J L, J E GOODRICH and W J KARI. *Asphalt composition tests: Their application and relation to field performance, Transportation Research Record 1096, pp 146–167, 1986.*

36 CRUMP G B. *Black but such as in esteem – the analytical chemistry of bitumen, Chairman's retiring address to the N W Region Analytical Division of the Royal Society of Chemistry, Jan 1981.*

but also as salts (e.g. of carboxylic acids), transition metal complexes and porphyrin-type complexes. An analysis of bitumens from various sources is shown in Table 3.2.

The predominant metals present in most fuel oils are sodium, vanadium, iron, nickel and chromium with most of the sodium present as sodium chloride. Vanadium and nickel are largely present as porphyrin structures, which also represent large numbers of different molecules depending upon ring substituents and structural isomerism. Bitumen and fuel oil are related products and Crump[37] surmises that similar elements will be present in bituminous compounds.

3.2 Bitumen structure

Bitumen is traditionally regarded as a colloidal system[38] consisting of high molecular weight asphaltene micelles dispersed or dissolved in a lower molecular weight oily medium (maltenes). The micelles are considered to be asphaltenes together with an absorbed sheath of high molecular weight aromatic resins which act as a stabilising solvating layer. Away from the centre of the micelle, there is a gradual transition to less polar aromatic resins, these layers extending outwards to the less aromatic oily dispersion medium.

In the presence of sufficient quantities of resins and aromatics of adequate solvating power, the asphaltenes are fully peptised and the resulting micelles have good mobility within the bitumen. These are known as 'SOL' type bitumens and are illustrated in Fig. 3.5.

If the aromatic/resin fraction is not present in sufficient quantities to peptise the micelles, or has insufficient solvating power, the asphaltenes can associate together further. This can lead to an irregular open packed structure of linked micelles in which the internal voids are filled with an intermicellar fluid of mixed constitution. These bitumens are known as 'GEL' types, as depicted in Fig. 3.6, the best examples being the oxidised grades used for roofing purposes. In practice, most bitumens are of intermediate character.

The colloidal behaviour of the asphaltenes in bitumens results from aggregation and solvation. The degree to which they are peptised will have a considerable influence on the resultant viscosity of the system. Such effects decrease with increasing temperature and the GEL character

37 CRUMP G B. *Black but such as in esteem – the analytical chemistry of bitumen, Chairman's retiring address to the N W Region Analytical Division of the Royal Society of Chemistry, Jan 1981.*

38 GIRDLER R B. *Proc Assoc Asph Pav Tech, vol 34, p 45, 1965.* Association Asphalt Paving Technologists, Seattle.

Asphaltenes

High molecular weight aromatic hydrocarbon

Low molecular weight aromatic hydrocarbon

○ Aromatic/naphthenic hydrocarbons

⌒ Naphthenic/aliphatic hydrocarbons

— Saturated hydrocarbons

Fig. 3.5 Schematic representation of a SOL type bitumen

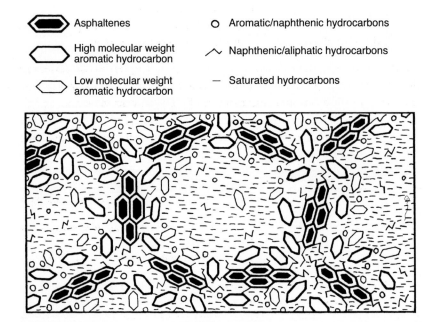

Asphaltenes

High molecular weight aromatic hydrocarbon

Low molecular weight aromatic hydrocarbon

○ Aromatic/naphthenic hydrocarbons

⌒ Naphthenic/aliphatic hydrocarbons

— Saturated hydrocarbons

Fig. 3.6 Schematic representation of a GEL type bitumen

of certain bitumens may be lost when they are heated to high temperatures. The viscosities of the saturates, aromatics and resins depend on the molecular weight distribution. The higher the molecular weight, the higher the viscosity. The viscosity of the continuous phase, i.e. the maltenes, imparts an inherent viscosity to the bitumen that is increased by the presence of the dispersed phase, i.e. the asphaltenes. The saturates fraction decreases the ability of the maltenes to solvate the asphaltenes because high saturate contents can lead to marked agglomeration of the asphaltenes. Accordingly, an increase in GEL character and a lower temperature dependence for bitumens results not only from the asphaltene content but also from the saturates content.

3.3 The relationship between constitution and rheology

Systematic blending of saturates, aromatics, resins and asphaltene fractions separated from bitumen has demonstrated the effect that constitution has on rheology[39–43]. By holding the asphaltene content constant and varying the concentration of the other three fractions, it has been demonstrated that:

- increasing the aromatics content at a constant saturates to resins ratio has little effect on rheology other than a marginal reduction in shear susceptibility;
- maintaining a constant ratio of resins to aromatics and increasing the saturates content softens the bitumen; and
- the addition of resins hardens the bitumen, reduces the penetration index and shear susceptibility but increases the viscosity.

It has also been shown that the rheological properties of bitumens depend strongly on the asphaltene content. At constant temperature, the viscosity of a bitumen increases as the concentration of the asphaltenes blended into the parent maltenes is increased. However, the increase in viscosity is substantially greater than would be expected if

39 MCKAY J R *et al. Petroleum asphaltenes: Chemistry and composition, ACS advances in chemistry series no 170, Paper 9.*

40 REERINK H. *Size and shape of asphaltene particles in relationship to high temperature viscosity, Ind Eng Chem Prod Res Deve, vol 12, no 1, 1973.*

41 GRIFFIN R L and T K MILES. *Relationship of asphalt properties to chemical constitution, J Chem Eng Data, vol 6, Jul 1961.*

42 GRIFFIN R L *et al. Influence of composition of paving asphalt on viscosity, Viscosity/ temperature, susceptibility and durability, J Chem Eng Data, vol 4, no 4, Oct 1959.*

43 AMERICAN SOCIETY for TESTING and MATERIALS. *The influence of asphalt composition on rheology, Papers on road and paving materials (bituminous), ASTM Special technical publication no 294, 1960.* ASTM, Philadelphia.

the asphaltenes were spherical, non-solvated entities. This suggests that the asphaltenes can interact with each other and/or the solvating medium. Even in a dilute toluene solution, the viscosity increase observed with increasing asphaltenes corresponds to a concentration of non-solvated spheres, some five times higher than the amount of asphaltenes used. Bitumen asphaltenes are believed to be stacks of plate-like sheets formed of aromatic/naphthenic ring structures. The viscosity of a solution, in particular a dilute solution, depends on the shape of the asphaltene particles. Size is important only if shape changes significantly as size increases. At high temperatures, the hydrogen bonds holding the sheets/stacks together are broken, resulting in a change in both the size and shape of the asphaltenes. Dissociation of the asphaltene entities continues until the limiting moiety, the unit sheet of condensed aromatic and naphthenic rings, is reached. Consequently, viscosity falls as temperature increases. However, as a hot bitumen cools, associations between asphaltenes occur to produce extended sheets. These, in turn, interact with other chemical types present (aromatics and resins) as well as stacking together to form discrete asphaltene particles.

The marked increase in non-Newtonian behaviour as a bitumen cools is a consequence of the inter-molecular and intra-molecular attractions between asphaltenes and other entities. Under shear, these extended associations will deform or even dissociate in a way that is not adequately described by classical Newtonian concepts. Consequently, at ambient and intermediate temperatures, it is reasonable to conclude that the rheology of bitumens is dominated by the degree of association of asphaltene particles and the relative amount of other species present in the system to stabilise these associations.

3.4 The relationship between broad chemical composition and physical properties

Atmospheric and vacuum distillation removes the lighter components from the bitumen feedstock. The loss of distillates leads to the preferential removal of saturates and concentration of asphaltenes. Air-blowing of bitumens from a given vacuum residue or fluxed vacuum residue results in a considerable increase in the asphaltene content and decrease in the aromatics content. Saturate and resin contents remain substantially of the same order as before blowing; this is shown graphically in Fig. 3.7. Table 3.3 shows a comparison of the chemical compositions between bitumens derived by distillation and blowing of the same feedstock.

Long-term full-scale road trials have been undertaken to determine if the chemical composition of bitumen changes with time using a full range of practical conditions, i.e. different mixture types, aggregates

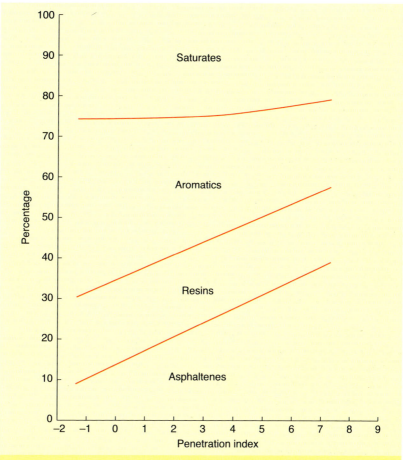

Fig. 3.7 Relationship between broad chemical composition and penetration index[44]

and bitumen contents[45]. The results of the studies are shown in Fig. 3.8 in terms of the ageing index (ratio of the viscosity of recovered bitumen, η_r, to the viscosity of the original bitumen, η_o, at 25°C) and also the chemical components. The major changes in viscosity are associated with the mixing and laying process. Changes in the viscosity of the binder are small over time. For the chemical composition, the asphaltene

44 LUBBERS H E. *Bitumen in de weg- en waterbouw, April 1985*. Nederlands Adviesbureau voor Bitumentoepassingen, Gouda.
45 CHIPPERFIELD E H, J L DUTHIE and R B GIRDLER. *Asphalt characteristics in relation to road performance, Proc Am Assoc Asph Pav Tech, vol 39, p 575, 1970.* Association Asphalt Paving Technologists, Seattle.

Table 3.3 Comparison of the broad chemical compositions of distilled and blown bitumens manufactured from a single short residue[46]

	Vacuum residue	Distillation				Blowing		
Penetration at 25 °C: dmm	285	185	99	44	12	84	46	9
Asphaltenes: %w	9·1	9·9	10·5	11·3	12·5	15·2	17·3	22·9
Resins: %w	18·6	16·7	18·2	17·7	21·3	21·0	22·1	21·5
Aromatics: %w	51·2	53·0	52·4	58·4	53·8	47·6	45·0	40·5
Saturates: %w	16·2	15·1	14·1	11·2	9·4	16·2	15·6	15·1

Note: The recovery of components is incomplete because of the techniques used.

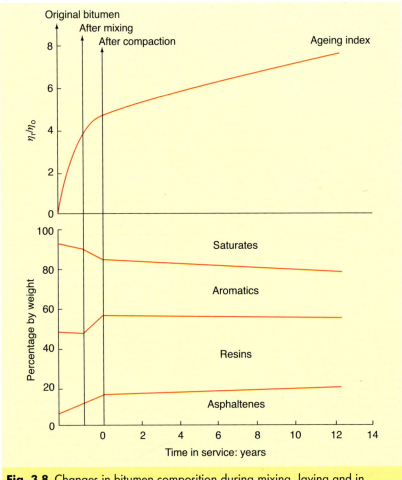

Fig. 3.8 Changes in bitumen composition during mixing, laying and in service[47]

content increases with mixing and shows a gradual increase with time. The resin and aromatic content decrease with time. Although little change was expected in the saturates content, some increase was noted probably due to oil spillage from vehicles on the road. Overall changes after mixing are very small even though the initial void contents of the mixtures studied were relatively high (5 to 8%). Recovered bitumen samples were obtained from the top 3 mm of cores extracted from the test sections and service temperatures were high.

Whilst chemical composition can be related to physical properties given specific components, it should be noted that bitumens of very different chemical compositions can have very similar physical properties if derived from different crude oils[48]. It is, therefore, impossible to describe bitumens generally in terms of chemical component concentrations and defining individual components (such as a minimum asphaltene content) has little, if any, relevance.

46 CHIPPERFIELD E H. *Bitumen production properties and uses in relation to occupational exposures, IARC Review on bitumen carcinogenicity, Institute of Petroleum Report IP 84-006, 1984.* IP, London.

47 CHIPPERFIELD E H, J L DUTHIE and R B GIRDLER. *Asphalt characteristics in relation to road performance, Proc Am Assoc Asph Pav Tech, vol 39, pp 575–613, 1970.* Association Asphalt Paving Technologists, Seattle.

48 CHIPPERFIELD E H, J L DUTHIE and R B GIRDLER. *Research on road bitumen, J Inst Mun Eng, vol 95, no 7, pp 216–222, Jul 1968.*

Chapter 4

Specifications and quality of bitumens

In the UK, bitumens are manufactured to three standards; BS EN 12591: 2000[49] (which applies throughout Europe), BS 3690-2: 1989[50] and BS 3690-3: 1983[51]. BS EN 12591: 2000 covers bitumens for road purposes, BS 3690-2: 1989 covers bitumens for industrial uses and BS 3690-3: 1983 covers blends of bitumen with coal tar, lake asphalt and pitch.

Four types of bitumen are manufactured by Shell Bitumen which are characterised by a combination of penetration, softening point and viscosity, namely

- penetration grades, characterised by penetration and softening point
- oxidised and hard grades, characterised by softening point and penetration
- cutback grades, characterised by viscosity.

In addition to the tests specified in the Standards cited above, Shell Bitumen carries out a further range of laboratory tests to ensure that bitumens are manufactured to a consistent standard maintaining suitability for purpose.

4.1 Penetration grade bitumens

Penetration grade bitumens are specified by the penetration[52] and

---

49 BRITISH STANDARDS INSTITUTION. *Bitumen and bituminous binders, Specifications for paving grade bitumens, BS EN 12591: 2000*. BSI, London.

50 BRITISH STANDARDS INSTITUTION. *Bitumens for building and civil engineering, Specification for bitumens for industrial purposes, BS 3690-2: 1989*. BSI, London.

51 BRITISH STANDARDS INSTITUTION. *Bitumens for building and civil engineering, Specification for bitumen mixtures, BS 3690-3: 1983*. BSI, London.

52 BRITISH STANDARDS INSTITUTION. *Methods of test for petroleum and its products, Bitumen and bituminous binders, Determination of needle penetration, BS EN 1426: 2000*. BSI, London.

The Shell Bitumen Handbook

Table 4.1 Specifications for paving grade bitumens with penetrations from 20 to 330 dmm[53]

	Unit	Test method	Grade designation										
			20/30	30/45	35/50	40/60	50/70	70/100	100/150	160/220	250/330
Penetration at 25°C	×0·1 mm	EN 1426	20–30	30–45	35–50	40–60	50–70	70–100	100–150	160–220	250–330
Softening point	°C	EN 1427	55–63	52–60	50–58	48–56	46–54	43–51	39–47	35–43	30–38
Resistance to hardening, at 163°C		EN 12607-1 or EN 12607-3									
– change of mass, maximum, ±	%		0·5	0·5	0·5	0·5	0·5	0·8	0·8	1·0	1·0
– retained penetration, minimum	%		55	53	53	50	50	46	43	37	35
– softening point after hardening, minimum	°C	EN 1427	57	54	52	49	48	45	41	37	32
Flash point, minimum	°C	EN 22592	240	240	240	230	230	230	230	220	220
Solubility, minimum	% (m/m)	EN 12592	99·0	99·0	99·0	99·0	99·0	99·0	99·0	99·0	99·0

44

softening point[54] tests. Designation is by penetration range only, e.g. 40/60 pen bitumen has a penetration that ranges from 40 to 60 inclusive and a softening point of 48 to 56°C. The unit of penetration is given as decimillimetre (dmm). This is the unit that is measured in the penetration test (discussed, along with the softening point test, in detail in Chapter 7). Notwithstanding, penetration grade bitumens are usually referred to without stating units. Table 1 of BS EN 12591: 2000 gives details of key performance parameters. It is summarised here as Table 4.1. It refers to a number of European Standards which are relevant to the performance of paving grade bitumens[55–58].

The majority of penetration grade bitumens produced are used in road construction. The tendency over the last decade of the twentieth century was to adopt harder bitumens that produce asphalts which have superior properties to those that are manufactured using the softer grades.

4.2 Oxidised bitumens

Oxidised bitumens are used almost entirely for industrial applications, e.g. roofing, flooring, mastics, pipe coatings, paints, etc. They are specified and designated by reference to both softening point and penetration tests, e.g. 85/40 is an oxidised grade bitumen with a softening point of 85 ± 5°C and a penetration of 40 ± 5 dmm. Oxidised bitumens also have to comply with solubility and loss on heating criteria. Table 1 of BS 3690-2: 1989 details the specification requirements for the six oxidised grade bitumens specified in the UK, reproduced here as Table 4.2. The softening points of oxidised grades of bitumen are much higher than those of the corresponding penetration grade bitumen and therefore

53 BRITISH STANDARDS INSTITUTION. *Bitumen and bituminous binders, Specifications for paving grade bitumens, BS EN 12591: 2000. BSI, London.*

54 BRITISH STANDARDS INSTITUTION. *Methods of test for petroleum and its products, Bitumen and bituminous binders, Determination of softening point, Ring and ball method, BS EN 1427: 2000. BSI, London.*

55 BRITISH STANDARDS INSTITUTION. *Bitumen and bituminous binders, Determination of the resistance to hardening under influence of heat and air, RTFOT method, BS EN 12607-1: 2000. BSI, London.*

56 BRITISH STANDARDS INSTITUTION. *Bitumen and bituminous binders, Determination of the resistance to hardening under influence of heat and air, RFT method, BS EN 12607-3: 2001. BSI, London.*

57 BRITISH STANDARDS INSTITUTION. *Methods of test for petroleum and its products, Petroleum products, Determination of flash and fire points, Cleveland open cup method, BS EN 22592: 1994. BSI, London.*

58 BRITISH STANDARDS INSTITUTION. *Methods of test for petroleum and its products, Bitumen and bituminous binders, Determination of solubility, BS EN 12592: 2000. BSI, London.*

Table 4.2 Specification for oxidised bitumens[59]

Property	Test method	Grade of bitumen					
		75/30	85/25	85/40	95/25	105/35	115/15
Softening point, °C	BS 2000: Part 58	75 ± 5	85 ± 5	85 ± 5	95 ± 5	105 ± 5	115 ± 5
Penetration at 25°C, dmm	BS 2000: Part 49	30 ± 5	25 ± 5	40 ± 5	25 ± 5	35 ± 5	15 ± 5
Loss on heating for 5 h at 163°C, loss by mass %, max	BS 2000: Part 45	0·2	0·2	0·5	0·2	0·5	0·2
Solubility in trichloroethylene, % by mass, min	BS 2000: Part 47	99·5	99·5	99·5	99·5	99·5	99·5

Note: At the time of publication of this book, no normalised European Standard existed for oxidised bitumens. Accordingly, BS 3690-2: 1989 remained applicable and all test methods referred to remain valid.

Table 4.3 Specification for hard bitumens[59]

Property	Test method	Grade of bitumen	
		H80/90	H100/120
Softening point, °C	BS 2000: Part 58	85 ± 5	110 ± 10
Penetration at 25 °C, dmm	BS 2000: Part 49	9 ± 3	6 ± 4
Loss on heating for 5 h at 163 °C, loss by mass %, max	BS 2000: Part 45	0·05	0·05
Solubility in trichloroethylene, % by mass, min	BS 2000: Part 47	99·5	99·5

Note: At the time of publication of this book, no normalised European Standard existed for hard bitumens. Accordingly, BS 3690-2: 1989 remained applicable and all test methods referred to remain valid.

the temperature susceptibility, i.e. the penetration index, is much higher, from +2 to +8.

4.3 Hard bitumens

Hard bitumens are also used entirely for industrial applications, e.g. coal briquetting, paints, etc. They are also specified by reference to both softening point and penetration tests but are designated by a softening point range only and a prefix H, e.g. H80/90 is a hard grade bitumen with a softening point between 80 and 90°C. The penetration index varies from 0 to +2. Table 4 of BS 3690-2: 1989 details the specification requirements of hard grade bitumens, and is reproduced here as Table 4.3.

4.4 Cutback bitumens

In the UK, BS EN 12591: 2000 replaced BS 3690-1: 1989 on 1 January 2002 with the latter being declared obsolescent from that date. However, the new Standard makes no reference to cutback bitumens. Since the UK continues to use significant volumes of cutback bitumens, BS 3690-1: 1989 will continue to be in use for the foreseeable future.

59 BRITISH STANDARDS INSTITUTION. *Bitumens for building and civil engineering, Specification for bitumens for industrial purposes, BS 3690-2: 1989*. BSI, London.

Table 4.4 Specification for cutback bitumens[60]

Property	Test method	Grade of cutback		
		50 sec	100 sec	200 sec
Viscosity (STV) at 40 °C, 10 mm cup, secs	BS 2000: Part 72	50 ± 10	100 ± 20	200 ± 40
Distillation	BS 2000: Part 27			
(a) distillate to 225 °C, % by volume, max		1	1	1
360 °C, % by volume		10 ± 3	9 ± 3	7 ± 3
(b) penetration at 25 °C of residue from distillation to 360 °C, dmm	BS 2000: Part 49	100 to 350	100 to 350	100 to 350
Solubility in trichloroethylene, % by mass, min	BS 2000: Part 47	99·5	99·5	99·5

Note: At the time of publication of this book, no normalised European Standard existed for cutback bitumens. Accordingly, BS 3690-2: 1989 remained applicable and all test methods referred to remain valid.

Cutback bitumens are manufactured by blending either 70/100 pen or 160/220 pen bitumen with kerosene to comply with a viscosity specification. In the UK, cutback bitumens are specified and designated by the flow time (in seconds) through a standard tar viscometer (STV)[61] (see Fig. 7.7). Three grades are available: 50 sec, 100 sec and 200 sec. The majority of cutback bitumen is used in surface dressing (discussed in detail in Chapter 19) but a significant amount is also used for the manufacture of both standard and deferred set asphalts. Table 2 of BS 3690-1: 1989 details the specification requirements for cutback bitumens and these are shown here in Table 4.4.

In addition to STV tests and solubility, cutback bitumens have to comply with a distillation specification[62] and a penetration requirement on the residual bitumen. This ensures that during application and in service, the diluent will evaporate at a consistent and predictable rate and that the residual bitumen will have the appropriate properties in service.

The suffix X on the SHELPHALT range of cutback bitumens indicates that they have been doped with a specially formulated heat-stable passive adhesion agent. This additive assists 'wetting' of the aggregate and resists stripping of the binder from the aggregate in the presence of water.

4.5 Bitumen quality

Over many years, Shell has investigated the relationship between laboratory measured properties of penetration grade bitumens and

60 BRITISH STANDARDS INSTITUTION. *Bitumens for building and civil engineering, Specification for bitumens for roads and other paved areas, BS 3690-1: 1989.* BSI, London.
61 BRITISH STANDARDS INSTITUTION. *Methods of test for petroleum and its products, Determination of viscosity of cutback bitumen, BS 2000-72: 1993.* BSI, London.
62 BRITISH STANDARDS INSTITUTION. *Methods of test for petroleum and its products, Determination of distillation characteristics of cutback bitumen, BS 2000-27: 1993.* BSI, London.

their performance in asphalt mixtures on the road. The ability to predict the long-term behaviour of asphalts becomes more important as traffic loading increases and performance requirements become ever more demanding. Performance on the road depends on many factors including the design, the nature of the application and the quality of the individual components. In volumetric terms, bitumen is a relatively minor component of an asphalt but it has a crucial role acting both as a durable binder and conferring visco-elastic properties on the asphalt.

Essentially, satisfactory performance of a bitumen on the road can be ensured if four properties are controlled:

- rheology
- cohesion
- adhesion
- durability.

The rheology of the bitumen at service temperatures is adequately characterised by the values of penetration and penetration index[63].

By studying the correlation between field performance and measured properties of experimental and commercial bitumens, Shell Global Solutions has developed a set of laboratory tests to assess the quality of a bitumen. The set of tests includes six on the bitumen and three on the asphalt. For ease of assessment, these nine results are presented in the form of a regular polygon, called the bitumen QUALAGON©, depicted in Fig. 4.1 and explained further in Section 4.5.3. There are tests within the QUALAGON© that cover the three remaining key performance elements:

- cohesion – low-temperature ductility
- adhesion – retained Marshall
- durability – rolling thin-film oven test (oxidation stability)
  true boiling point – gas liquid chromatography (volatility)
  exudation droplet test (homogeneity)
  field trials (oxidation stability).

4.5.1 Cohesion

The cohesive strength of bitumen is characterised by its ductility at low temperature, see Fig. 4.2. In this test, three 'dumb-bells' of bitumen are immersed in a water bath and stretched at a constant rate of 50 mm per minute until fracture occurs. The distance the specimen is stretched

---

63 DE BATS F TH and G VAN GOOSWILLIGEN. *Practical rheological characterisation of paving grade bitumens, Proc 4th Eurobitume Symp, Madrid, pp 304–310, Oct 1989.*

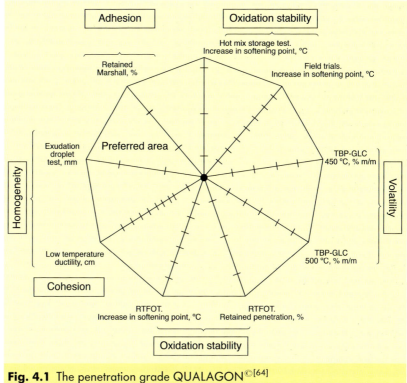

Fig. 4.1 The penetration grade QUALAGON©[64]

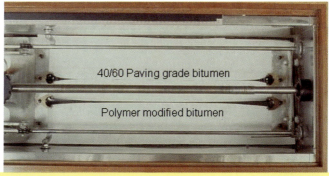

Fig. 4.2 The ductility test[65]

64 VAN GOOSWILLIGEN G, F TH DE BATS and T HARRISON. *Quality of paving grade bitumen – a practical approach in terms of functional tests, Proc 4th Eurobitume Symp, Madrid, pp 290–297, Oct 1989.*

65 AMERICAN SOCIETY for TESTING and MATERIALS. *Standard test method for ductility of bituminous materials, D113-99, 1 Nov 1999.* ASTM, Philadelphia.

before failure is reported as the ductility. The test temperature is adjusted, depending on the penetration of the bitumen under test, e.g. 10°C for 80 to 100 pen, 13°C for 60 to 70 pen and 17°C for 40 to 50 pen. Under these conditions, the test has been found to discriminate between bitumens of different cohesive strengths.

4.5.2 Adhesion

In the QUALAGON©, the adhesion characteristics of a bitumen are assessed by a retained Marshall test[66]. In the test, eight Marshall specimens are manufactured using a prescribed aggregate, aggregate grading and bitumen content. The eight specimens are divided into two groups of four such that the average void content of the two groups are equal. One group of four is tested immediately using the standard Marshall test and the remaining four specimens are vacuum treated under water at a temperature of between 0 and 1°C to saturate, as far as possible, the pore volume of the mixture with water. Subsequently, the specimens are stored in a water bath at 60°C for 48 hours. The Marshall stability of these four specimens is then determined. The ratio of the Marshall stability of the treated specimens to the initial Marshall stability is termed the 'retained Marshall stability'.

Although a relatively wide scatter in absolute stability values and poor reproducibility are inherent in Marshall testing, it has been found that consideration of the percentage relative to the result from the standard procedure reduces the differences between results from different laboratories.

The results of laboratory tests together with observations of performance in practice have identified key links between functional properties and the constitution of the bitumen. This work has indicated that if the molecular weight distribution and chemical constitution of the bitumen is unbalanced, it can exhibit inhomogeneity which may adversely affect both the cohesive and adhesive properties of the bitumen[67].

4.5.3 Durability

Durability can be defined as the ability to maintain satisfactory rheology, cohesion and adhesion in long-term service. As a part of the quality criteria for bitumen, the following have been identified as the prime

---

66 MARSHALL CONSULTING & TESTING LABORATORY. *The Marshall method for the design and control of bituminous paving mixtures, 1949*. Marshall Consulting and Testing Laboratory, Mississippi.

67 VAN GOOSWILLIGEN G, F TH DE BATS and T HARRISON. *Quality of paving grade bitumen – a practical approach in terms of functional tests, Proc 4th Eurobitume Symp, Madrid, pp 290–297, Oct 1989*.

Fig. 4.3 The rolling thin-film oven test[68]

durability factors:

- oxidative hardening
- evaporative hardening
- exudative hardening.

Oxidative and evaporative hardening
Hardening due to oxidation has long been held to be the main cause of ageing. In mixtures where the finely structured voids are interconnected, hardening in situ by oxidation is important. However, under such circumstances, hardening due to evaporation of the more volatile components of the bitumen can also be significant.

Tests on the bitumen   The rolling thin-film oven test[68] (RTFOT) is an ageing test and measures hardening by both oxidation and evaporation. The apparatus for this test is shown in Fig. 4.3. In this test, a thin film of bitumen is continuously rotated around the inner surface of a glass jar at

68 AMERICAN SOCIETY FOR TESTING and MATERIALS. *Standard test method for effect of heat and air on a moving film of asphalt (Rolling thin-film oven test)*, *D2872-97, 1 Nov 1997*. ASTM, Philadelphia.

163°C for 75 minutes, with an injection of hot air into the jar every 3 to 4 seconds. The amount of bitumen hardening during the test, i.e. drop in penetration and increase in softening point, correlates well with that observed during the manufacture of an asphalt. The oxidation stability of bitumens is plotted on the QUALAGON[©] as the percentage retained penetration and the increase in softening point.

Evaporative hardening can be assessed by measuring the front-end volatility of the bitumen by true boiling point gas liquid chromatography (TBP-GLC). In this test, a small sample of bitumen (150 mg) is dissolved in carbon disulphide and is separated on two chromatographic columns. The first column separates the heavy end components such as asphaltenes, heavy polar aromatics, etc. from the bitumen. The hydrocarbons eluted from this column are then separated on a second column.

TBP-GLC is a rapid and accurate method of front-end volatility analysis and is plotted on the QUALAGON[©] as the percentage by mass recovered at 450 and 500°C, thus taking account of the shape of the volatility curve. The loss of 0.2% of the mass in the RTFOT correlates well with the TBP-GLC recovery limits at 450 and 500°C[69].

Tests on the mixture   The hardening due to oxidation and evaporation of a thin film of bitumen in contact with aggregate is assessed by two mixture tests: the hot mixture storage test[69] and the change in the softening point of the bitumen during commercial asphalt manufacture. The hot mixture storage test simulates ageing conditions during mixing and hot storage. A prescribed mixture is manufactured in the laboratory and a specified quantity of this mixture is stored for 16 hours at 160°C in a sealed tin. Thus, the volume of air entrained in the sample is known and is constant from test to test. The bitumen is recovered from both the mixed and stored material and the penetration and softening points are determined from samples of bitumen recovered from these two materials. The ageing of the bitumen during mixing and storage is expressed as the difference between the softening point of the bitumen after storage and that of the original bitumen.

At laboratory scale, this is a very severe test and the change in softening point is very much larger than that which would be found during actual bulk storage. Nevertheless, the test correlates with the hardening tendency of a mixture at high temperature when in prolonged contact with air.

---

69 VAN GOOSWILLIGEN G, F TH DE BATS and T HARRISON. *Quality of paving grade bitumen – a practical approach in terms of functional tests, Proc 4th Eurobitume Symp, Madrid, pp 290–297, Oct 1989.*

Exudative hardening

If the constitution of a bitumen is unbalanced, it may, when in contact with a porous aggregate, exude an oily component into the surface pores of the aggregate. This may result in a hardening of the bitumen film remaining on the surface of the aggregate. Exudation is primarily a function of the amount of low molecular weight paraffinic components present in the bitumen relative to the amount and type of asphaltenes.

Shell Global Solutions has developed the exudation droplet test[70] to quantitatively measure the exudation tendency of a bitumen. In this test, bitumen droplets are applied to the recesses in custom-made white marble plates. The plates are stored at 60°C for 4 days under a nitrogen blanket. During this period, oily rings develop around the bitumen droplet that can be measured under ultraviolet light using a microscope. Ring widths vary from a few tenths of a millimetre for a balanced bitumen to several millimetres for an unbalanced bitumen.

Hardening in service as a result of exudation can be substantial and depends not only on the exudation tendency of the bitumen but also on the porosity of the aggregate. If the aggregate possesses low porosity, the quantity of exudate absorbed is negligible, irrespective of the exudation tendency of the bitumen. Similarly, if the exudation tendency of the bitumen is low, the quantity of exudate absorbed will be negligible, irrespective of the porosity of the aggregate.

4.5.4 Verification of the QUALAGON© tests and test criteria

Many years' experience using the QUALAGON© in conjunction with rheological data have confirmed that the set of criteria upon which it is based provides a satisfactory quantitative measure of bitumen quality and its performance in service.

Good-quality bitumens generally have properties within the 'preferred' area of the QUALAGON©. However, this does not mean that a bitumen which is partially outside the preferred area is necessarily a bad bitumen or one of poor quality. The QUALAGON© limits are not pass/fail criteria and the whole figure must be interpreted with care and judgement. For this reason, the QUALAGON© is not suitable for specification purposes but it is an invaluable assessment tool providing an excellent guide to performance in service.

---

70 VAN GOOSWILLIGEN G, F TH DE BATS and T HARRISON. *Quality of paving grade bitumen – a practical approach in terms of functional tests, Proc 4th Eurobitume Symp, Madrid, pp 290–297, Oct 1989.*

4.6 The CEN bitumen specifications

Harmonisation of European Standards for petroleum products was set as a target by the 'Comité Européen de Normalisation' (CEN) during the mid 1980s to eliminate barriers to trade within the member states of the European Union. One of the first steps in achieving this aim was the publication of the 'Construction Products Directive' (CPD) in December 1988. The CPD requires that construction products used in member states must be fit for their intended use, satisfying certain essential requirements:

- mechanical resistance and stability
- safety in case of fire
- hygiene, health and environmental criteria
- safety in use
- protection against noise
- energy, economy and heat retention.

These essential requirements have to be taken into account when drafting European Standards (often described as 'ENs'). The task of producing ENs has been entrusted to the CEN and, in turn, to the CEN technical committees (TCs) and working groups (WGs). It was the CEN working group dealing with paving grade bitumens, TC227 WG1, which proposed a pan-European specification that has resulted in the publication of BS EN 12591: 2000[71]. This is based on the traditional tests that are used for characterising bitumens. The specification, shown in Table 4.5, includes a series of mandatory tests and properties that must be adopted by all members of the EC. It also includes optional properties that can be adopted or rejected by each national body (for the UK this is BSI Subcommittee B/510/19, Bitumen and related products). Of the optional tests, the UK has adopted kinematic viscosity at 135°C as a mandatory test.

4.7 The SHRP/SUPERPAVE bitumen specification

The Strategic Highways Research Program (SHRP), initiated in the USA in 1987, was a coordinated effort to produce rational specifications for bitumens and asphalts based on performance parameters. The motivation was to produce pavements which performed well in service. These pavements were subsequently called 'SUPERPAVE' (SUperior PERforming PAVEments).

One of the results of this work was the 'Superpave asphalt binder specification' which categorises grades of bitumen according to their

---

71 BRITISH STANDARDS INSTITUTION. *Bitumen and bituminous binders, Specifications for paving grade bitumens, BS EN 12591: 2000.* BSI, London.

Table 4.5 CEN specification for paving grade bitumens[71]

	Grade								
	20/30	30/45	35/50	40/60	50/70	70/100	100/150	160/220	250/330
MANDATORY PROPERTIES									
Penetration at 25°C, ×0·1 mm	20–30	30–45	35–50	40–60	50–70	70–100	100–150	160–220	250–330
Softening point, °C	55–63	52–60	50–58	48–56	46–54	43–51	39–47	35–43	30–38
Resistance to hardening:									
Change of mass, % maximum	±0·5	±0·5	±0·5	±0·5	±0·5	±0·8	±0·8	±1·0	±1·0
Retained penetration, % minimum	55	53	53	50	50	46	43	37	35
Softening point after hardening, °C, minimum	57	54	52	49	48	45	41	37	32
Flash point, °C, minimum	240	240	240	230	230	230	230	220	220
Solubility, % minimum	99·0	99·0	99·0	99·0	99·0	99·0	99·0	99·0	99·0
OPTIONAL PROPERTIES									
Paraffin wax content, % maximum	2·2 / 4·5	2·2 / 4·5	2·2 / 4·5	2·2 / 4·5	2·2 / 4·5	2·2 / 4·5	2·2 / 4·5	2·2 / 4·5	2·2 / 4·5
Dynamic viscosity at 60°C, Pa s, minimum	440	260	225	175	145	90	55	30	18
Kinematic viscosity at 135°C, mm²/s, minimum	530	400	370	325	295	230	175	135	100
Fraass breaking point, °C, maximum	−5	−5	−5	−7	−8	−10	−12	−15	−16
Resistance to hardening:									
(Note one of the following alternatives may be chosen)									
(1) Increase in softening point, °C, maximum	8	8	8	9	9	9	10	11	11
(2) Increase in softening point, °C, maximum	10	11	11	11	11	11	12	12	12
and Fraass breaking point, °C, maximum	−5	−5	−5	−7	−8	−10	−12	−15	−16
(3) Increase in softening point, °C, maximum	10	11	11	11	11	11	12	12	12
and penetration index	−1·5 min +0·7 max	−1·5 min +0·7 max	−1·5 min +0·7 max	−1·5 min +0·7 max	−1·5 min +0·7 max	−1·5 min +0·7 max	−1·5 min +0·7 max	−1·5 min +0·7 max	−1·5 min +0·7 max
(4) Increase in softening point, °C, maximum	10	11	11	11	11	11	12	12	12

Table 4.6 The complete SUPERPAVE binder specification[72]

	Performance grade																	
	PG 52							**PG 58**					**PG 64**					
	−10	−16	−22	−28	−34	−40	−46	−16	−22	−28	−34	−40	−10	−16	−22	−28	−34	−40
Average 7-day maximum pavement design temperature, °C	<52							<58					<64					
Minimum pavement design temperature, °C	>−10	>−16	>−22	>−34	>−34	>−40	>−46	>−16	>−22	>−28	>−34	>−40	>−10	>−16	>−22	>−28	>−34	>−40
Original binder																		
Minimum flash point, °C	230																	
Viscosity, ASTM D 4402 Max 3 Pa s, test temperature, °C	135																	
Dynamic shear TP5: G'/sin δ, minimum, 1·00 kPa Test temperature @ 10 rad/s, °C	52							58					64					
Binder after rolling thin-film oven test																		
Mass loss, maximum, %	1.00																	
Dynamic shear TP5: G'/sin δ, minimum, 2·20 kPa Test temperature @ 10 rad/s, °C	52							58					64					
Binder after pressure ageing vessel																		
PAV ageing temperature, °C	90							100					100					
Dynamic shear TP5: G'/sin δ, maximum, 5000 kPa Test temperature @ 10 rad/s, °C	25	22	19	16	13	10	7	25	22	19	16	13	31	28	25	22	19	16
Physical hardening	**Report**																	
Creep stiffness, TP1: S, maximum, 300 MPa m-value, maximum, 0·30 Test temperature @ 60s, °C	0	−6	−12	−18	−24	−30	−36	−6	−12	−18	−24	−30	0	−6	−12	−18	−24	−30
Direct tension, TP3: Failure strain, minimum, 1·0% Test temperature @ 1·0 mm/min, °C	0	−6	−12	−18	−24	−30	−36	−6	−12	−18	−24	−30	0	−6	−12	−18	−24	−30

performance characteristics in different environmental conditions. The specification was intended to limit the potential of a bitumen to contribute to permanent deformation, fatigue failure and low-temperature cracking of asphalt pavements.

Table 4.6 shows the complete SUPERPAVE binder specification. It is intended to control permanent deformation, low-temperature cracking and fatigue in asphalt pavements. This is achieved by controlling various physical properties measured with the equipment described above. In this specification, the physical properties remain constant for all grades but the temperature at which these properties must be achieved varies depending on the climate in which the binder is to be used. For example, a PG 52-40 grade is designed to be used in an environment where the average seven day maximum pavement temperature is 52°C and a minimum pavement design temperature of −40°C.

4.7.1 Rotational viscometer

The rotational viscometer is specified to ensure that the viscosity of the bitumen, at normal application temperatures, will be pumpable, coat the aggregate and enable satisfactory compaction of the asphalt.

4.7.2 The bending beam rheometer

The bending beam rheometer[73] (BBR) is a simple device which measures how much a beam of bitumen will deflect under a constant load at temperatures corresponding to its lowest pavement service temperature when bitumen behaves like an elastic solid. The creep load is intended to simulate the stresses that gradually increase in a pavement as the temperature falls.

Two parameters are determined in this test, the creep stiffness is a measure of the resistance of the bitumen to constant loading and the creep rate is a measure of how the asphalt stiffness changes as loads are applied.

If the creep stiffness is too high, the asphalt will behave in a brittle manner and cracking will be more likely. A high creep rate (sometimes denoted as the 'm' value) is desirable because, as the temperature changes and thermal stresses accumulate, the stiffness will change relatively quickly. A high value of creep rate indicates that the binder will tend

---

72 ASPHALT INSTITUTE. *Superpave performance graded asphalt binder specification and testing, Superpave Series No 1, pp 15–19, 1997.* AI, Kentucky.

73 BAHIA H U, D A ANDERSON and D W CHRISTENSEN. *The bending beam rheometer: a simple device for measuring low-temperature rheology of asphalt binders, Proc Assoc Asph Pav Tech, pp 117–135, 1991.* Association of Asphalt Paving Technologists, Seattle.

to disperse stresses that would otherwise accumulate to a level where low-temperature cracking could occur.

The dynamic shear rheometer (DSR)[74] (see Section 7.6.2) and BBR provide information relating to the stiffness behaviour of bitumen over a wide range of temperatures. Although stiffness can be used to estimate failure properties, for modified binders the relation between stiffness and failure is less well defined. Accordingly, an additional test to measure strength and strain at break has been included, the direct tension test.

4.7.3 The direct tension test

Bitumens that can be stretched long distances before failure are termed 'ductile' and those that break before significant stretching has occurred are termed 'brittle'. Unfortunately, the BBR is unable to fully characterise the ability of some bitumens to stretch before failure. The direct tension test[75] measures ultimate tensile strain of a bitumen at low temperatures (between 0 and $-36°C$). A dumb-bell specimen is loaded in tension at a constant rate and the failure strain, the change in length divided by the original length, is determined. In the test, failure is defined by the stress where the load on the specimen reaches its maximum value. At this failure stress, the minimum strain at failure must be 1%.

4.7.4 Bitumen ageing tests

In the SUPERPAVE specification, two ageing tests are carried out, the rolling thin-film oven test[76] (RTFOT) which simulates ageing during asphalt manufacture and the pressure ageing vessel[77] (PAV) which is believed to simulate in-service ageing.

The PAV test was developed to evaluate long-term in-service ageing. The device uses pressure and temperature to compress time so that very long-term ageing can be simulated in only 20 hours. A sample of the bitumen under test is aged in the RTFOT and then placed in the PAV for further ageing. In the PAV test, bitumen in a thin film on a tray is placed in a sample rack in the PAV and the material is aged at a pressure

74 ASPHALT INSTITUTE. *Superpave performance graded asphalt binder specification and testing, Superpave series no 1, pp 15–19, 1997*. AI, Kentucky.

75 AMERICAN ASSOCIATION OF STATE HIGHWAY OFFICIALS. *Standard test method for determining the fracture properties of asphalt binder in direct tension, D6723-02*. AASHTO, Washington DC.

76 AMERICAN SOCIETY FOR TESTING and MATERIALS. *Standard test method for effect of heat and air on a moving film of asphalt (Rolling thin-film oven test), D2872-97, 1 Nov 1997*. ASTM, Philadelphia.

77 AMERICAN SOCIETY FOR TESTING and MATERIALS. *Standard practice for accelerated aging of asphalt binder using a pressurized aging vessel (PAV), D6521-00, 1 Nov 2000*. ASTM, Philadelphia.

of 2070 kPa at a temperature of 90, 100 or 110°C. (Within the industry there is considerable concern that ageing materials at temperatures in this range is too far removed from the temperature experienced in service and that such an approach may give misleading results.) After the PAV test, the residue is used for DSR, BBR and direct tension tests.

Chapter 5

Polymer modified and special bitumens

On the majority of roads, conventional asphalts perform well. However, demands made upon roads increase year by year. Ever increasing numbers of commercial vehicles with super single tyres and increased axle loads take their toll and it is clear that this trend will continue in the future. To assist the Highway Engineer to meet this growing challenge, there now exists a wide range of proprietary asphalts made with polymer modified bitumens and a range of polymer modified bitumens for generic asphalts, all of which have been proven in service.

Good basic design, stage construction and timely maintenance enable asphalt roads not only to withstand all the demands that are made of them but also to provide the high standards of safety and comfort that have become synonymous with bitumen-bound road materials. However, certain areas of a road are much more highly stressed than others. Accordingly, if the entire network is to perform satisfactorily during the design life, these areas may require special attention. There is a general awareness of the need for special materials on bridge decks, for example, but the needs of other critical areas can often be overlooked during construction or maintenance. It is clearly advantageous that maintenance or strengthening work be required along the entire length at a given point in time, rather than attention being repeatedly required at certain locations at shorter intervals. Disruption to traffic flow during maintenance work constitutes a significant proportion of the cost of undertaking roadworks. The use of modified bitumens offers a solution to reducing the frequency of maintenance required at particular locations and provides a much longer service life for maintenance treatments at difficult sites.

The degree of improvement required, and hence the cost, will depend upon the needs of the site to be treated. Modest improvements in deformation resistance can be achieved using reclaimed tyre rubber. Sulphur and thermoplastic polymers can improve both the workability of the asphalt during compaction and its deformation resistance in service and specially processed binders such as Multiphalte offer substantially improved deformation resistance. Moving further up the performance curve thermoplastic elastomers improve both deformation resistance and fatigue characteristics and thermosetting systems offer the ultimate performance of exceptional deformation resistance coupled with outstanding flexibility.

5.1 The role of bitumen modifiers in asphalt

As the bitumen is responsible for the visco-elastic behaviour characteristic of asphalt, it plays a large part in determining many aspects of road performance, particularly resistance to permanent deformation and cracking. In general, the proportion of any induced strain in asphalt that is attributable to viscous flow, i.e. non-recoverable, increases with both loading time and temperature. The effect of this is illustrated in Figs. 5.1(a) and (b). Figure 5.1(a) shows the response of an asphalt sample in a simple creep test (this is discussed further in Chapter 16). The strain resulting from the applied loading shows an instantaneous elastic response followed by a gradual increase in strain with time until the load is removed. The change in strain with time is caused by the viscous behaviour of the material. On removal of the load, the elastic strain is recovered instantaneously and some additional recovery occurs with time. This is known as 'delayed elasticity'. Ultimately, a permanent residual strain remains, which is irrecoverable and is directly caused by viscous behaviour.

The response to a load pulse induced in an element of an asphalt due to moving traffic loads is shown in Fig. 5.1(b). Here it is not possible to distinguish between the two components of elastic response but the small permanent strain and larger elastic strain are shown. Although the permanent strain illustrated in Fig. 5.1(b) is small for a single pulse load, when many millions of load applications are applied to a pavement, a large accumulation will develop. It is this component that results in surface deformation. From the above, it is clear why more deformation occurs at high ambient temperatures and where traffic is slow moving or stationary.

One of the prime roles of a bitumen modifier is to increase the resistance of the asphalt to permanent deformation at high road temperatures without adversely affecting the properties of the bitumen or asphalt at other temperatures. This is achieved by one of two methods, both of which result in a reduction in permanent strain. The first approach is to stiffen the bitumen so that the total visco-elastic response of the asphalt is reduced. The second option is to increase the elastic component of the bitumen thereby reducing the viscous component.

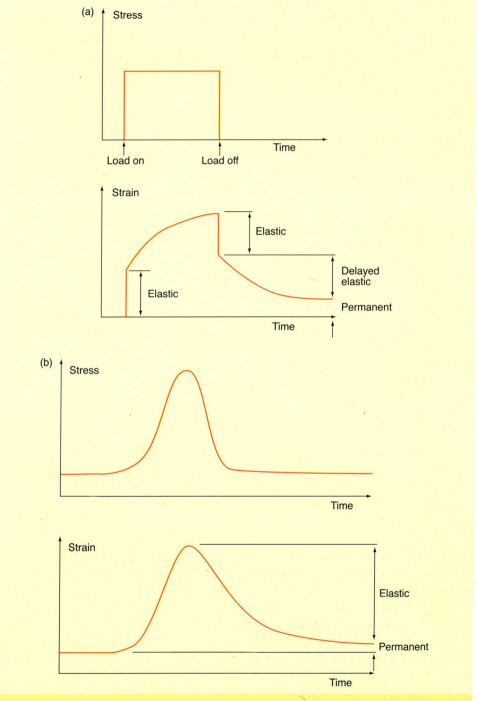

Fig. 5.1 Visco-elastic responses of an asphalt under (a) a static load and (b) a moving wheel load

Increasing the stiffness of the bitumen is also likely to increase the dynamic stiffness of the asphalt. This will improve the load-spreading capability of the material, increase the structural strength and lengthen the expected design life of the pavement. Alternatively, it may be possible to achieve the same structural strength but with a thinner layer. Increasing the elastic component of the bitumen will improve the flexibility of the asphalt; this is important where high tensile strains are induced.

5.2 The modification of bitumen

For many years, researchers and development chemists have experimented with modified bitumens, mainly for industrial uses, adding asbestos, special fillers, mineral fibres and rubbers. In the last thirty years many researchers have looked at a wide spectrum of modifying materials for bitumens used in road construction. Table 5.1 details the majority of bitumen modifiers and additives that have been examined.

Table 5.2 contains a summary of some of the modifiers used with bitumens and asphalts in service situations and (✓) indicates where they enhance the performance of the asphalt.

For the modifier to be effective and for its use to be both practicable and economic, it must[78]:

- be readily available
- resist degradation at asphalt mixing temperatures
- blend with bitumen
- improve resistance to flow at high road temperatures without making the bitumen too viscous at mixing and laying temperatures or too stiff or brittle at low road temperatures
- be cost effective.

The modifier, when blended with bitumen, should:

- maintain its premium properties during storage, application and in service
- be capable of being processed by conventional equipment
- be physically and chemically stable during storage, application and in service
- achieve a coating or spraying viscosity at normal application temperatures.

---

78 DENNING J H and J CARSWELL. *Improvements in rolled asphalt surfacings by the addition of organic polymers, Laboratory Report 989, 1981.* Transport and Road Research Laboratory, Crowthorne.

Table 5.1 Some additives used to modify bitumen

Type of modifier	Example
Thermoplastic elastomers	Styrene–butadiene–styrene (SBS)
	Styrene–butadiene–rubber (SBR)
	Styrene–isoprene–styrene (SIS)
	Styrene–ethylene–butadiene–styrene (SEBS)
	Ethylene–propylene–diene terpolymer (EPDM)
	Isobutene–isoprene copolymer (IIR)
	Natural rubber
	Crumb tyre rubber
	Polybutadiene (PBD)
	Polyisoprene
Thermoplastic polymers	Ethylene vinyl acetate (EVA)
	Ethylene methyl acrylate (EMA)
	Ethylene butyl acrylate (EBA)
	Atactic polypropylene (APP)
	Polyethylene (PE)
	Polypropylene (PP)
	Polyvinyl chloride (PVC)
	Polystyrene (PS)
Thermosetting polymers	Epoxy resin
	Polyurethane resin
	Acrylic resin
	Phenolic resin
Chemical modifiers	Organo-metallic compounds
	Sulphur
	Lignin
Fibres	Cellulose
	Alumino-magnesium silicate
	Glass fibre
	Asbestos
	Polyester
	Polypropylene
Adhesion improvers	Organic amines
	Amides
Antioxidants	Amines
	Phenols
	Organo-zinc/organo-lead compounds
Natural asphalts	Trinidad Lake Asphalt (TLA)
	Gilsonite
	Rock asphalt
Fillers	Carbon black
	Hydrated lime
	Lime
	Fly ash

Table 5.2 Benefits of different types of modifiers[79]

Modifier	Permanent deformation	Thermal cracking	Fatigue cracking	Moisture damage	Ageing
Elastomers[80–82]	✓	✓	✓		✓
Plastomers[83,84]	✓				
Tyre rubber[85]		✓	✓		
Carbon black[86,87]	✓				✓
Lime[88,89]				✓	✓
Sulphur[90,91]	✓				
Chemical modifiers	✓				
Antioxidants					✓
Adhesion improvers				✓	✓
Hydrated lime[92]				✓	✓

79 BROWN S F, R D ROWLETT and J L BOUCHER. *Asphalt modification, Highway research: sharing the benefits. Proceedings of the conference: The United States strategic highway research program, pp 181–203, 1990.* Thomas Telford, London.

80 LITTLE D N, J W BUTTON, R M WHITE, E K ENSLEY, Y KIM and S J AHMED. *Investigation of asphalt additives, FHWA/RD-87-001, Nov 1986.* Texas Transportation Institute, Texas A&M University, College Station, Texas,

81 CALTABIANO M A. *Reflection cracking in asphalt overlays, M Phil Thesis, University of Nottingham, 1990.*

82 MONISMITH C L and A A TAYEBALI. *Behaviour of mixes containing a conventional and a polymer modified (styrelf) asphalt, University of California at Berkeley, Feb 1988.*

83 TREMELIN S S. *Neoprene modified asphalt, E I Du Pont de Nemours & Company Inc, 121st Meeting, Philadelphia, Pennsylvania, May 1982.*

84 SHULER T S and R D PAVLOVICH. *Characterization of polymer modified binders, New Mexico State Highway Department, Research Report 52001-1F, Jan 1987.*

85 SCHNORMEIER R H. *Time proves asphalt–rubber seals cost effective, National seminar on asphalt–rubber, San Antonio, Texas, 1981.*

86 ROSTLER F S R, R M WHITE and E M DANNENBURG. *Carbon black as a reinforcing agent for asphalt, Proc Assoc Asph Pav Tech, vol 55, pp 376–401, 1986.* Association of Asphalt Paving Technologists, Seattle.

87 YAO Z and C L MONISMITH. *Behaviour of asphalt mixtures with carbon black reinforcement, Proc Assoc Asph Pav Tech, vol 55, pp 564–585, 1986.* Association of Asphalt Paving Technologists, Seattle.

88 KIM O K, J MONTALVO, C A BELL, R G HICKS, and J E WILSON. *Effect of moisture and aging on asphalt pavement life: Part 1 – Effect of moisture, Oregon State University, Department of Civil Engineering, Corvallis, Oregon, Report no FHWA-OR-RD-86-01, Jan 1986.*

89 PLANCHER H, E L GREEN, and J C PETERSEN. *Reduction to oxidative hardening of asphalts by treatment with hydrated lime, a mechanistic study, Proc Assoc Asph Pav Tech, vol 45, pp 1–24, 1976.* Association of Asphalt Paving Technologists, Seattle.

90 DENNING J H and J CARSWELL. *Improvements in rolled asphalt surfacings by the addition of sulphur, Laboratory Report 963, 1981.* Transport and Road Research Laboratory, Crowthorne.

91 DEME I. *Shell sulphur asphalt products and processes, International Road Federation Symp on sulphur asphalt in road construction, Bordeaux, France, Feb 1981.*

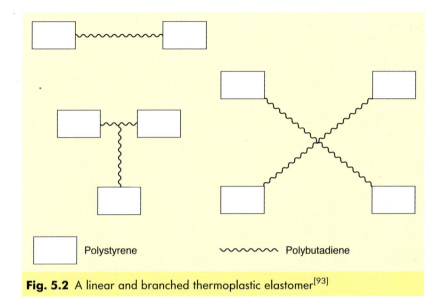

Polystyrene   Polybutadiene

Fig. 5.2 A linear and branched thermoplastic elastomer[93]

5.2.1 The modification of bitumen using thermoplastic elastomers

Of the four main groups of thermoplastic elastomers – polyurethane, polyether–polyester copolymers, olefinic copolymers and styrenic block copolymers – it is the styrenic block copolymers that have proved to present the greatest potential when blended with bitumen[93–95].

Styrenic block copolymers, also termed thermoplastic rubbers or thermoplastic elastomers, may be produced by a sequential operation of successive polymerisation of styrene–butadiene–styrene (SBS) or styrene–isoprene–styrene (SIS). Alternatively, a di-block precursor can be produced by successive polymerisation of styrene and mid-block monomer, followed by a reaction with a coupling agent. Thus, not only linear copolymers but also multi-armed copolymers can be produced; these are often referred to as radial or branched copolymers, as shown in Fig. 5.2.

92 KENNEDY T W, F L ROBERTS and K W LEE. *Evaluation of asphalt mixtures using the Texas freeze–thaw pedestal test, Proc Assoc Asph Pav Tech, vol 51, pp 327–341, 1982.* Association of Asphalt Paving Technologists, Seattle.

93 BULL A L and W C VONK. *Thermoplastic rubber/bitumen blends for roof and road, Shell Chemicals Technical Manual TR 8.15, 1984.* Shell International Petroleum Company Ltd, London.

94 HENDERSON J F and M SZWARC. *The use of living polymers in the preparation of polymer structures of controlled architecture, Rev Macromolecular Chem, 1968, vol 3, pp 317–401.*

95 BONEMAZZI *et al. Correlation between the properties of polymers and polymer modified bitumens, Eurobitume and Eurasphalt Conf, Strasbourg, Brussels, 1996.*

Thermoplastic elastomers derive their strength and elasticity from a physical cross-linking of the molecules into a three-dimensional network. This is achieved by the agglomeration of the polystyrene end-blocks into separate domains, as shown schematically in Fig. 5.3, providing the physical cross-links for a three-dimensional polybutadiene or polyisoprene

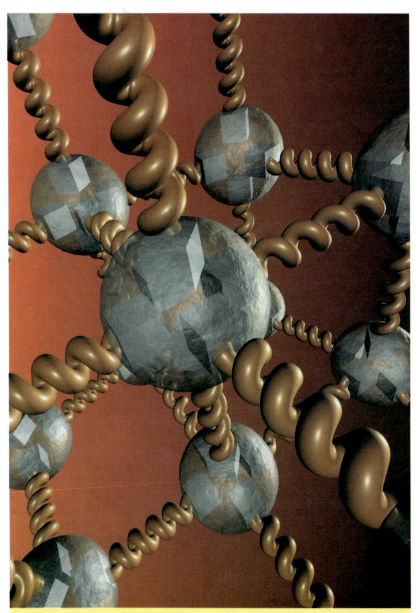

Fig. 5.3 Schematic structure of thermoplastic elastomers at ambient temperature

rubbery matrix. It is the polystyrene end-blocks that impart strength to the polymer and the mid-block that gives the material its exceptional elasticity[96]. At temperatures above the glass transition point of polystyrene (100°C) the polystyrene softens as the domains weaken and will even dissociate under stress, thus allowing easy processing. Upon cooling, the domains re-associate and the strength and elasticity is restored, i.e. the material is thermoplastic.

The quality of the polymer dispersion achieved is influenced by a number of factors:

- the constitution of the bitumen
- the type and concentration of the polymer
- the shear rate applied by the mixer[97].

When the polymer is added to the hot bitumen, the bitumen immediately starts to penetrate the polymer particles causing the styrene domains of the polymer to become solvated and swollen. Once this has occurred, the level of shear exerted on the swollen particles is critical if a satisfactory dispersion is to be achieved within a realistic blending time. Thus, medium or, preferably, high shear mixers are required to adequately disperse thermoplastic elastomers into the bitumen.

As discussed in Chapter 3, bitumens are complex mixtures that can be subdivided into groups of molecules that have common structures:

- saturates
- aromatics
- resins
- asphaltenes.

Saturates and aromatics can be viewed as carriers for the polar aromatics, i.e. the resins and asphaltenes. The polar aromatics are responsible for the visco-elastic properties of the bitumen at ambient temperatures. This is due to association of the polar molecules that leads to large structures, in some cases even to three-dimensional networks, i.e. 'GEL' type bitumen. The degree to which this association takes place depends on the temperature, the molecular weight distribution, the concentration of the polar aromatics and on the solvency power of the saturates and aromatics in the maltenes phase. If the concentration and molecular weight of the asphaltenes is relatively low, the result will be a 'SOL' type bitumen.

96 VONK W C and G VAN GOOSWILLIGEN. *Improvement of paving grade bitumens with SBS polymers, Proc 4th Eurobitume Symp, Madrid, pp 298–303, 1989.*

97 COLLINS J H, M G BOULDIN, R GELLES and A BERKER. *Improved performance of paving asphalt by polymer modification, Proc Assoc Asph Pav Tech, vol 60, pp 43–79, 1991.* Association of Asphalt Paving Technologists, Seattle.

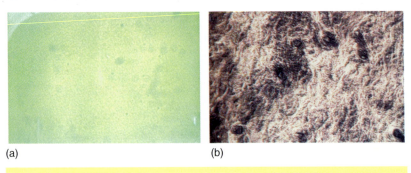

(a)   (b)

Fig. 5.4 The microscopic structures of (a) compatible and (b) incompatible bitumen/polymer systems

important the compatibility

   The addition of thermoplastic elastomers with a molecular weight similar to or higher than that of the asphaltenes disturbs the phase equilibrium. The polymer and the asphaltene 'compete' for the solvency power of the maltenes phase and if insufficient maltenes are available, phase separation may occur. Figure 5.4 shows the microscopic structures of (a) 'compatible' and (b) 'incompatible' bitumen/polymer systems. The compatible system has a homogeneous 'sponge-like' structure whereas the incompatible system has a coarse discontinuous structure. Phase separation or incompatibility can be demonstrated by a simple hot storage test[98] in which a sample of the polymer modified bitumen is placed in a cylindrical container and stored at elevated temperature, usually 150°C, in an oven for up to seven days. At the end of the storage period, the top and bottom of the sample are separated and tested. Incompatibility is usually assessed by the difference in softening point between the top and bottom samples – if the difference is less than 5°C the binder is considered to be storage stable.

   The main factors influencing storage stability and binder properties are:

- the amount and molecular weight of the asphaltenes
- the aromaticity of the maltene phase
- the amount of polymer present
- the molecular weight and structure of the polymer
- the storage temperature.

Shell Bitumen has developed a number of compatible SBS modified binders for a variety of surfacing applications including deformation-resistant hot rolled asphalt, stone mastic asphalt and thin surfacings.

---

98 EUROPEAN COMMITTEE FOR STANDARDIZATION. *Bitumen and bituminous binders, Determination of storage stability of modified bitumen, EN 13399, 1998.* CEN, Brussels.

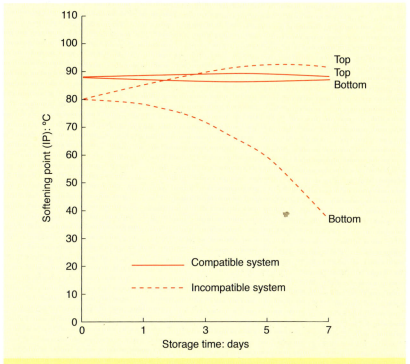

Fig. 5.5 Effect of the compatibility of the system on the storage stability of bitumen/thermoplastic elastomer blends

The philosophy of producing a compatible system is based on the practicalities and user friendliness of the bitumen to facilitate the medium-term storage that is often required. The effect of compatibility on storage stability is demonstrated in Fig. 5.5. It shows the numerical difference in softening point of the top and bottom samples of two bitumens containing 7% of SBS thermoplastic rubber after 1, 3, 5 and 7 days' storage at 140°C. The results clearly show that the compatible system is extremely stable whereas the incompatible system has separated dramatically with virtually no polymer remaining in the bottom sample after 7 days' storage. The implications of this for storage are obvious. In practice, it is possible to use polymer modified binders that are not storage stable. However, they must be handled with great care and stored in tanks with stirrers or extended circulation to prevent separation of the polymer.

The polymer content of a sample of a polymer modified bitumen can be determined using an infrared spectrophotometer of the type shown in Fig. 5.6. The accuracy of the determination depends on a number of factors but usually the polymer content can be determined within ±0·5%.

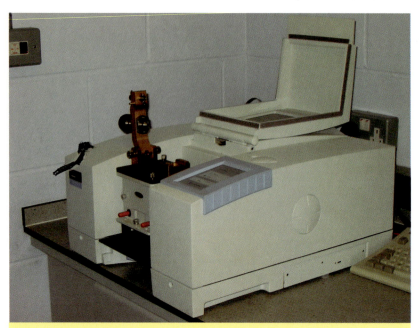

Fig. 5.6 An infrared spectrophotometer used to determine the polymer content of bitumen/polymer systems

Cariphalte DM, a high polymer content bitumen containing 7% SBS, was developed in the 1980s to enhance the overall performance of hot rolled asphalts. The effect of the addition of 7% SBS on the rheological profile of the binder compared to standard 40/60 bitumen is shown in Fig. 5.7. This figure clearly shows that, at high road surface temperatures of 55 to 60°C, Cariphalte DM has a much higher viscosity than 40/60 bitumen and is therefore much stiffer. Coupled with the elastic behaviour of the polymer network, asphalts containing Cariphalte DM exhibit a much greater resistance to deformation than standard materials. At service temperatures below 0°C, Cariphalte DM is more flexible than 40/60 bitumen and hence offers a reduced susceptibility to cracking. Similar behaviour is observed with other members of the Cariphalte family of binders.

The degree of improvement in deformation resistance in polymer modified bitumens is well documented[99,100] and has been confirmed

99 DENNING J H and J CARSWELL. *Improvements in rolled asphalt surfacings by the addition of organic polymers, Laboratory Report 989, 1981.* Transport and Road Research Laboratory, Crowthorne.

100 DOWNES M J W, R C KOOLE, E A MULDER and W E GRAHAM. *Some proven new binders and their cost effectiveness, 7th Australian Asph Pavement Assoc Conf, Brisbane, Australia, Aug 1988.*

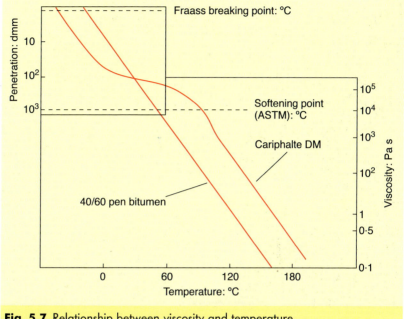

Fig. 5.7 Relationship between viscosity and temperature

through many commercial contracts which included the rigorous requirements which are prescribed for sites accorded Classification No 2 in Clause 943 of the Specification for Highway Works[101]. Figures 5.8(a) and (b) demonstrate that the addition of Cariphalte DM improves the deformation resistance of hot rolled asphalt. These photographs were taken on the slow lane of the A14 Trunk Road at Bury St Edmunds and are at locations within 100 m of each other. The conventional material (a) has developed ruts in excess of 40 mm deep over a five year period compared to the modified asphalt (b) that exhibits only minor deformation of about 7 mm and clearly retains good texture[102].

The degree of improvement in the flexibility of asphalt has been quantified by constant strain fatigue tests again carried out by both the then Transport and Road Research Laboratory[103] and Shell Global

101 HIGHWAYS AGENCY *et al. Manual of contract documents for highway works, Specification for highway works, vol 1, cl 943, 2002.* The Stationery Office, London.

102 PRESTON J N. *Shell Cariphalte DM, The A14 – a case history, Shell Bitumen Technical Brochure, 1996.* Shell International Petroleum Company Ltd, London.

103 DENNING J H and J CARSWELL. *Improvements in rolled asphalt surfacings by the addition of organic polymers, Laboratory Report 989, 1981.* Transport and Road Research Laboratory, Crowthorne.

(a)

(b)

Fig. 5.8 Deformation in HRA surface course made with (a) 40/60 bitumen and (b) Cariphalte DM

Solutions[104]. Fatigue tests carried out on hot rolled asphalt at 5°C and at a frequency of 50 Hz show that, over a wide range of applied strain levels, Cariphalte DM improves fatigue life by a factor of at least three as shown in Fig. 5.9. Thus, this material is eminently suitable for locations where repeated high tensile strains are present.

In the late 1970s, research at the Shell Global Solutions laboratory in Germany resulted in the development of the SBS modified binder Caribit. The very first section built with Caribit was laid in Leimen, a town near the city of Heidelberg in Germany in 1981. Caribit 65 was

104 DOWNES M J W, R C KOOLE, E A MULDER and W E GRAHAM. *Some proven new binders and their cost effectiveness, 7th Australian Asphalt Pavement Association Conf, Brisbane, Australia, Aug 1988.*

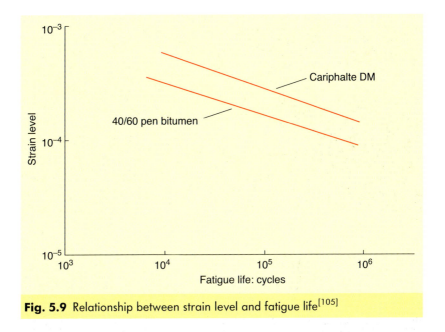

Fig. 5.9 Relationship between strain level and fatigue life[105]

used for the surface course, a 0/11 asphalt concrete. As this site was the first to use Caribit, there have been a number of examinations of the material over the past twenty years to assess the performance of the material in service. In particular, the elastic behaviour and the ageing of the binder were monitored. The susceptibility to ageing was measured by the increase in the softening point of the binder recovered from the asphalt and the elastic behaviour was measured by the elastic recovery of the recovered binder. At irregular intervals, cores were taken to show the performance of the binder and the road. Figures 5.10(a) and (b) show the ageing and elastic recovery of the asphalt over a 20 year period.

Figure 5.10(a) shows that, as expected, during mixing and laying of the asphalt, the softening point increased. During the following 18 years, there was no significant change in the softening point. One of the advantages of using polymer modified binders is the achievement of thick binder films. The thicker the binder film, the smaller the risk of oxidative hardening of the binder. Obviously, the aggregate has been coated with an adequate binder film thickness because the softening point remained at the same level for 18 years after mixing.

105 WHITEOAK C D. *Broadening the boundaries of binder technology, Highways, vol 56, no 1942, p 35, Oct 1988.* Alad Ltd, St Albans, Herts.

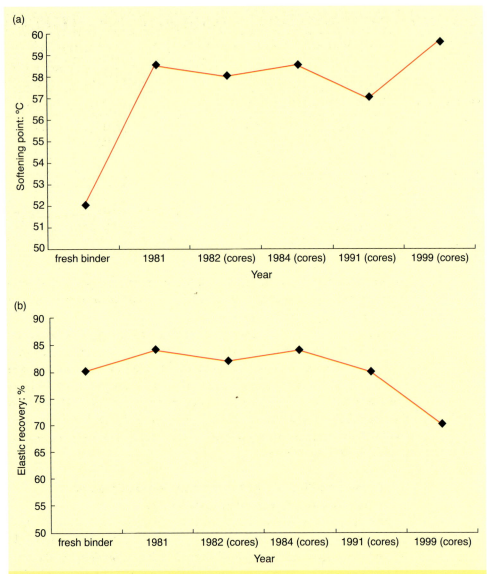

Fig. 5.10 Changes in (a) softening point and (b) elastic recovery of an asphalt made with Caribit over a 20 year period in service

In Germany, the minimum requirement for elastic recovery of the fresh binder is 50%. Figure 5.10(b) shows that both the fresh and aged binder comfortably fulfil this requirement, indicating significant ageing resistance of this binder.

In well compacted asphalt concrete manufactured using penetration grade bitumen, the softening point typically increases by 1°C per

Table 5.3 Properties of the Cariphalte DM and 40/60 bitumens recovered from the A38

Binder	Age of sample	Deformation: mm	Void content of cores: %	Binder properties	
				Penetration @ 25 °C: dmm	Softening point (IP): °C
Cariphalte DM	Initial binder properties	–	–	89	88·0
	after 1 year	–	4·0	79 (11%)	83·0
	after 4½ years	1·1	4·2	83 (7%)	81·0
	after 7 years	1·9	3·8	88 (1%)	80·0
	after 10 years	3·3	3·5	73 (18%)	80·0
40/60	Initial binder properties	–	–	52	50·5
	after 1 year	2·2	3·9	42 (19%)	53·5
	after 4½ years	4·1	3·6	36 (31%)	56·0
	after 7 years	5·1	3·2	34 (35%)	56·0
	after 10 years	6·3	3·4	35 (33%)	56·0

Note. Percentage change in parentheses.

annum. With the use of polymer modified bitumen, it is possible to sub-stantially reduce the ageing of the binder as a result of the thicker binder films on the aggregate.

In the UK, similar investigations have been carried out on the hot rolled asphalt surface course laid on the A38 at Burton-on-Trent as part of a national evaluation of modified binders[106]. The level of deformation and in situ binder properties were monitored over a period of 10 years and the results of this study are detailed in Table 5.3. After 18 years on the road, the hot rolled asphalt made with Cariphalte DM is in excellent condition and is markedly outperforming the material made with penetra-tion grade 40/60 bitumen.

5.2.2 The modification of bitumen by the addition of non-rubber thermoplastic polymers

Polyethylene, polypropylene, polyvinyl chloride, polystyrene and ethyl-ene vinyl acetate (EVA) are the principal non-rubber thermoplastic polymers that have been examined in modified road binders. As thermo-plastics, they are characterised by softening on heating and hardening on cooling. These polymers tend to influence the penetration more than the softening point, which is the opposite tendency of thermoplastic elastomers.

Thermoplastic polymers, when mixed with bitumen, associate below certain temperatures increasing the viscosity of the bitumen. However, they do not significantly increase the elasticity of the bitumen and,

106 DAINES M E. *Trials of porous asphalt and rolled asphalt on the A38 at Burton, Research Report 323, 1992.* Transport and Road Research Laboratory, Crowthorne.

when heated, they can separate which may give rise to a coarse dispersion on cooling.

EVA copolymers are thermoplastic materials with a random structure produced by the copolymerisation of ethylene with vinyl acetate. Copolymers with low vinyl acetate content possess properties similar to low density polyethylene. As the level of vinyl acetate increases, so the properties of the copolymer change. The properties of EVA copolymers are classified by molecular weight and vinyl acetate content as follows.

- *Molecular weight.* The molecular weights of many polymers are defined in terms of an alternative property; standard practice for EVAs is to measure the melt flow index (MFI), a viscosity test that is inversely related to molecular weight; the higher the MFI, the lower the molecular weight and viscosity; this is analogous to the penetration test for bitumen; the higher the penetration, the lower the average molecular weight and viscosity of the bitumen;

- *Vinyl acetate content.* In order to appreciate the main effects of vinyl acetate on the properties of the modified binder, it is useful to consider a simple illustration of its structure as shown in Fig. 5.11; this shows how the regular polyethylene segments of the chain can pack closely together and form what are known as 'crystalline regions'; it also shows how the bulky vinyl acetate groups disrupt this closely packed arrangement to give non-crystalline or amorphous rubbery regions; the crystalline regions are relatively

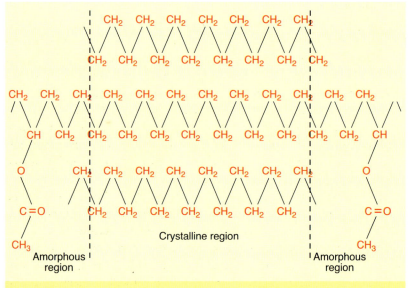

Fig. 5.11 Packing of polyethylene in EVA

stiff and give a considerable reinforcing effect whereas the amorphous regions are rubbery; quite obviously, the more vinyl acetate groups (or the higher the vinyl acetate content) the higher the proportion of rubbery regions and, conversely, the lower the proportion of crystalline regions.

A wide range of EVA copolymers is available, specified by both MFI and vinyl acetate content. EVA copolymers are easily dispersed in, and have good compatibility with, bitumen. They are thermally stable at the temperatures at which asphalt is normally mixed. However, during static storage, some separation may occur and it is therefore recommended that the blended product should be thoroughly circulated before use.

5.2.3 *The modification of bitumen by the addition of rubbers*
Polybutadiene, polyisoprene, natural rubber, butyl rubber, chloroprene, random styrene–butadiene–rubber amongst others have all been used with bitumen but their effect is mainly to increase viscosity. In some instances, the rubbers have been used in a vulcanised (cross-linked) state, e.g. reclaimed tyre crumb, but this is difficult to disperse in bitumen. Successful dispersion requires high temperatures and long digestion times and can result in a heterogeneous binder with the rubber acting mainly as a flexible filler.

5.2.4 *The modification of bitumen by the addition of viscosity reducers*
Several performance properties of asphalt can be improved by adding low molecular weight polyethylene or paraffin wax to bitumen. At temperatures above the melting point of the additive, the viscosity of the bitumen is significantly reduced compared with that of bitumen modified with sulphur (see Section 5.2.5). This, in turn, allows for a reduction in asphalt mixing and laying temperatures by up to 30°C, thereby saving energy and reducing fume emissions. Also, critical applications where hand laying may be required or laying material at low ambient temperatures may be facilitated as a result of the longer time period available between asphalt mixing and compaction.

At the crystallisation temperature of the additive, the viscosity of the modified bitumen rises sharply which in turn may lead to enhanced asphalt stiffness thereby reducing the permanent deformation of the carriageway. However, the low-temperature properties may be adversely affected. Figure 5.12 shows typical viscosity/stiffness–temperature relationship of a penetration grade bitumen modified with a selected paraffin wax.

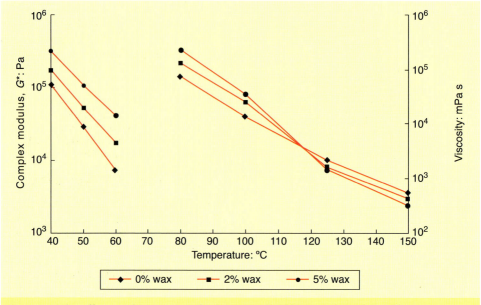

Fig. 5.12 The effect on the viscosity of a bitumen of adding a selected paraffin wax

5.2.5 *The modification of bitumen by the addition of sulphur*

Sulphur is used to modify both bitumen and asphalt. In relatively low concentrations, 2 to 3%, it reduces the high-temperature viscosity of bitumen improving the workability of asphalt when hot and the deformation resistance when cold[107]. It is also used in some polymer modified bitumens to cross-link the bitumen and the polymer to improve storage stability.

At higher concentrations, the sulphur substantially reduces the viscosity of asphalt making it virtually self-compacting. This makes it possible to lay this material through a paver without the need for roller compaction. Alternatively, it can be made sufficiently free-flowing that it can be used for filling potholes and levelling by hand with a trowel. As the material cools, the excess sulphur partially fills and conforms to the shape of the voids in the compacted material keying together the aggregate particles. When this material is cold, the resulting friction between particles within the asphalt makes it very resistant to deformation and very durable. The major weakness of this material is that above 150°C hydrogen sulphide is emitted and this requires the use of specialist equipment[107].

107 DEME I. *Shell sulphur asphalt products and processes, International Road Federation Symp on sulphur asphalt in road construction, Bordeaux, France, Feb 1981.*

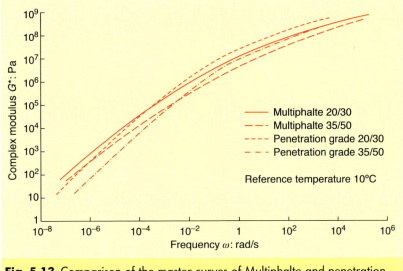

Fig. 5.13 Comparison of the master curves of Multiphalte and penetration grade bitumens

5.3 Multigrade bitumens

The ideal road binder would have uniform stiffness and creep behaviour throughout the whole range of service temperatures. Multigrade bitumens are a step in this direction as they are less temperature susceptible than penetration grade bitumens.

In the mid 1980s, Shell introduced Multiphalte, a range of multigrade bitumens. These are manufactured by means of a special refining process[108,109]. Multiphalte bitumens have low-temperature susceptibility resulting in improved performance at high temperatures and better low-temperature characteristics than hard grades, making them ideal for use in surface courses.

Figure 5.13 illustrates the Multiphalte bitumen concept. This shows the complex modulus of four bitumens at 10°C. In the high frequency (or low temperature) range, the stiffness of the Multiphalte bitumen is less than that of penetration grade bitumen of the same grade. In the low frequency (or high temperature) zone, the Multiphalte has a higher modulus than the penetration grade bitumen. These results clearly show that Multiphalte bitumens possess better rutting resistance than

108 KOOLE R, K VALKERING and D LANÇON. *Development of a multigrade bitumen to alleviate permanent deformation, Australian Pacific Rim AAPA Asphalt Conf, Sydney, Australia, Nov 1991.*

109 MAIA A F. *Multiphalte: A cost effective bitumen to combat rutting, 1st Malaysian Road Conf, Kuala Lumpur, Malaysia, Jun 1994.*

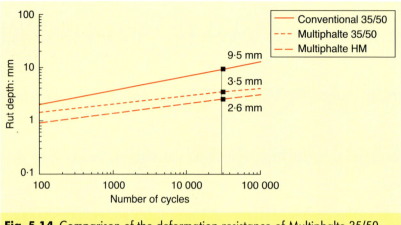

Fig. 5.14 Comparison of the deformation resistance of Multiphalte 35/50, Multiphalte HM and conventional 35/50 bitumens

penetration grade bitumens while still retaining good low-temperature behaviour.

Shell Bitumen launched Multiphalte HM (High Modulus) in France in the mid 1990s. This 20/30 grade performs better at low temperature than penetration grade hard bitumen thus minimising the risk of thermal fatigue cracking. Multiphalte HM also has excellent anti-rutting and structural properties.

Several studies have also been conducted in the laboratory to characterise Multiphalte bitumens and asphalts in order to predict their performance on the road in terms of both their rutting resistance and low-temperature cracking. Most deformation-resistance tests with Multiphalte bitumens have involved use of the French Laboratoires Central des Ponts et Chaussées (LCPC) wheel-track tester. This test continuously measures the rut depth generated by a loaded wheel at 60°C. Figure 5.14 shows the deformation in this test on Multiphalte 35/50, Multiphalte HM and a penetration grade 35/50 bitumen. The deformation after 30 000 cycles is more than halved using the Multiphalte binders.

Along with the laboratory tests, two rutting trials were conducted in 1992 and 1993 on the LCPC circular test track in Nantes, France. Figures 5.15(a) and (b) show the rut depth as a function of the total number of axle passes for the two trials[110].

These two trials show that the asphalt mixture based on Multiphalte 60/80 is twice as resistant to rutting as the asphalt made with the

110 CORTE J F, J P KERZREHO, Y BROSSEAUD and A SPERNOL. *Study of rutting of wearing courses on the LCPC test track, Proc 8th Int Conf on Asphalt Pavements, Seattle, WA, USA, Article no. 191, Aug 1997.*

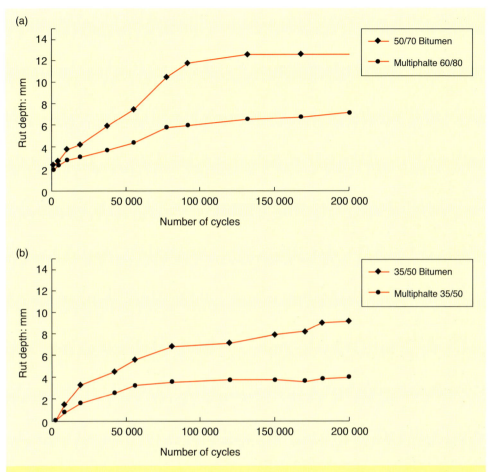

Fig. 5.15 Changes in rut depth for asphalts used on the LCPC circular test track. (a) Penetration grade 50/70 and Multiphalte 60/80 bitumens. (b) Penetration grade 35/50 and Multiphalte 35/50 bitumens

penetration grade 50/70 bitumen. These results are in line with those found in both the laboratory and field trials. This is supported by field data in the UK. After 10 years in service, hot rolled asphalt surface course made with Multiphalte 35/50 laid on the A38 in 1984 had developed a rut depth of 3·1 mm compared to a rut depth of 6·3 mm for an asphalt manufactured using penetration grade 40/60 bitumen in another part of the carriageway[111].

111 STRICKLAND D. *Evaluation of the performance of Multiphalte 35/50 on the A38 during the first 10 years service, Shell Bitumen Technical Brochure, 1998.* Shell Bitumen, Wythenshawe, UK.

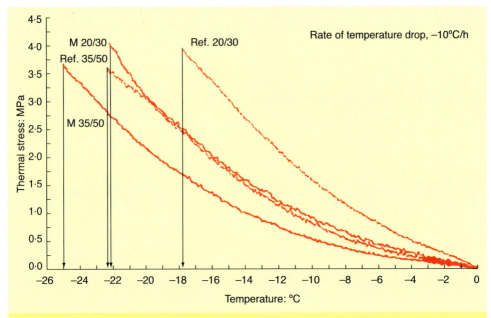

Fig. 5.16 Comparison of penetration grade and Multiphalte bitumens in a restrained cooling test

In the laboratory, the cracking resistance of Multiphalte has been determined in a restrained cooling test where an asphalt specimen is maintained at a constant length in the test rig and is subjected to decreasing temperature[112]. The stress resulting from contraction of the material until specimen fracture is measured. Figure 5.16 shows a comparison of restrained cooling results for several bitumens. Even though greater thermal stresses are generated in Multiphalte HM, its results are similar to those for the penetration grade 35/50. In addition, Multiphalte 35/50 has a lower fracture temperature than the penetration grade bitumen.

5.4 Pigmentable binders

Section 20.8 describes the available means of achieving a coloured surfacing. However, one important way of producing a coloured surface is to use a pigmentable binder. The majority of coloured asphalts are manufactured by adding pigment, usually iron oxide, during the mixing process. Appropriately coloured aggregates are used to ensure that the overall appearance of the material is maintained when aggregate is exposed after trafficking. The main drawbacks to colouring mixtures using conventional bitumens are:

---

112 ACHIMASTOS L, P-J CÉRINO, and Y MARCIANO. *Low temperature behaviour of the new generation of anti-rutting binders, Eurobitume, 2000.*

- the only acceptable colour that can be achieved is red; and
- the quantity of iron oxide required to achieve an acceptable red is fairly high, which significantly increases the cost of the mixture.

To enable asphalt to be pigmented in colours other than red, Shell developed Mexphalte C. This synthetic binder was produced specifically for this purpose and now polymer modified versions of Mexphalte C are available. Mexphalte C is a 'synthetic' binder which, because it contains no asphaltenes, is readily pigmented to virtually any colour[113,114]. It possesses rheological and mechanical properties similar to penetration grade bitumen with the exception that the high-temperature equi-viscous temperature of this binder is about 15°C lower than the corresponding penetration grade. Coloured asphalts manufactured using this binder require 1 to 2% of pigment to achieve a satisfactory colour, whereas only red can be achieved with penetration grade bitumen using up to 5% or more of iron oxide in the mixture. Figures 5.17 and 5.18 illustrate the dramatic results that can be achieved using Mexphalte C.

There are a number of novel applications for coloured asphalt including the following.

- The application of red pigmented asphalt, usually sand carpet, as a protective layer over waterproof membranes laid on bridge decks. When the surfacing on the bridge deck is to be replaced, contractors planing off the old material will only mill down to the red sand carpet, thereby avoiding damage to the expensive waterproof membrane.
- When surfacing tunnels, a light coloured asphalt will improve driver visibility and reduce the electrical energy required to illuminate the tunnel[115].
- For bus lanes and other lanes requiring demarcation[116].
- For recreational and other areas to improve visual impact[117,118].

113 SCHELLEKENS J C A and J KORENSTRA. *Shell Mexphalte C brings colour to asphalt pavements, Shell Bitumen Review 62, pp 22–25, Jun 1987*. Shell International Petroleum Company Ltd, London.

114 LE COROLLER A and R HERMENT. *Shell Mexphalte C and Colas Colclair, A range of products for all road surfacing processes, Shell Bitumen Review 64, pp 6–10, Sep 1989*. Shell International Petroleum Company Ltd, London.

115 GENARDINI C. *Mexphalte C – For road surfacing in tunnels, Shell Bitumen Review 67, pp 20–21, Feb 1994*. Shell International Petroleum Company Ltd, London.

116 O'CONNOR G. *Mexphalte C – Bus lane differentiation in Sydney, Shell Bitumen Review 69, pp 12–13, Feb 1999*. Shell International Petroleum Company Ltd, London.

117 LOHAN G. *Mexphalte C – Adding colour to Ireland's history, Shell Bitumen Review 69, p 11, Feb 1999*. Shell International Petroleum Company Ltd, London.

118 GUSTAFSSEN P. *Shell MEXPHALTE C, Liseburg amusement park in Gothenburg, Sweden opened for the 1987 season in Shell colours, Shell Bitumen Review 63, pp 12–14, Apr 1988*. Shell International Petroleum Company Ltd, London.

Fig. 5.17 The Zenith in Toulouse, France[119]

5.5 Fuel-resisting binders

Mexphalte and Cariphalte Fuelsafe are two of the most recent additions to the Shell family of polymer modified binders. The binders improve both the low- and high-temperature performance of asphalt but it is the resistance to the effects of contact with fuels and lubricants that is dramatically improved. Cariphalte Fuelsafe is a further development on Mexphalte Fuelsafe that was developed in the mid 1990s. Comparative immersion tests with A1 jet fuel show that the weight loss during immersion is an order of magnitude less with Cariphalte Fuelsafe than with Mexphalte Fuelsafe.

---

119 CERINO P-J. *Mexphalte C reaches its zenith in the "Pink City"*, *Shell Bitumen Review 70, p 19, 2001.* Shell International Petroleum Company Ltd, London.

Fig. 5.18 Parvis de Notre-Dame-de Fourvières in France

It has been shown by use of 28 day diesel immersion tests that Cariphalte Fuelsafe reduces material loss by up to 60 times compared with the equivalent penetration grade binder. The lorry parking area at Donnington Services at junction 24 on the M1 in England was surfaced with stone mastic asphalt made with Mexphalte Fuelsafe in 1999[120]. After 4 years' service, although there is considerable diesel and lubricant deposits on the asphalt surface, this material is performing excellently.

5.6 Thermosetting binders

Thermosetting polymers are produced by blending two liquid components, one containing a resin and the other a hardener, that react chemically to form a strong three-dimensional structure. Two-component epoxy resins blended with bitumen[121] display the properties of modified thermosetting resins rather than those of bitumen. The family of two-component thermosetting resins was developed around thirty years ago and has found wide application as surface coatings and adhesives. The principal differences between bitumen (a thermoplastic) and thermosetting binders are as follows.

120 VIA BITUMINA. *The secret of Mexphalte Fuelsafe, Via Bitumina 1, p 2, Jun 2000.* Shell International Petroleum Company Ltd, London.

121 DINNEN A. *Epoxy bitumen binders for critical road conditions, Proc 2nd Eurobitume Symp, pp 294–296, Cannes, Oct 1981.*

Table 5.4 Comparison of typical properties of Shell epoxy asphalt and conventional asphalt[122]

Test parameter	Shell epoxy asphalt	Conventional asphalt
Marshall stability: kN	45	7·5
Marshall flow: mm	4·0	4·0
Marshall quotient: kN/mm	11·2	1·9
Wheel-tracking rate at 45°C: mm/h	0	3·2
Stiffness modulus: N/m^2 (3 point blending, frequency 10 Hz, strain $1·8 \times 10^{-4}$):		
0°C	$2·0 \times 10^{10}$	$1·5 \times 10^{10}$
20°C	$1·2 \times 10^{10}$	$3·0 \times 10^{9}$
40°C	$3·3 \times 10^{9}$	$4·0 \times 10^{8}$
60°C	$9·5 \times 10^{8}$	–
Flexural fatigue resistance, cycles to failure (Composite specimen: 50 mm asphalt on 10 mm steel plate, constant deflection 1·0 mm, 25°C, 5 Hz haversine loading):		
6% binder	$1·0 \times 10^{6}$	$3·0 \times 10^{4}$
7% binder	$>2·0 \times 10^{6}$	$2·0 \times 10^{5}$

- When the two components in a thermosetting binder are mixed the usable life of the binder is limited.
- The amount of time available is strongly dependent on temperature – the higher the temperature the shorter the usable life.
- After a thermoset product is applied it continues to cure and increase in strength, cf. organo-manganese compounds.
- The rate of curing on the road is dependent on the ambient temperature.
- As temperature increases bitumen softens and flows.
- Thermosetting binders are much less temperature susceptible and are virtually unaffected by the temperature changes experienced on a road.
- The cured thermosetting binder is an elastic material and does not exhibit viscous flow.
- When cured, thermosetting binders are very resistant to attack by chemicals, including solvents, fuels and oils.

Well-known examples of thermosetting systems are high-strength surfaces such as Shell Grip, Spray Grip and Shell Epoxy Asphalt.

High-strength surfacings are discussed in Chapter 19 but it is Shell Epoxy Asphalt, in particular, that most clearly demonstrates the combination of strength and flexibility that can be achieved using thermosetting binders. Table 5.4 gives a comparison of the properties of two

122 DINNEN A. *Epoxy bitumen binders for critical road conditions, Proc 2nd Eurobitume Symp, pp 294–296, Cannes, Oct 1981.*

hot rolled asphalt mixtures, one containing bitumen and the other containing the epoxy asphalt binder. The wheel tracking results show no measurable rutting on the epoxy asphalt, demonstrating its superb resistance to permanent deformation and the Marshall stability of the epoxy asphalt can be up to ten times greater than that of conventional asphalt. The dynamic stiffness of the epoxy asphalt is significantly higher than the conventional material, particularly at high ambient temperatures. Additionally, flexural fatigue tests using asphalt bonded to a steel plate (simulating flexing of an orthotropic bridge deck) showed that in this test the fatigue life of the epoxy asphalt was at least an order of magnitude greater than that of conventional asphalt.

Since its introduction in the 1960s, Shell Epoxy Asphalt has been used in a variety of situations all over the world[123]. In 1986, a full-scale road trial using the epoxy binder in a hot rolled asphalt mixture was laid on the M6 at Keele, Staffordshire, UK. The evaluation of this trial is continuing and to date the material is performing well. More recently, the Erskine Bridge in Scotland was surfaced with epoxy asphalt[124].

5.7 Cost–performance relationships for modified binders

For any premium product to gain credence with Highway Engineers, it must be shown to be cost effective. Total life or life cycle costing is an economic evaluation technique that is increasingly being used to justify new construction or maintenance expenditure. In this technique, all the costs are considered in terms of discounted costs and benefits over the period of time that the road is required (net present cost concept).

Downes *et al.*[125] considered it particularly appropriate to apply the principles of life cycle costing to the consideration of the cost effectiveness of polymer modified bitumens vis-à-vis penetration grade bitumens. Cariphalte DM was one of the modified materials evaluated and it was shown that at a discount rate of 7%, the net saving using Cariphalte DM was marginally above 20%. If user costs and accidents were taken into account then the savings increased to 45%.

Although much has been done already there is, perhaps, still more to be done. That the majority of polymer modified bitumens provide technically sound solutions to a number of road construction problems is

123 REBBECHI J J. *Epoxy asphalt surfacing of the Westgate Bridge, Proc, Australian Road Research Board, vol 10, pp 136–146, 1980.*

124 CLUETT C. *Epoxy asphalt, Erskine Bridge refurbishment 1994, Shell Bitumen Review 68, pp 20–23, 1996.* Shell International Petroleum Company Ltd, London.

125 DOWNES M J W, R C KOOLE, E A MULDER and W E GRAHAM. *Some proven new binders and their cost effectiveness, 7th Australian Asphalt Pavement Association Conf, Brisbane, Australia, Aug 1988.*

proven. That most of them achieve this with only minimal changes to established working practices is equally sure. The next step must be for all sectors of the industry to work together to develop a greater understanding of the potential of these materials in order to provide better, more cost effective roads for the benefit of both the industry and the community as a whole.

Chapter 6

Bitumen emulsions

Whatever the end use, application conditions usually require bitumen to behave as a mobile liquid. In principle, there are three ways to make a highly viscous bitumen into a low-viscosity liquid:

- heat it
- dissolve it in solvents
- emulsify it.

Bitumen emulsions are two-phase systems consisting of bitumen, water and one or more additives to assist in formation and stabilisation and to modify the properties of the emulsion. The bitumen is dispersed throughout the water phase in the form of discrete globules, typically 0·1 to 50 μm in diameter, which are held in suspension by electrostatic charges stabilised by an emulsifier.

Bitumen emulsions can be divided into four classes. The first two are, by far, the most widely used:

- cationic emulsions
- anionic emulsions
- non-ionic emulsions
- clay-stabilised emulsions.

The terms anionic and cationic stem from the electrical charges on the bitumen globules. This identification system originates from one of the fundamental laws of electricity – like charges repel, unlike charges attract. If an electrical potential is applied between two electrodes immersed in an emulsion containing negatively charged particles of bitumen, they will migrate to the anode. In that case, the emulsion is described as 'anionic'. Conversely, in a system containing positively charged particles of bitumen, they will move to the cathode and the emulsion is described as 'cationic'. The bitumen particles in a non-ionic emulsion are neutral and, therefore, will not migrate to either pole. These types of emulsion are rarely used on highways.

Clay-stabilised emulsions are used for industrial rather than for road applications. In these materials, the emulsifiers are fine powders, often natural or processed clays and bentonites, with a particle size much less than that of the bitumen particles in the emulsion. Although the bitumen particles may carry a weak electrical charge, the prime mechanism that inhibits their agglomeration is the mechanical protection of the surface of the bitumen by the powder together with the thixotropic structure of the emulsion which hinders movement of the bitumen particles. Clay-stabilised emulsions are discussed in more detail in the *Shell Bitumen Industrial Handbook*[126].

In 1906, the first patent covering the application of bituminous dispersions in water for road building[127] was taken out. Initially, efforts were made to form emulsions by purely mechanical means. However, it rapidly became apparent that mechanical action alone was insufficient and, since these pioneering days, emulsifiers have been used[128]. At first, the naturally occurring organic acids in the bitumen were utilised by adding sodium or potassium hydroxide to the aqueous phase. The subsequent reaction formed an anionic soap that stabilised the dispersion[129]. A great variety of acidic chemicals have been used to promote the stability of anionic bitumen emulsions. These include residues from fatty acid distillation, rosin acids, hydroxystearic acid and lignin sulphonates, all of which may be blended with the bitumen prior to emulsification or dissolved in the alkaline aqueous phase.

Since the early 1950s, cationic emulsifiers have become increasingly popular because of their affinity to many solid surfaces. This is an important property in road construction because good adhesion of bitumen to different types of mineral aggregate is essential. The cationic emulsifiers most widely used are amines, amidoamines and imidazolines[130].

6.1 Emulsifiers

The emulsifier performs several functions within the bitumen emulsion. It:

- makes emulsification easier by reducing the interfacial tension between bitumen and water

---

126 MORGAN P and A MULDER. *The Shell Bitumen Industrial Handbook, 1995*. Shell Bitumen, Wythenshawe, UK.
127 VAN WESTRUM S. *German Patent 173.6391, 1906*.
128 ALBERT K and L BEREND. *Chem. Fabrik. Austrian Patent 72451, 1916*.
129 BRADSHAW L C. *Paint Technol, vol 24, pp 19–23, 1960*.
130 SCHWITZER M K. *Chem Ind, vol 21, pp 822–831, 1972*.

- determines whether the emulsion formed is water-in-oil or oil-in-water type
- stabilises the emulsion by preventing coalescence of droplets
- dictates the performance characteristics of the emulsion such as setting rate and adhesion.

The emulsifier is the single most important constituent of any bitumen-in-water emulsion. In order to be effective, the emulsifier must be water soluble and possess the correct balance between hydrophilic and lipophilic properties. Emulsifiers can be used singly or in combination to provide special properties.

In the emulsion, the ionic portion of the emulsifier is located at the surface of the bitumen droplet whilst the hydrocarbon chain orientates itself on the surface of the bitumen and is firmly bound to it. This is illustrated in Fig. 6.1.

Emulsions may also contain unbound emulsifier which can influence the final properties of the emulsion. The ionic portion of the emulsifier imparts a charge to the droplets themselves and counter-ions like sodium or chloride diffuse into the water phase.

In anionic surfactants, the electrovalent and polar head group is negatively charged and imparts a negative charge to the surface of the bitumen droplets.

$$R-COO^-Na^+$$

In cationic emulsifiers, the electrovalent and polar head group is positively charged and imparts a positive charge to the surface of the bitumen droplets.

$$R-NH_3^+Cl^-$$

In non-ionic emulsifiers, the hydrophilic head group is covalent, polar and dissolves without ionisation. Any charge on the bitumen emulsion droplets is derived from ionic species in the bitumen itself.

$$R-COO(CH_2CH_2O)_xH$$

In simple emulsifiers of the above chemical structures, R represents the hydrophobic portion of the emulsifier and is usually a long chain hydrocarbon consisting of 8 to 22 carbon atoms derived from naturals fats and oils such as tallow or from petroleum such as alkylbenzenes. The hydrophilic head group can variously contain amines, sulphonates, carboxylates, ether and alcohol groups. Emulsifiers with polyfunctional head groups containing more than one of these types are widely used.

Complex wood-derived emulsifiers include Vinsol Resin®, tannins and ligninsulphonates which contain polycyclic hydrophobic portions and

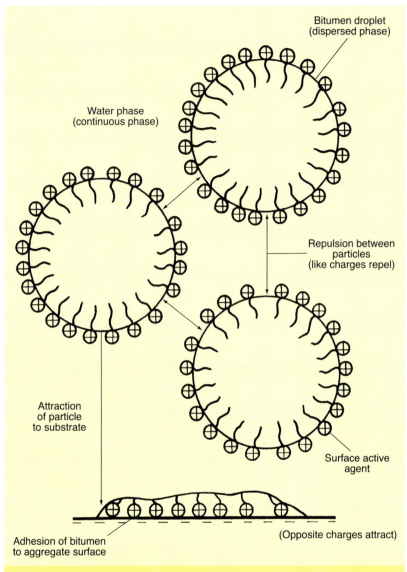

Fig. 6.1 Schematic diagram of charges on bitumen droplets

several hydrophilic centres. Proteinaceous materials such as blood and casein have also been used in bitumen emulsifiers. In general, the structure of these molecules is complex, as can be seen by the structure for lignosulphonate illustrated in Fig. 6.2.

Cationic emulsions constitute the largest volume of emulsions produced worldwide, they are produced from the following types or mixtures: monoamines, diamines, quaternary ammonium compounds,

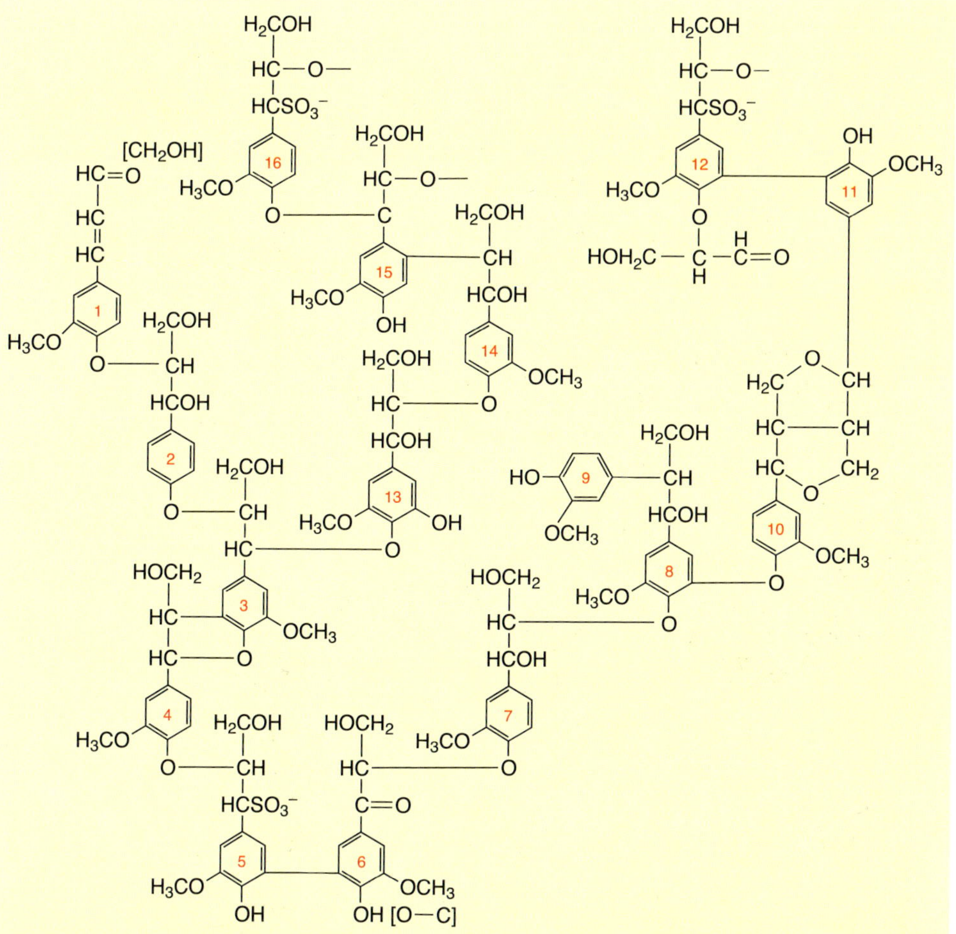

Fig. 6.2 The structure of lignosulphonate

alkoxylated amines, amidoamines, and can be represented structurally as follows.

Example of chemical structure	Chemical type
$(R = C_8 - C_{22})$	
$R-NH_2$	Monoamine
$R-NH_2CH_2CH_2CH_2NH_3^{2+} 2Cl^-$	Diamine
$RCONHCH_2CH_2CH_2NH(CH_3)_2^+ Cl^-$	Amidoamine
$\begin{array}{c} CH_3 \\ | \\ R-N^+-CH_3Cl^- \\ | \\ CH_3 \end{array}$	Quaternary ammonium compound

Example of chemical structure	Chemical type
$(CH_2CH_2O)_xH$	Alkoxylated amine

$$R-N\begin{cases} (CH_2CH_2O)_xH \\ (CH_2CH_2O)_yH \end{cases}$$

Most of these are supplied in neutral basic form and need to be reacted with an acid to become water soluble and cationic in nature. Therefore, cationic emulsions are generally acidic with a pH <7. Hydrochloric acid is normally used which reacts with the nitrogen atom to form an ammonium ion, this reaction can be represented as follows:

$$R-NH_2 + HCl \longrightarrow R-NH_3^+Cl^-$$

Amine + Acid $\longrightarrow$ Alkylammonium chloride

Quaternary ammonium compounds do not require any reaction with acids as they are already salts and water soluble but the water phase pH can be adjusted with acid if required to modify the performance of the emulsion.

Anionic emulsifiers constitute the second largest volume of emulsion produced worldwide and are usually stabilised with fatty acids or sulphonate emulsifiers.

Fatty acids are insoluble in water and are made soluble by reacting with an alkali, normally sodium or potassium hydroxide, so that anionic emulsions are alkaline with a pH >7.

$$\underset{\text{Fatty acid}}{R-COOH} + \underset{\text{Sodium hydroxide}}{NaOH} \longrightarrow \underset{\text{Soap}}{R-COO^-Na^+} + \underset{\text{Water}}{H_2O}$$

Sulphonates are usually supplied as water soluble sodium salts. Further neutralisation is not required but an excess of sodium hydroxide is used in order to keep the pH of the emulsion higher than seven and also to ionise the natural acids contained in the bitumen.

Non-ionic emulsifiers are not produced in significant quantities and are normally only used to modify both anionic or cationic emulsions. Typical non-ionic emulsifiers include nonylphenolethoxylates and ethoxylated fatty acids.

6.2 The manufacture of bitumen emulsions

Most bitumen emulsions are manufactured by a continuous process using a colloid mill. This equipment consists of a high-speed rotor which revolves at 1000 to 6000 revs/min in a stator. The clearance between the rotor and the stator is typically 0·25 to 0·50 mm and is usually adjustable.

Hot bitumen and emulsifier solutions are fed separately but simultaneously into the colloid mill, the temperatures of the two components being critical to the process. The viscosity of the bitumen entering the colloid mill should not exceed 0·2 Pa s (2 poise). Bitumen temperatures in the range 100 to 140°C are used in order to achieve this viscosity with the penetration grade bitumens that are normally used in emulsions. To avoid boiling the water, the temperature of the water phase is adjusted so that the temperature of the resultant emulsion is less than 90°C. As the bitumen and emulsifier solutions enter the colloid mill they are subjected to intense shearing forces that cause the bitumen to break into small globules. The individual globules become coated with the emulsifier which gives the surface of the droplets an electrical charge. The resulting electrostatic forces prevent the globules from coalescing.

When the bitumen is not a soft penetration grade bitumen, the process is more difficult. Higher temperatures are needed to allow the bitumen to be pumped to and dispersed in the mill; dispersion of the bitumen requires more power input to the mill which further increases the product temperature. Pressurised mills are used with bitumens having a high viscosity at normal emulsification temperatures and to allow higher throughput with normal bitumens. Emulsions with temperatures up to 130°C are produced under high pressure and the emulsion output must be cooled below 100°C before being discharged into normal storage tanks.

As an alternative to a colloid mill, a static mixer may be used. This contains no moving parts. The high shear necessary to produce an emulsion is generated by pumping the input materials at high speed through a series of baffles designed to produce highly turbulent flow. The benefits of having no moving parts and no shaft seals are obvious; additional benefits claimed are closer control of the bitumen particle size in the emulsion produced with consequent closer control of critical emulsion properties (discussed in Section 6.5).

A batch process can be used for the production of small volumes of emulsion. The type of mixer which is used is chosen to suit the consistency of the end product; it may be a high-speed propeller for low-viscosity road emulsions or a slow Z-blade mixer for paste-like industrial emulsions.

Schematic diagrams of continuous and batch emulsion manufacturing facilities are shown in Figs. 6.3(a) and (b).

6.3 Properties of bitumen emulsions

Surface dressing applications are, by far, the largest use for bitumen emulsions. Hence, emulsion properties discussed here are with particular reference to surface dressing requirements. Notwithstanding, the principles have general application. The most important properties of bitumen

97

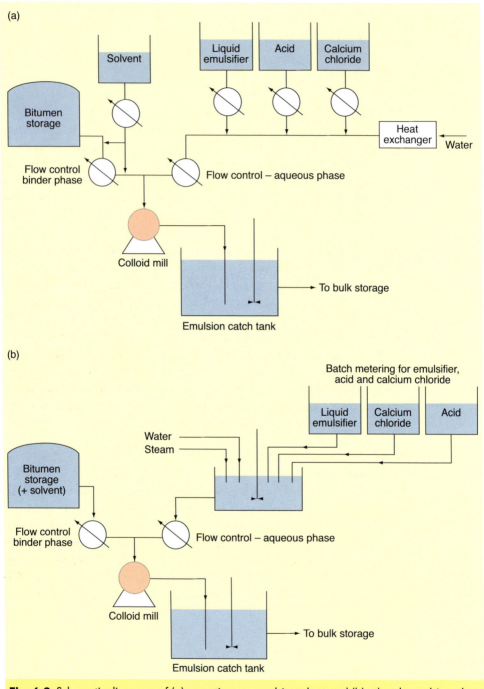

Fig. 6.3 Schematic diagrams of (a) a continuous emulsion plant and (b) a batch emulsion plant

emulsions are:

- stability
- viscosity (or, more accurately, rheology)
- breaking
- adhesivity.

There are conflicting requirements for the properties of bitumen emulsions. The ideal emulsion would be stable under storage, transport and application conditions but would break rapidly very soon after application leaving a binder having the properties of the original bitumen adhering strongly to the road and the chippings. It would have a low viscosity for ease of handling and application and would flow to minimise irregular spraying but would not flow due to road irregularities, camber or gradient.

It is generally assumed that the bitumen produced when an emulsion breaks is the same as the bitumen that was used to produce the emulsion, but there are exceptions. The emulsifier used may modify the recovered bitumen, particularly properties that are surface dependent such as adhesion. Clay-stabilised emulsions can be formulated to produce bitumen films that have non-flow properties even at very high temperatures. These are useful for roofing and insulating applications. Modifiers in the form of an emulsion can be added to a bitumen emulsion. The two disperse phases remain separate until the emulsion breaks after which the coagulated material is an intimate mixture at the micron level of the two previously emulsified materials.

6.3.1 Emulsion stability

Settlement

Emulsions, particularly those having low bitumen content and low viscosity, are prone to settlement. At ambient temperatures, the grades of bitumen normally used in emulsions have a density that is slightly greater than that of the aqueous phase of the emulsion. Consequently, the bitumen particles tend to fall through the aqueous phase, resulting in a bitumen rich lower layer and a bitumen deficient upper layer. The velocity of the downward movement of the particles can be estimated using Stokes' law[131]:

$$v = \frac{2}{9} \frac{gr^2(\rho_1 - \rho_2)}{\eta}$$

131 STOKES G G. *On the effect of internal friction of fluids on the motion of pendulums, Trans Cambridge Phil Soc, vol 9, pp 8–106, 1851.*

where g = gravitational force, r = particle radius, ρ_1 = specific gravity of the bitumen, ρ_2 = specific gravity of the aqueous phase and η = viscosity of the aqueous phase.

However, Stokes' law applies to particles that are free to move and have no inter-particular forces, conditions that are frequently not met in a bitumen emulsion.

Settlement can be reduced by equalising the densities of the two phases. One way of achieving this is to add calcium chloride to the aqueous phase. However, because the coefficients of thermal expansion of bitumen and the aqueous phase are not the same, their densities can be made equal only at one specific temperature. Since large particles settle more rapidly than do small ones, settlement can be abated by reducing either the mean particle size or the range of particle sizes that are present. Increasing the viscosity of the aqueous phase will also reduce the rate of settlement. Indeed, if the aqueous phase behaviour can be made non-Newtonian by introducing a yield value, settlement can be eliminated completely. In addition to gravity, there are repulsive and attractive forces between the bitumen droplets caused by the layers of emulsifier on the droplets that impede or accelerate settlement.

Coalescence follows settlement in two stages. Firstly, bitumen droplets agglomerate into clumps; this reversible phenomenon is called flocculation. Secondly, the resultant flocks fuse together to form larger globules; this irreversible process is called coalescence. This can be spontaneous or it can be induced by mechanical action.

Pumping, heating and transport stability

Two bitumen particles in an emulsion will coalesce if they come into contact. Contact is prevented by electric charge repulsion and the mechanical protection offered by the emulsifier. Any effect that overcomes these forces will induce flocculation and coalescence. Flow of the emulsion, caused by pumping, heating (convection currents) or transport is one such effect. Some emulsifiers have a tendency to foam which is, itself, a potential cause of coalescence since bitumen particles in the thin film of a bubble are subjected to the forces of surface tension.

6.3.2 Emulsion viscosity

Since surface dressing emulsions are almost always applied by spray, their viscosity under spraying conditions is of prime importance. Unfortunately, it is almost impossible to measure. Spraying of K1-70 emulsion involves high shear rates and temperatures up to 90°C, which initiate the emulsion breaking process. This is good from a practical point of view but a major problem from the rheologist's standpoint. Emulsion viscosity is usually measured with an orifice viscometer (Engler, Redwood II or standard tar viscometer, see Section 7.3) in which the shear rate is relatively low. However, any given portion of emulsion passes through the region of shear once only and effects caused by the evaporation of water from the surface of the emulsion (in extreme cases, skinning) do not affect the emulsion flow from the bottom of the viscometer.

Emulsions having a high concentration of disperse phase (bitumen) rarely have Newtonian viscosity characteristics, i.e. the apparent viscosity changes with the shear rate at which the viscosity is measured. In addition, the rate of change of viscosity with temperature is not the same for different emulsions. When comparing two different emulsions, it is possible for one of them to have a lower viscosity and better spray distribution at the spraying temperature of 85°C while also showing higher viscosity and less run-off from the road at a road temperature of 30°C. Single-point viscosity measurements can, therefore, be misleading although more data are difficult to obtain.

In principle, there are three ways of increasing the viscosity of an emulsion:

• by increasing the concentration of the disperse phase (bitumen)
• by increasing the viscosity of the continuous phase
• by reducing the particle size distribution range.

The converse changes will, similarly, decrease the viscosity of an emulsion. Emulsion viscosity is almost independent of the viscosity of the disperse phase (bitumen). It is possible to produce emulsions of hard bitumen (<10 pen) which are readily pourable at 10°C.

Increasing the bitumen content
At low bitumen contents, the effect is small. At high bitumen contents, a small increase in concentration can induce a dramatic change in viscosity which may be uncontrollable.

Modification of the aqueous phase
The viscosity of a bitumen emulsion is highly dependent on the aqueous phase composition. It has been shown that, in the case of conventional cationic road emulsions, viscosity can be increased by decreasing the acid content or increasing the emulsifier content. Additives intended specifically as viscosity modifiers can also be used.

Increasing the flow rate though the mill
By increasing the flow rate through the mill, the particle size distribution of the emulsion will be changed. At bitumen contents less than 65%, the viscosity of the emulsion is virtually independent of flow rate. However, at bitumen contents greater than 65% where the globules of bitumen are

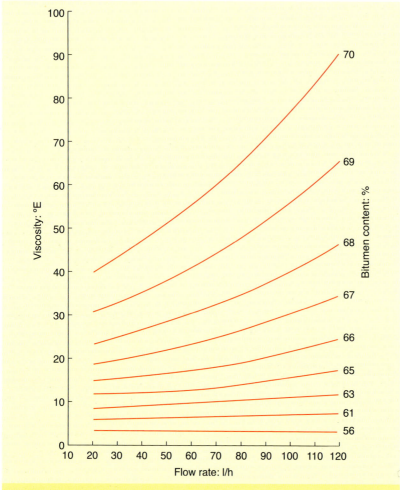

Fig. 6.4 Emulsion viscosity (in degrees Engler) as a function of flow rate for different bitumen contents

packed relatively closely together, inducing a change in the particle size distribution by changing the flow rate has a marked effect on viscosity as shown in Fig. 6.4.

Decreasing the viscosity of the bitumen in the mill
If the viscosity of the bitumen entering the colloid mill is reduced, the particle size of the emulsion will be reduced which, in turn, tends to increase the viscosity of the emulsion.

6.3.3 *Breaking of emulsions*
There are, in principle, six parameters that can be used to change the breaking properties of emulsions:

- bitumen content
- aqueous phase composition
- particle size distribution
- environmental conditions
- chippings
- the use of breaking agents.

Bitumen content
At high bitumen contents, the bitumen particles are more likely to come into contact with each other resulting in an increase in the rate of break.

Aqueous phase composition
The breaking rate of a bitumen emulsion has been shown to be increased by reducing the acid content, increasing the emulsifier content or by decreasing the ratio between the acid and emulsifier contents.

Particle size distribution
The smaller the size of the bitumen particles, the finer will be the dispersion, resulting in a slower breaking rate of the emulsion.

Environmental conditions
Evaporation of water is influenced by wind velocity, humidity and temperature in that order. Temperature and humidity are related: as the air temperature falls, relative humidity increases. Working at night with emulsions can therefore be difficult with low temperature and relative humidity reaching 100% causing the loss of water to cease entirely.

At higher temperatures, the bitumen particles in the emulsion are more mobile and the bitumen is softer. In such circumstances, particles are more likely to come into contact and, therefore, more likely to coalesce.

Chippings
As stated above, the spraying conditions for K1-70 emulsion initiate the emulsion breaking process. Therefore, it is imperative that chippings are applied very soon after the emulsion has been applied to the surface of the road. This is necessary to ensure that the emulsion is still capable of wetting the chippings. When the chippings are applied, breaking is accelerated by absorption of emulsifier onto the aggregate and evaporation of water. The former can be completely inhibited by the use of coated chippings. Conversely, dust can cause rapid breaking of the emulsion onto the dust with no adhesion of bitumen to the chippings. Within these extremes, the surface area of the chippings (i.e. their size and shape) has a considerable influence on the rate of break of the emulsion.

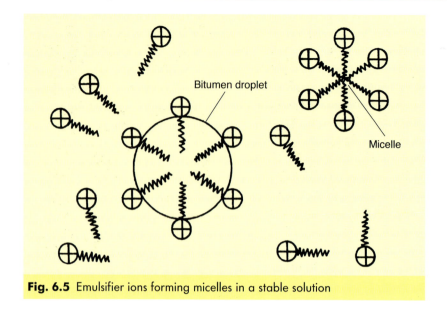

Fig. 6.5 Emulsifier ions forming micelles in a stable solution

The emulsion contains emulsifier molecules in both the water and on the surface of the droplets. Some of the emulsifier ions form micelles and, in a stable emulsion, an equilibrium exists as shown in Fig. 6.5. If some of the emulsifier ions are removed from the solution, the balance is restored by ions from the micelles and the surface of the droplets. This occurs when an emulsion comes into contact with a mineral aggregate. The negatively charged aggregate surface rapidly absorbs some of the ions from the solution weakening the charge on the surface of the droplets. This initiates the breaking process as shown in Fig. 6.6. A point is reached where the charge on the surface of the droplets is so depleted that rapid coalescence takes place. The aggregate is now covered in hydrocarbon chains and, therefore, the liberated bitumen adheres strongly to its surface.

The use of breaking agents
The use of breaking agents can accelerate the breaking of an emulsion. For surface dressing emulsions, it is possible to spray a chemical breaking agent either simultaneously with the emulsion or just after the emulsion has been applied to the road. Care is required in the use of breaking agents. Applying too little will have no effect, applying too much may break the emulsion but adversely affect its adhesivity. Poor distribution of a breaking agent can have similar effects.

6.3.4 Emulsion adhesivity
It is very important in all applications where bitumen is used as an adhesive between solid surfaces that the bitumen 'wets' the surface to

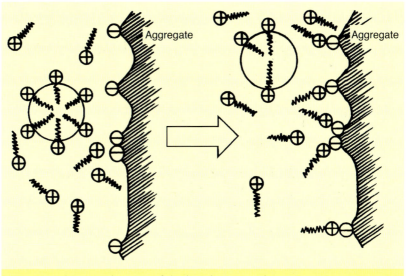

Fig. 6.6 Schematic diagram of the breaking process

create the maximum contact area. With dry substrates, the 'critical surface tension of wetting' of the aggregate must be high enough to ensure that the bitumen spreads easily over the surface. The resultant adhesion generally exceeds the cohesion of the bitumen. However, when the surface of an aggregate is covered with water, the wetting of the aggregate becomes a three-phase phenomenon that can occur only if the balance of the interfacial energies favours wetting by the bitumen. Cationic emulsifiers are particularly efficient at reducing free surface energy of a polar aggregate forming a thermodynamically stable condition of minimum surface energy resulting from the emulsifier being attracted to the aggregate surface[132,133].

Most cationic emulsifiers are also anti-stripping agents; consequently initial bonding is assured. However, the quality of the bond between the bitumen and the aggregate depends on a number of factors:

- the type and amount of emulsifier
- the bitumen grade and constitution
- the pH of the emulsifier solution
- the particle size distribution of the emulsion
- the aggregate type.

132 HIGHWAYS RESEARCH BOARD. *Effect of water on bitumen–aggregate mixtures, Special Report no 98, 1968*. Highway Research Board, Washington.

133 HEUKELOM W and P W O WIJGA. *Bitumen testing, 1973*. Koninklijke/Shell Laboratorium, Amsterdam.

6.4 Classification and specification of bitumen emulsions
6.4.1 BS 434-1: 1984

Bitumen emulsions are classified in BS 434-1: 1984[134] by a three-part alphanumeric designation. The first part, either A or K, indicates anionic or cationic emulsion. The second, 1 to 4, indicates the breaking rate or stability (the higher the number, the greater the stability or the slower the break). The third part of the code indicates the bitumen content of the emulsion as a percentage of the total. For example:

K1-70 is a cationic emulsion, rapid breaking with a bitumen content of 70%

A2-50 is an anionic emulsion, semi-stable with a bitumen content of 50%.

BS 434 specifies bitumen emulsions in terms of:

- particle charge;
- sieve residue (coarse sieve, 710 μm);
- sieve residue (fine sieve, 150 μm);
- viscosity as an efflux time from a flow cup (degrees Engler or Redwood II seconds);
- 'bitumen content' – the term used is in quotation marks because the test method specified in fact measures water content with 'bitumen content' defined as the difference between water content and 100%, thus 'bitumen content' includes emulsifier and solvent;
- coagulation at low temperature;
- storage stability (short period test);
- storage stability (long period test);
- stability of mixing with coarse aggregate; and
- stability of mixing with cement.

However, only particle charge, binder content and viscosity are specified for K1-70 emulsion, the grade most widely used for surface dressing. The viscosity limits are not mandatory provided the emulsion gives satisfactory performance in a sprayer, i.e. meets transverse distribution requirements of BS 1707: 1989[135].

In common with many other facets of highway specifications, there is a trend towards performance specifications for surface dressing work rather than the recipe specifications that have historically predominated. This does not remove the need for laboratory test methods – these are still required for quality control and quality assurance purposes.

134 BRITISH STANDARDS INSTITUTION. *Bitumen road emulsions (anionic and cationic), Specification for bitumen road emulsions, BS 434-1: 1984*. BSI, London.
135 BRITISH STANDARDS INSTITUTION. *Specification for hot binder distribution for road surface dressing, BS 1707: 1989*. BSI, London.

6.4.2 *European specifications*

A European specification for the properties of bitumen emulsions is currently being drafted based upon harmonisation of the existing national standards of member countries. While this is being developed, BS 434, the current specification, will remain unchanged.

6.4.3 *Specifications for emulsions containing polymer modified bitumens*

There is no British Standard specifically covering emulsions which have been manufactured using polymer modified bitumens. Given that these emulsions are normally used in the same way and for the same purposes as emulsions made with normal bitumens, no special properties are required for the emulsions. Such emulsions are generally used on sites that have more difficult service conditions where there would be a higher probability of failure if a normal bitumen was used.

It would be reasonable to require some form of evidence that emulsion supplied to a particular contract truly contains bitumen that has been modified. Since many binder tests cannot be applied directly to emulsions, a binder recovery procedure is required. Natural drying, even for 1 mm films, requires a period of several days. Alternatives are accelerated drying at high temperature with forced air flow or distillation.

6.5 Modification of bitumen emulsion properties
6.5.1 *Particle size distribution*

This is one property of bitumen emulsions for which there is no British Standard and no standard test method. The user does not see the property directly but it influences many of the emulsion properties that are critical to success in application and service.

Mean particle diameter can be determined, laboriously, using a Thoma cell and microscope. However, automatic equipment that can measure both mean particle size and particle size distribution of emulsions is available, but bitumen emulsions can present difficulties because they usually contain a few particles that are very large and may contain some that are very small.

The strong influence of particle size distribution on the properties of bitumen emulsions is due to the facts that the surface area of a spherical particle is proportional to the square of its diameter and its mass is proportional to the cube of its diameter. Many performance properties of an emulsion are influenced by the amount of 'free' emulsifier in the aqueous phase, i.e. the amount of emulsifier that has not been absorbed onto the bitumen particles. The amount of emulsifier that is absorbed onto the bitumen particles depends on the total surface area of those particles. Even a small mass proportion of bitumen present as submicron particles can create a large surface area.

The distribution of emulsion droplet size is dependent on the interfacial tension between the bitumen and the aqueous phase (the lower the interfacial tension, the easier the bitumen disperses) and on the energy used in dispersing the bitumen. For a given mechanical energy input, harder bitumens will produce coarser emulsions and high penetration or cutback bitumens will produce finer emulsions. It is possible to influence the particle size and distribution by modifying the materials and process used to make an emulsion.

The addition of acid to the bitumen

The addition of naphthenic acids to a non-acidic bitumen is important for the production of anionic emulsions. The acids react with the alkaline aqueous phase to form soaps that are surface active and which stabilise the dispersion. The addition of naphthenic acids causes a decrease in the mean particle size of the emulsion without changing its size distribution.

Manufacturing conditions

Manufacturing conditions substantially influence the particle size distribution of the emulsion:

- *Temperature.* Increasing the temperature of either the aqueous phase or the bitumen normally increases the mean particle size of the emulsion.
- *Bitumen content.* Increasing the bitumen content increases the mean particle size and tends to reduce the range of particle sizes.
- *Composition of the aqueous phase.* For cationic emulsions manufactured using hydrochloric acid and an amine emulsifier, the particle size can be decreased by increasing either the acid or the emulsifier content; if the ratio of acid to amine is kept constant, the particle size can also be reduced by increasing the amine/acid content; the size distribution does not appear to be related to the concentration of these two components.
- *Operating conditions of the colloid mill.* The gap and rotational speed of the colloid mill strongly influence the particle size and distribution of the emulsion; a small gap will result in a small particle size with a relatively narrow range of sizes; high rotational speed will produce a small particle size.
- *Increasing the flow rate through the mill.* By increasing the flow rate through the mill the particle size distribution of the emulsion will be changed; at bitumen contents less than 65%, the viscosity of the emulsion is virtually independent of flow rate; however, at

bitumen contents greater than 65% where globules of bitumen are packed relatively closely together, inducing a change in the particle size distribution by changing the flow rate has a marked effect on viscosity as shown in Fig. 6.4.

- *Decreasing the viscosity of the bitumen.* If the viscosity of the bitumen entering the colloid mill is lowered, the particle size of the emulsion will be reduced which will tend to increase the viscosity of the emulsion.

6.5.2 *Effects of bitumen properties*
Electrolyte content of the bitumen
Bitumens usually contain a small amount of electrolyte, principally sodium chloride. The amount varies from a few parts per million up to a few hundred parts per million, i.e. less than 0·1%. It is present as very small crystals thoroughly dispersed in the bitumen. This small amount of electrolyte can exert considerable influence on the viscosity of emulsions produced from the bitumen. At levels up to 20 ppm, the electrolyte has little or no effect. As the amount increases, the viscosity of emulsions produced from the bitumen rises (for a given bitumen content) to a maximum, then at an electrolyte level of about 300 ppm the emulsion viscosity suddenly falls again. The effect is due to osmosis. Although bitumen is an excellent waterproofing agent, films of bitumen are not completely impermeable to water when they are less than a micron in thickness, often the case in bitumen emulsions. At such a thickness, a film of bitumen acts as a semi-permeable membrane. In the bitumen emulsion, some particles contain a crystal of electrolyte that causes a reduction in osmotic pressure. Water diffuses into the particle to equalise the internal pressure with the external pressure. This has the effect of increasing the volume of the bitumen particle and reducing the amount of water in the continuous phase. If large numbers of particles are affected, the result is a significant increase in emulsion viscosity.

At high electrolyte concentrations, the process of water diffusion continues to the point where the water-filled bitumen particle bursts, releasing all of the trapped water. This is where the emulsion viscosity suddenly falls, as the true continuous phase volume content returns to the nominal value. This is not desirable as the bitumen from the burst particles can initiate coagulation of other particles.

Moderate salt content is usually considered beneficial by emulsion manufacturers since it allows production of emulsions at a given viscosity with a slightly lower bitumen content. The electrolyte content of bitumens, having a naturally low level, can be artificially increased by adding electrolyte to the bitumen. The electrolyte must be present in the bitumen as very small well dispersed crystals to have any useful effect. The effect of a high natural electrolyte content can be reduced by adding electrolyte (e.g. calcium chloride) to the aqueous phase of the emulsion.

Bitumen density
High bitumen density can cause rapid settlement in emulsions, leading to coagulation during static storage. The problem can be alleviated by adding a high boiling point solvent (e.g. kerosene) to bitumen which is a grade harder than would normally be used, or by increasing the specific gravity of the aqueous phase (e.g. by adding calcium chloride or by using an emulsifier which imparts a yield value to the aqueous phase).

Acid value
The presence of natural naphthenic acids in the bitumen is beneficial to the production of most anionic emulsions. The acids react with the excess of alkali in the aqueous phase during emulsification, acting as an efficient natural emulsifier. However, some industrial emulsions cannot be made with acidic bitumens; the naphthenic acids take preference at the particle surface over the emulsifiers in the aqueous phase and give an emulsion which is unstable.

6.5.3 Emulsions containing polymer modified bitumen

Manufacture
One of the main reasons for using polymer modified binders is improved high-temperature performance. Thus, it is inevitable that the manufacture of emulsions from these binders will prove more difficult than from unmodified bitumens. The processes require higher temperatures and greater power input. Fortunately, polymer levels are usually less than would be used for an asphalt binder which mitigates the problem to some extent.

Emulsion properties
There is one emulsion property that changes significantly when a polymer modified bitumen is used, the bitumen content needed to achieve a given viscosity. K1-70 emulsion, based on penetration grade bitumen, is generally made at a bitumen content of 67 to 70%, at which it has satisfactory performance in most spraying equipment. It is difficult to achieve a higher bitumen content without increasing the viscosity of the emulsion beyond the point where adequate distribution from typical spray bars can be obtained. However, emulsions of polymer modified bitumens with bitumen contents up to 80% can be manufactured whose performance in spraying is as good as that of a standard K1-70 emulsion.

This effect is apparent even at low levels of modification, down to 1% of polymer in bitumen for some polymers. It is a benefit in its own right (independent of any performance benefit that may be obtained from the modified binder) since it represents a 30% reduction in the amount of water that must be transported with the required bitumen to evaporate from the road surface after application. The reason for the effect is the change in particle size distribution of the emulsion containing modified binder compared with that in a standard binder. The former often has a much wider spread of particle size distribution than the latter.

Other emulsion properties are generally similar to those of standard bitumen emulsion.

Method of use
Emulsions of polymer modified bitumen are generally used in exactly the same way as their unmodified equivalents. It must not be assumed that the use of polymer modified products provides a safety margin for poor contracting practices. On the contrary, the modified products usually require greater care in their use and are less tolerant of unsuitable site and weather conditions. Their benefits are in the performance levels that can be obtained under high traffic stress conditions from a well-executed application.

6.5.4 *Manufacturing variables and emulsion properties*
Previous sections have given an overview of the factors that influence the properties of emulsions and a number of alternative approaches are available to the emulsion manufacturer to adjust emulsion properties. However, it is clear that it is virtually impossible to adjust one property of the emulsion without influencing others. This interdependence is illustrated in Fig. 6.7.

6.6 Uses of bitumen emulsions[136]
The vast majority of bitumen emulsions are used in surface dressing applications and are discussed in detail in Chapter 19. However, their versatility makes them suitable for a wide variety of applications. The Road Emulsion Association Ltd publishes a range of data sheets on applications for emulsions.

6.6.1 *Slurry seals*
Slurry seals are a mixture of bitumen emulsion and graded aggregate of pourable consistency. In practice, the materials are mixed and screeded using a purpose-built vehicle. The process was first developed over

---

136 ROAD EMULSION ASSOCIATION LTD. *Bitumen road emulsion, Technical Data Sheet no 1, Mar 1998.* REAL, Clacton-On-Sea.

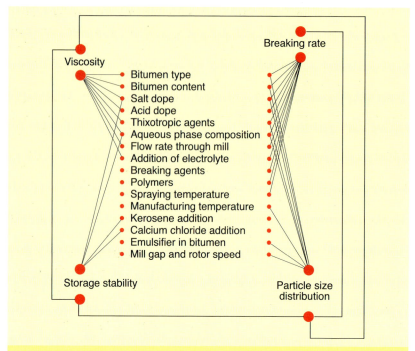

Fig. 6.7 Interrelationships between manufacturing variables and properties for bitumen emulsions

fifty years ago as a maintenance treatment for airfield runways. At the time, semi-stable anionic emulsions were used, the applied thickness was 3 mm maximum and the curing time was many hours. Proprietary formulations and cationic emulsions have reduced the curing times to minutes, thicker layers can be applied and a degree of surface texture can be obtained making the process suitable for road maintenance.

6.6.2 *Bitumen emulsions in road mixtures*[137]

Growing concern about energy conservation and environmental protection has generated an interest in the production of cold asphalts. The UK Environmental Protection Act of 1990 continues to have an increasing effect on the highway industry in terms of tightening controls on emissions from asphalt manufacturing plants. Similar legislation is likely to become enforced in most, if not all, developed countries. Although the idea of widescale production of cold materials is relatively recent in countries like the UK, elsewhere, the manufacture and utilisation of 'emulsified asphalts' has been commonplace for a number of years.

137 BRADSHAW L C. *Bitumen emulsion in road mixes, Shell Bitumen Review no 45, pp 8–11, 1974.* Shell International Petroleum Company Ltd, London.

France has been using cold materials since the 1960s for strengthening and reprofiling lightly trafficked roads and collaborative work between contractors and authorities led to the development of a material called 'grâve émulsion'. Although traditionally a continuously-graded 20 mm material, modern grâve émulsion typically comprises a 14 mm gradation with a bitumen content of 4 to 4·5%. Grâve émulsion can be stockpiled, laid using conventional paving equipment or by a blade grader and compacted at ambient temperatures. Although the material was originally used for minor maintenance works, more recently it has been applied in structural layers of moderately trafficked roads. A national standard for grâve emulsion was introduced in France in 1993[138].

Similar developments have taken place in the USA where environmental issues and the remoteness of some sites from asphalt plants provided the impetus for using cold mixtures. The Asphalt Institute published *A Basic Emulsion Manual, MS-19* in 1986[139] and a third edition of its *Asphalt Cold Mix Manual, MS-14* in 1989[140] and included cold mixtures called 'Emulsified Asphalt Materials' (EAMs) in its *Thickness Design Manual MS-1* in 1991[141]. In 1992, approximately 50 million tonnes of cold mixed asphalts were laid in the USA representing about 10% of the total asphalt production.

Cold mixture technology presents a new set of challenges to engineers who have traditionally relied upon hot asphalts. Whereas hot mixtures rely on the visco-elastic properties of bitumen, emulsion mixtures introduce a new series of conditions that must be met in order that such materials can be successfully produced and laid. The surface chemistry of the aggregate begins to have an important role and emulsions must be tailored to the mineralogy of different rock types.

Cold mixture theory

The classical concept of how cold mixtures work is that the emulsion breaks, either during mixing or compaction, coating the aggregate after which there is an increase in strength over time. The strength development is a result of the expulsion of water from the aggregate matrix and the coalescence and subsequent cohesion of the bitumen particles.

138 ASSOCIATION FRANÇAISE de NORMALISATION. *Assises de chaussées, Graves-émulsion, NF P 98-121, Nov 1993.* AFNOR, Paris.

139 ASPHALT INSTITUTE. *A basic emulsion manual, Manual Series no 19 (MS-19), 1986.* AI, Kentucky.

140 ASPHALT INSTITUTE. *Asphalt cold mix manual, Manual Series no 14 (MS-14), 3rd ed, 1989.* AI, Kentucky.

141 ASPHALT INSTITUTE. *Thickness design, Asphalt pavements for highways & streets, Manual Series no 1 (MS-1), 1991.* AI, Kentucky.

However, the characteristics of initial workability or being able to stock-pile the material and the subsequent development of mechanical strength in situ form conflicting requirements. By tailoring the emulsion to produce a mixture that will remain workable for days or weeks, the development of cohesion in the compacted mixture and, hence, the strength gain of the matrix will also be retarded. If the emulsion is tailored to produce a rapid break then the mixture will quickly exhibit developing stiffness and, hence, will only have a brief workability window. Therefore, there exists the potential for innovative technology to address such problems and develop cold mixtures which perform in a manner which is similar to that of hot mixed products.

The HAUC specification

Cold lay macadams have been used in the UK since the early part of the twentieth century for the surfacing of roads and the reinstatement of carriageway openings. Such materials were produced by fluxing the bitumen with light oils causing a reduction in the viscosity of the binder. This results in materials remaining workable for indefinite periods depending on the quantity of flux oil that is added to the mixture. Shortcomings in the performance of such cutback materials restricted their use primarily to the trench reinstatement market where fluxed macadams were laid as temporary materials prior to being replaced with conventional hot mixtures when the lower layers of the trench had fully consolidated.

The Highway Authorities and Utilities Committee (HAUC) introduced a specification in 1992[142] that required the immediate permanent reinstatement of carriageway excavations. Due to the relatively small quantities of materials required for such works and the remoteness of some sites, it is frequently impossible to use hot mixed materials. Therefore, HAUC has created demand for the use of permanent cold lay surfacing materials (PCSMs) and specified their use in terms of performance criteria based on the characteristics of hot mixtures.

Hot mixture equivalence is characterised by measuring the stiffness and resistance to permanent deformation of 150 mm diameter cores extracted from trials of the proposed mixtures. As PCSMs are materials that cure over time, hot mixture equivalence is not expected to be achieved until after a period of two years. The criteria that must be met by PCSMs at the end of such time are given in Tables 6.1 and 6.2.

A general 'condition indicator' of the material is also given by measuring the residual void content of the cores. Materials suitable for use on

---

142 DEPARTMENT OF TRANSPORT. *Specifications for the reinstatement of openings in highways: A code of practice, Highway Authority and Utilities Committee (HAUC), 1992.* HMSO, London.

Table 6.1 PCSM requirements for stiffness

Permanent cold lay surfacing material	Minimum property requirement at 20°C for equivalence to:		
	50 pen hot laid stiffness: MPa	100 pen hot laid stiffness: MPa	200 pen hot laid stiffness: MPa
20 mm nom. size binder course	4600	2400	900
10 mm nom. size binder course	3800	1900	800
6 mm nom. size binder course	2800	1400	600

Table 6.2 PCSM requirements for resistance to permanent deformation

All materials	Resistance to permanent deformation at 30°C
	Repeated load uniaxial creep resistance when tested in accordance with BS DD 185* shall not exceed: either: 16 000 microstrain for 1800 load applications or: 20 000 microstrain for 3600 load applications

*Note: this is now BS 598-111: 1995.

carriageways must exhibit void contents of between 2% and 10% at the end of the two year assessment period.

The introduction of the HAUC specification has provided an impetus to the development of cold mixtures in the UK which hitherto did not exist. One result of the development work carried out will be a broadening of the technology that will allow the knowledge and experience gained to be transferred into other areas such as bases, binder courses and surface courses. Ultimately, it may be possible to build road pavements using *only* cold mixed asphalts.

6.6.3 *Other road uses*
Crack filling
Bitumen emulsion, preferably containing rubber, is a relatively inexpensive and effective method of sealing cracks in asphalts to stop the ingress of water into the structural layers of a pavement. It is important that cracks are treated as soon as practicable to limit the damage. This is particularly true in areas of high rainfall or during winter when freeze/thaw cycles of water in cracks can result in further, often serious, deterioration.

Grouting
Grouting[143] is a process involving the construction or stabilisation of road surfacings and footpaths. Emulsion is applied to compacted dry

---

143 ROAD EMULSION ASSOCIATION LTD. *Patching and grouting, Technical Data Sheet no 8, Mar 1998.* REAL, Clacton-On-Sea.

or damp aggregate, the low viscosity of the emulsion permitting in-depth penetration through the void structure of the aggregate. The technique involves the construction of a combined base/surface course from a thickness of between 50 and 75 mm in one course or up to 100 mm in two courses, as may be specified. Suitable emulsion types for different applications are given below.

Surface dressing	K1-70, K1-60
Bond coats	K1-40, K2-40
Slurry seals	K3-60
Open graded macadams	K2-60, K2-70
Retreading	K2-60
Mist spraying	K1-40, K2-40
Concrete curing	K1-40, K2-40

Tack/bond coats

This is an established technique for ensuring that an adhesive and cohesive bond develops between the layers in a road pavement[144]. The use of such coats may be essential in some surfacing operations to ensure success of the process. Tack coats are often used where a base or binder course has been subjected to traffic, perhaps on a construction site. In addition, their use may contribute significantly to the assessed value of 'residual life' returned by deflection techniques.

6.6.4 Miscellaneous uses of bitumen emulsions

Details of miscellaneous applications for emulsions have been published by the Road Emulsion Association[145]. Industrial applications of bitumen emulsions are discussed in the *Shell Bitumen Industrial Handbook*[146]. Bitumen emulsions are also used in other civil engineering works, horticultural and agricultural applications. Some examples are given below.

Soil stabilisation[147]

Fresh topsoil on embankments or ploughed agricultural land is susceptible to surface erosion and slippage on embankments. It is essential,

144 ROAD EMULSION ASSOCIATION LTD. *Tack coating, Technical Data Sheet no 5, Mar 1998*. REAL, Clacton-On-Sea.

145 ROAD EMULSION ASSOCIATION LTD. *Miscellaneous uses of bitumen road emulsions, Technical Data Sheet no 12, Mar 1998*. REAL, Clacton-On-Sea.

146 MORGAN P and A MULDER. *The Shell Bitumen Industrial Handbook, 1995*. Shell Bitumen, Wythenshawe, UK.

147 BUTCHTA H. *Terrafix in agricultural applications, Shell Bitumen Review no 61, pp 7–8, Jun 1986*. Shell International Petroleum Company Ltd, London.

therefore, to either stabilise the surface with a binding agent or establish a vegetation root system. Bitumen emulsion sprayed on the surface of the topsoil binds the surface together and assists seed germination by retaining moisture in the soil, improving thermal insulation and protecting seeds from the elements and birds.

Slip layers and concrete curing

Bitumen emulsions are used to create a membrane between layers of concrete, the objective being to retain the strength of the upper layer by preventing water seepage into the lower layers. This avoids rigid adhesion between layers of different ages and strengths and helps to produce a stronger upper layer by preventing water absorption into the lower layers. Bitumen emulsion is also sprayed onto the top surface of freshly laid concrete to prevent the evaporation of water.

Protective coats

Bitumen emulsions are used for protecting buried concrete, pipelines and ironwork. To enhance the adhesive and cohesive characteristics of the cured binder film, a polymer modified emulsion is normally used.

6.7 Bibliography

For further information on the subject of bitumen emulsions, see *Bitumen Emulsions – General Information and Applications*[148] which is available in both French and English.

---

148 SYNDICAT DES FABRICANTS D'ÉMULSIONS ROUTIÈRES DE BITUMES. *Bitumen emulsions – General information and applications, 1991.* SFÉRB, Paris.

Chapter 7

Mechanical testing and properties of bitumens

Bitumen is a complex material with a complex response to stress. The response of a bitumen to stress is dependent on both temperature and loading time. Thus, the nature of any bitumen test and what it indicates about the properties of a bitumen must be interpreted in relation to the nature of the material.

A wide range of tests are performed on bitumens, from specification tests to more fundamental rheological and mechanical tests. A number of these were described in Chapter 4. This chapter considers categorisation tests and important properties of bitumens.

7.1 Standard specification tests for bitumens

Since a wide variety of bitumens are manufactured, it is necessary to have tests to characterise different grades. The two tests used in the UK to specify different grades of bitumen are penetration and softening point tests. Although they are arbitrary empirical tests, it is possible to estimate important engineering properties from the results, including high-temperature viscosity and low-temperature stiffness. The use of the penetration test for characterising the consistency of bitumen dates from the late nineteenth century[149].

As the penetration and softening point tests are empirically derived, it is essential that they are always carried out under exactly the same conditions. The Institute of Petroleum (IP)[150], the American Society for Testing and Materials (ASTM)[151,152] and British Standards Institution

149 BAVEN H C. *School of Mines Quart, vol 10, p 297, 1889.*

150 INSTITUTE OF PETROLEUM. *Methods for analysis and testing, vol 1, part 1.*

151 AMERICAN SOCIETY for TESTING and MATERIALS. *Standard test method for penetration of bituminous materials, D5-97, 1 Nov 1997.* ASTM, Philadelphia.

152 AMERICAN SOCIETY for TESTING and MATERIALS. *Standard test method for softening point of bitumen (Ring and ball method), D36-95, 1 Nov 1997.* ASTM, Philadelphia.

(BSI)[153,154] publish standard methods of testing bitumen. In many cases, the methods are identical and, therefore, methods are published jointly. However, some methods differ in detail, for example the IP and ASTM softening point method, and in these cases a correction factor is provided to relate test results obtained using these two test methods.

The majority of the methods quote limits for assessing the acceptability of test results. Limits of variability of results obtained by a single operator (repeatability) and by different operators in different laboratories (reproducibility) are specified. Thus, tolerance is given to allow for differences between operators and equipment at different locations.

7.1.1 The penetration test

The consistency of a penetration grade or oxidised bitumen is measured by the penetration test[153,155]. In this test, a needle of specified dimensions is allowed to penetrate a sample of bitumen, under a known load (100 g), at a fixed temperature (25°C), for a known time (5 s). The test apparatus is shown in Fig. 7.1.

The penetration is defined as the distance travelled by the needle into the bitumen. It is measured in tenths of a millimetre (decimillimetre, dmm). The lower the value of penetration, the harder the bitumen. Conversely, the higher the value of penetration, the softer the bitumen. This test is the basis upon which penetration grade bitumens are classified into standard penetration ranges.

It is essential that the test methods are followed precisely as even a slight variation can cause large differences in the result. The most common errors are:

- poor sampling and sample preparation;
- badly maintained apparatus and needles; and
- incorrect temperature and timing.

Temperature control is critical, control to ±0·1°C being necessary. Needles must be checked regularly for straightness, correctness of profile and cleanliness. Automatic timing devices are also necessary for accuracy and these must be checked regularly. Penetrations less than 2 dmm and greater than 500 dmm cannot be determined with accuracy.

153 BRITISH STANDARDS INSTITUTION. *Methods of test for petroleum and its products, Bitumen and bituminous binders. Determination of needle penetration, BS EN 1426: 2000, BS 2000-49: 2000.* BSI, London.

154 BRITISH STANDARDS INSTITUTION. *Methods of test for petroleum and its products, Bitumen and bituminous binders. Determination of softening point. Ring and ball method, BS EN 1427: 2000, BS 2000-58: 2001.* BSI, London.

155 AMERICAN SOCIETY for TESTING and MATERIALS. *Standard test method for penetration of bituminous materials, D5-97, 1 Nov 1997.* ASTM, Philadelphia.

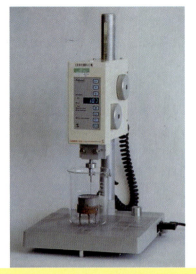

<mark>**Fig. 7.1** The penetration test</mark>

For each test, three individual measurements of penetration are made. The average of the three values is recorded to the nearest whole unit. The recorded penetration is reported if the difference between the individual three measurements does not exceed a specified limit.

The acceptable repeatability and reproducibility quoted in BS 1426: 2000[156] for the penetration test are as follows:

- *Repeatability*
  - If the penetration <50   2 dmm
  - If the penetration ≥50   4% of the mean of the two results
- *Reproducibility*
  - If the penetration <50   4 dmm
  - If the penetration ≥50   8% of the mean of the two results

7.1.2 *The softening point test*

The consistency of a penetration grade or oxidised bitumen can also be measured by determining its softening point[157,158]. In this test, a steel

156 BRITISH STANDARDS INSTITUTION. *Methods of test for petroleum and its products, Bitumen and bituminous binders. Determination of needle penetration, BS EN 1426: 2000, BS 2000-49: 2000.* BSI, London.

157 BRITISH STANDARDS INSTITUTION. *Methods of test for petroleum and its products, Bitumen and bituminous binders. Determination of softening point. Ring and ball method, BS EN 1427: 2000, BS 2000-58: 2001.* BSI, London.

158 AMERICAN SOCIETY for TESTING and MATERIALS. *Standard test method for softening point of bitumen (Ring and ball method), D36-95, 1 Nov 1997.* ASTM, Philadelphia.

Fig. 7.2 The softening point test

ball (weight 3·5 g) is placed on a sample of bitumen contained in a brass ring that is then suspended in a water or glycerine bath. The apparatus is shown in Fig. 7.2.

Water is used for bitumen with a softening point of 80°C or below and glycerine is used for softening points greater than 80°C. The bath temperature is raised at 5°C per minute, the bitumen softens and eventually deforms slowly with the ball through the ring. At the moment the bitumen and steel ball touch a base plate 25 mm below the ring, the temperature of the water is recorded. The test is performed twice and the mean of the two measured temperatures is reported to the nearest 0·2°C for a penetration grade bitumen and 0·5°C for an oxidised bitumen. If the difference between the two results exceeds 1·0°C, the test must be repeated. The reported temperature is designated the softening point of the bitumen and represents an equi-viscous temperature. In the ASTM version of the softening point test, the bath is not stirred whereas in the IP version the water or glycerine is stirred. Consequently, the softening points determined by these two methods differ. The ASTM results are generally 1·5°C higher than for the IP or BS method[159].

As with the penetration test, the procedure for carrying out the softening point test must be followed precisely to obtain accurate results. Sample preparation, rate of heating and accuracy of temperature measurement are all critical. For example, it has been shown[160] that by varying the heating rate from 4·5 to 5·5°C per minute, the tolerance permitted by the IP 58 test method, can result in a difference of 1·6°C in the measured softening point. Automatic softening point machines are available and these ensure close temperature control and automatically record the result at the end of the test.

The acceptable repeatability and reproducibility limits quoted in BS EN 1427: 2000[161] for the softening point test are as follows:

- *Repeatability*
  - For unmodified bitumens in water: 1·0°C
  - For modified bitumens in water: 1·5°C
  - For oxidised bitumens in glycerol: 1·5°C
- *Reproducibility*
  - For unmodified bitumens in water: 2·0°C
  - For modified bitumens in water: 3·5°C
  - For oxidised bitumens in glycerol: 5·5°C

The consistency of bitumen at the softening point temperature has been measured by Pfeiffer and Van Doormaal[162] in terms of penetration value. Using a specifically prepared, extra-long penetration needle they found a value of 800 pen for many but not all bitumens. The exact value was found to vary with penetration index and wax content. It has also been demonstrated by direct measurement that the viscosity at the softening point temperature of the majority of bitumens is 1200 Pa s (12 000 poise).

159 KROM C J. *Determination of the ring and ball softening point of asphaltic bitumens with and without stirring. J Inst Pet, vol 36, 1950.* Institute of Petroleum, London.
160 PFEIFFER J P H (Ed). *The properties of asphaltic bitumen, vol 4, Elsevier's Polymer Series, 1950.* Elsevier Science, Amsterdam.
161 BRITISH STANDARDS INSTITUTION. *Methods of test for petroleum and its products, Bitumen and bituminous binders, Determination of softening point, Ring and ball method, BS EN 1427: 2000.* BSI, London.
162 PFEIFFER J P H and VAN DOORMAAL P M. *The rheological properties of asphaltic bitumens, J Inst Pet, vol 22, pp 414–440, 1936.* Institute of Petroleum, London.

Fig. 7.3 The Fraass breaking point test

7.2 The Fraass breaking point test

The Fraass breaking point test[163] is one of very few tests that can be used to describe the behaviour of bitumens at very low temperatures (as low as −30°C). The test was developed by Fraass[164] in 1937. It is essentially a research tool that determines the temperature at which bitumen reaches a critical stiffness and cracks. A number of countries with very low winter temperatures, for example Canada, Finland, Norway and Sweden, have maximum allowable Fraass temperatures for individual grades of bitumen.

In the Fraass test, shown in Fig. 7.3, a steel plaque 41 mm × 20 mm coated with 0·5 mm of bitumen is slowly flexed and released. The temperature of the plaque is reduced at 1°C per minute until the bitumen reaches a critical stiffness and cracks. The temperature at which the sample cracks is termed the breaking point and represents an equi-viscous or equi-stiffness

163 BRITISH STANDARDS INSTITUTION. *Methods of test for petroleum and its products, Bitumen and bituminous binders. Determination of the Fraass breaking point, BS EN 12593: 2000, BS 2000-8: 2001.* BSI, London.

164 FRAASS A. *Test methods for bitumen and bituminous mixture with specific reference to low temperature, Bitumen, pp 152–155, 1937.*

124

temperature. It has been shown that, at fracture, the bitumen has a stiffness of 2.1×10^9 Pa[165] which is approaching the maximum stiffness of 2.7×10^9 Pa (discussed further in Section 7.6.4). The Fraass temperature can be predicted from penetration and softening point for penetration grade bitumens because it is equivalent to the temperature at which the bitumen has a penetration of 1·25.

7.3 Viscosity

This is a fundamental characteristic of a bitumen as it determines how the material will behave at a given temperature and over a temperature range. The basic unit of viscosity is the pascal second (Pa s). The absolute or dynamic viscosity of a bitumen measured in pascal seconds is the shear stress applied to a sample of bitumen in pascals divided by the shear rate per second; 1 Pa s $= 10$ P (poise). The absolute viscosity of a bitumen can be measured using a sliding plate viscometer.

Viscosity can also be measured in units of m^2/s, or more commonly mm^2/s (1 $mm^2/s = 1$ centistoke (cSt)). These units relate to kinematic viscosity. Kinematic viscosity is measured using a capillary tube viscometer. Kinematic viscosity is related to dynamic viscosity by the equation:

$$\text{kinematic viscosity} = \frac{\text{dynamic viscosity}}{\text{density}}$$

For many purposes, it is usual to determine the viscosity of a bitumen by measuring the time required for a fixed quantity of material to flow through a standard orifice. These methods are useful for specification and comparative purposes and, if required, the results can be converted into more fundamental units of viscosity.

7.3.1 Sliding plate viscometer

A fundamental method of measuring viscosity is the sliding plate viscometer. This apparatus applies the definition of absolute or dynamic viscosity, i.e. it takes the shear stress (Pa) applied to a film of bitumen (5 to 50 μm thick) sandwiched between two flat plates and measures the resulting rate of strain (s^{-1}). The viscosity in pascal seconds (Pa s) is given by the shear stress divided by the rate of strain. The apparatus comprises a loading system that applies a uniform shear stress during a measurement and a device to produce a record of the flow as a function of time. Generally, the records show a curved toe due to visco-elastic behaviour

165 THENOUX G, G LEES and C A BELL. *Laboratory investigation of the Fraass brittle test, Asphalt Technol, no 39, pp 34–46, July 1987.* Institute of Asphalt Technology, Dorking.

passing into a linear form whose slope is a direct measure of viscosity. Depending on the load and the size of the specimen, viscosities in the range of 10^5 to 10^9 Pa s can be measured. A special feature is that the shear stress is the same throughout the specimen making this instrument particularly suitable for investigating the phenomenon of shear stress dependence. Shear stress dependence is the variation in measured viscosity with the magnitude of the applied shear stress.

7.3.2 Capillary viscometer

Capillary viscometers are essentially narrow glass tubes through which the bitumen flows. The tube has narrow and wide sections and is provided with two or more marks to indicate a particular volume or flow as shown in Figs. 7.4 and 7.5.

The value of kinematic viscosity is measured by timing the flow of bitumen through a glass capillary viscometer at a given temperature. Each viscometer is calibrated and the product of flow time and viscometer calibration factor gives the kinematic viscosity in mm^2/s. Temperature–viscosity curves are obtained by measuring kinematic viscosity at a number of different temperatures, and plotting the viscosity on a logarithmic scale against temperature on a linear scale (discussed further in Section 7.4).

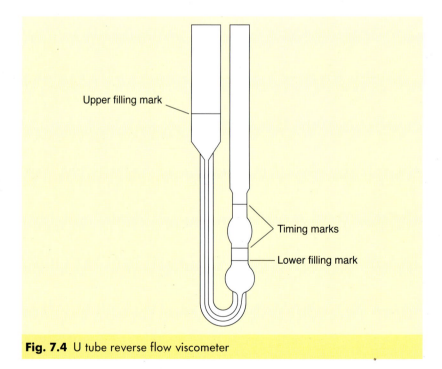

Upper filling mark

Timing marks

Lower filling mark

Fig. 7.4 U tube reverse flow viscometer

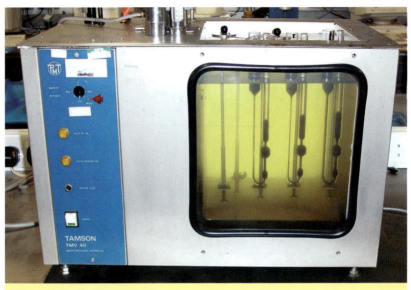

Fig. 7.5 Kinematic viscometer

7.3.3 *Rotational viscometer*

Rotational viscometers are normally used to determine, and in some cases specify, the viscosity of bitumens at application temperatures. An example of a Brookfield rotational viscometer and Thermocel[166] is shown in Fig. 7.6. Essentially, the device consists of a thermostatically controlled chamber containing a sample of hot bitumen. The spindle is lowered into the bitumen and rotated. The torque required to rotate the spindle at a pre-specified shear rate is measured and converted into the viscosity of the bitumen. A direct reading of viscosity in Pa s is given on the display.

The rotational viscosity of bitumen is usually determined at 150°C but with this type of apparatus the viscosity can be determined over a relatively wide range of temperatures, between 120°C and 180°C, and shear rates.

7.3.4 *Cup viscometer*

Cutback bitumens and emulsions are both specified by viscosity. Cutback bitumens are specified using the standard tar viscometer[167] (STV,

166 AMERICAN SOCIETY for TESTING and MATERIALS. *Standard test method for viscosity determination of asphalt at elevated temperatures using a rotational viscometer, D4402-02, 1 Jul 2002.* ASTM, Philadelphia.
167 BRITISH STANDARDS INSTITUTION. *Methods of test for petroleum and its products, Determination of viscosity of cutback bitumen, BS 2000-72: 1993.* BSI, London.

Fig. 7.6 Rotational viscometer

shown in Fig. 7.7). The viscosity of emulsions can be determined using either the Engler viscometer or the Redwood II viscometer[168]. In these tests, a metal cup is filled with cutback bitumen or emulsion at a standard temperature and the time is recorded in seconds for a standard volume of material to flow out through the orifice in the bottom of the cup.

There are several cup-type viscometers available which differ mainly in the size of opening through which the bitumen is drained. Since the stress is provided by gravity, the absolute viscosity, η, in Pa s is given by:

$$\eta = \text{flow time (in s)} \times \text{density (in g/ml)} \times \text{constant}$$

In the above equation, the value of the constant depends on the instrument used; these are listed in Table 7.1. The test results may also be

---

168 BRITISH STANDARDS INSTITUTION. *Bitumen road emulsions (anionic and cationic), Specification for bitumen road emulsions, BS 434-1: 1984.* BSI, London.

Fig. 7.7 Standard tar viscometer

Table 7.1 Constants used according to type of cup viscometer	
Viscometer	Constant
Saybolt Universal	0·000 218
Redwood I	0·000 247
Saybolt Furol	0·002 18
Redwood II	0·002 47
Engler	0·007 58
Standard tar viscometer (4 mm)	0·013 2
Standard tar viscometer (10 mm)	0·400

expressed as the kinematic viscosity, ν, in m^2/s, or more commonly mm^2/s, defined as:

$$\nu = \text{flow time} \times \text{constant}$$

7.4 The bitumen test data chart

In the late 1960s, Heukelom developed a system that permitted penetration, softening point, Fraass breaking point and viscosity data to be described as a function of temperature on one chart[169]. This is known

169 HEUKELOM W. *A bitumen test data chart for showing the effect of temperature on the mechanical behaviour of asphaltic bitumens, J Inst Pet, vol 55, pp 404–417, 1969.* Institute of Petroleum, London.

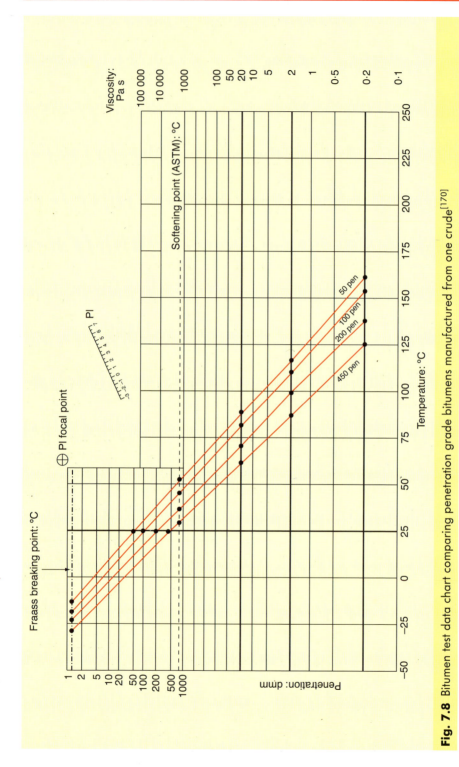

Fig. 7.8 Bitumen test data chart comparing penetration grade bitumens manufactured from one crude[170]

as the bitumen test data chart (BTDC). The chart consists of one horizontal scale for the temperature and two vertical scales for the penetration and viscosity. The temperature scale is linear, the penetration scale is logarithmic and the viscosity scale has been devised so that penetration grade bitumens with 'normal' temperature susceptibility or penetration indices (PIs, discussed in Section 7.5) give straight-line relationships. A typical bitumen test data chart is shown in Fig. 7.8.

The BTDC shows how the viscosity of a bitumen depends on temperature but it does not take account of the loading time. However, as the loading times for penetration, softening point and the Fraass breaking point tests are similar, these test data can be plotted with kinematic viscosity test data on the BTDC, as shown in Fig. 7.8. As the test results on this chart form a straight-line relationship, it is possible to predict the temperature–viscosity characteristics of a penetration grade bitumen over a wide range of temperatures using only the penetration and softening point.

During the manufacture and compaction of an asphalt there are optimum bitumen viscosities. This is illustrated in Fig. 7.9 for a dense bitumen macadam mixture manufactured using 200 pen bitumen. If the viscosity of the bitumen is too high during mixing, the aggregate will not be properly coated and if the viscosity is too low, the bitumen will coat the aggregate easily but may drain off the aggregate during storage or transportation. For satisfactory coating, the viscosity should be approximately 0·2 Pa s.

During compaction, if the viscosity is too low, the mixture will be excessively mobile, resulting in pushing of the material in front of the roller. High viscosities will significantly reduce the workability of the mixture and little additional compaction will be achieved. It is widely recognised that the optimal bitumen viscosity for compaction is between 2 and 20 Pa s.

Thus, the BTDC is a useful tool for ensuring that appropriate operating temperatures are selected to achieve the appropriate viscosity for any grade of bitumen. Consideration of the viscosity requirements during manufacture and application has led to the operating temperatures given in Fig. 7.9. Particular mixtures or circumstances may dictate other operating temperatures.

The chart can also be used for comparing the temperature/viscosity characteristics of different types of bitumens. Three classes of bitumen

170 HEUKELOM W. *An improved method of characterizing asphaltic bitumens with the aid of their mechanical properties, Proc Assoc Asph Pav Tech, vol 42, pp 62–98, 1973.* Association of Asphalt Paving Technologists, Seattle.

131

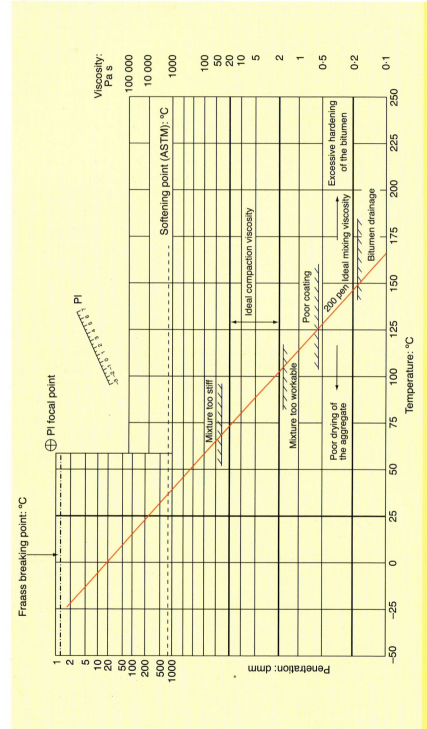

Fig. 7.9 Bitumen test data chart showing 'ideal' bitumen viscosities for optimal mixing and compaction of a dense bitumen macadam

can be distinguished using the BTDC: class S, class B and class W bitumens.

7.4.1 Class S bitumens

The test data for a large group of bitumens can be represented by straight lines on the BTDC, within the repeatability of the test. This group, which has been designated class S (S for straight line), comprises penetration bitumens of different origins with limited wax content. Figure 7.8 shows a chart with straight lines for a number of different penetration grade bitumens manufactured from one base crude. The lines move towards the left of the chart as the bitumens become softer. However, their slopes are equal, indicating that temperature susceptibilities are similar. Bitumens with the same penetration at 25°C but from different origins, are shown in Fig. 7.10. The origin may influence the temperature susceptibility which is reflected by the slope of the line. Accordingly, the temperature viscosity characteristics of S type bitumens may be determined from their penetration and softening point only.

7.4.2 Class B bitumens

The test data of class B (B for blown) bitumens give curves on the chart as shown in Fig. 7.11. The curves can be represented by two intersecting straight lines. The slope of the line in the high-temperature range is about equal to that of an unblown bitumen of the same origin, but the line in the lower temperature range is less steep. Physically, there is no transition point, but it is very convenient that they are still straight lines in the penetration and viscosity regions. Each of them can be characterised with two test values. Thus, in all, four tests are required for a complete description; penetration, softening point and two high-temperature viscosity measurements.

7.4.3 Class W bitumens

Class W (W for waxy) bitumens also give curves consisting of two straight lines which, however, are dissimilar to those of blown bitumens. The two branches of the curve give slopes that are similar but are not aligned. Figure 7.11 shows an example of an S type bitumen together with a curve for a similar bitumen with a wax content of 12 per cent. At low temperatures, when the wax is crystalline, there is hardly any difference between the two curves. At higher temperatures, where the wax is molten, the curve for the waxy bitumen is significantly lower down on the chart. Between the two straight branches, there is a transition range in which the test data are scattered because the thermal history of the sample influences the viscosity result obtained over this range of temperatures.

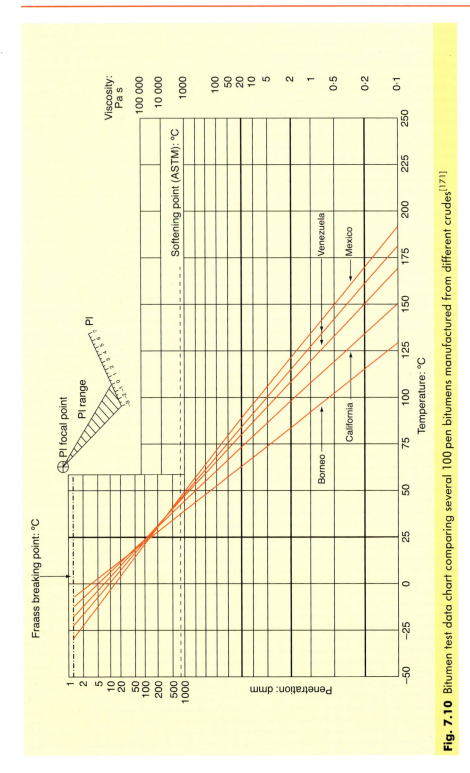

Fig. 7.10 Bitumen test data chart comparing several 100 pen bitumens manufactured from different crudes[171]

171 HEUKELOM W. *A bitumen test data chart for showing the effect of temperature on the mechanical behaviour of asphaltic bitumens, J Inst Pet, vol 55, pp 404–417, 1969.* Institute of Petroleum, London.

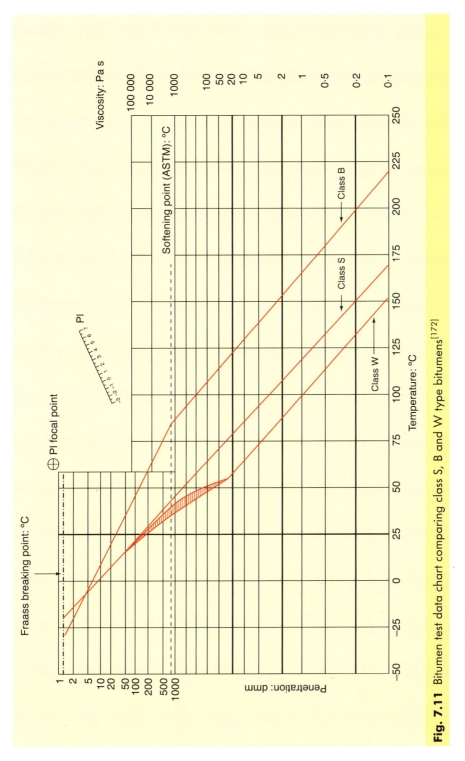

Fig. 7.11 Bitumen test data chart comparing class S, B and W type bitumens[172]

172 HEUKELOM W. *A bitumen test data chart for showing the effect of temperature on the mechanical behaviour of asphaltic bitumens, J Inst Pet, vol 55, pp 404–417, 1969.* Institute of Petroleum, London.

7.5 Temperature susceptibility – penetration index (PI)

All bitumens display thermoplastic properties, i.e. they become softer when heated and harden when cooled. Several equations exist that define the way that the viscosity (or consistency) changes with temperature. One of the best known is that developed by Pfeiffer and Van Doormaal[173]. If the logarithm of penetration, P, is plotted against temperature, T, a straight line is obtained such that:

$$\log P = AT + K$$

where A = the temperature susceptibility of the logarithm of the penetration and K = a constant.

The value of A varies from about 0·015 to 0·06 showing that there may be a considerable difference in temperature susceptibility. Pfeiffer and Van Doormaal developed an equation for the temperature susceptibility that assumes a value of about zero for road bitumens. For this reason, they defined the penetration index (PI) as:

$$\frac{20 - PI}{10 + PI} = 50A$$

or, explicitly,

$$PI = \frac{20(1 - 25A)}{1 + 50A}$$

The value of PI ranges from around -3 for highly temperature susceptible bitumens to around $+7$ for highly blown low-temperature susceptible (high PI) bitumens. The PI is an unequivocal function of A and hence it may be used for the same purpose. The values of A and PI can be derived from penetration measurements at two temperatures, T_1 and T_2, using the equation:

$$A = \frac{\log \text{ pen at } T_1 - \log \text{ pen at } T_2}{T_1 - T_2}$$

The consistency at the softening point can be expressed in terms of penetration, both by linear extrapolation of logarithm of pen versus temperature and by direct measurement with an extra-long penetration needle at the ASTM softening point temperature. Pfeiffer and Van Doormaal found that most bitumens had a penetration of about 800 dmm at the ASTM softening point temperature. Replacing T_2 in the above equation by the ASTM softening point temperature and the penetration at T_2 by

---

173 PFEIFFER J P H and P M VAN DOORMAAL. *The rheological properties of asphaltic bitumens, J Inst Pet, vol 22, pp 414–440, 1936.* Institute of Petroleum, London.

800 they obtained the equation:

$$A = \frac{\log \text{pen } T_1 - \log 800}{T_1 - \text{ASTM softening point}}$$

Substituting this equation in the equation for PI on the previous page and assuming a penetration test temperature of 25°C gives:

$$\text{PI} = \frac{1952 - 500 \log \text{pen} - 20 SP}{50 \log \text{pen} - SP - 120}$$

The assumption of penetration of 800 at the softening point temperature is not valid for all bitumens. It is therefore advisable to calculate the temperature susceptibility using the penetration at two temperatures, T_1 and T_2.

The nomographs shown in Figs. 7.12 and 7.13 enable the approximate value of PI to be deduced from either the penetration at 25°C and the softening point temperature, or the penetration of the bitumen at two different temperatures. Due to the spread of the actual value of penetration at the softening point temperature, the value of PI calculated from one penetration and one softening point may vary from the precise value calculated from two penetrations. However, since the form of properties is generally determined for control of bitumen in terms of the specification, it is normally the case that these figures are used.

One drawback of the PI system is that it uses the change in bitumen properties over a relatively small range of temperatures to characterise bitumen. Extrapolations to extremes of temperature can sometimes be misleading. The PI can be used to give a good approximation of the behaviour to be expected, but confirmation using stiffness or viscosity measurements is desirable.

7.6 Engineering properties of bitumen

Bitumen is widely used as a construction material in civil engineering, building and for industrial applications but its mechanical properties are more complex than materials such as steel or concrete. However, the mechanical properties of bitumens, which are of considerable value to both the civil engineer and the industrial bitumen user, can be defined in terms that are analogous to the elasticity moduli of solid materials.

7.6.1 The concept of stiffness

A viscous material is one that is semi-fluid in nature. When stressed, it will deform or tend to deform, any deformation being unrecovered when the loading is removed. Elastic materials also deform or tend to deform when stressed. However, when the loading is removed, any deformation is fully recovered. Bitumens are visco-elastic materials.

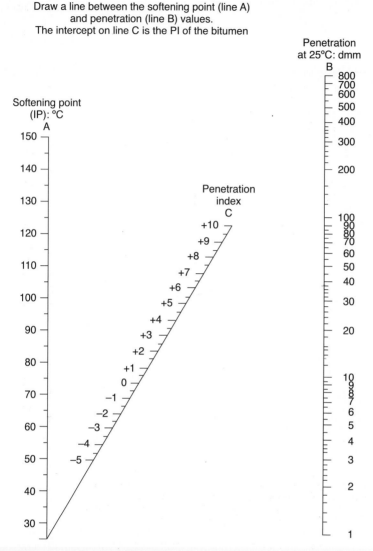

Draw a line between the softening point (line A)
and penetration (line B) values.
The intercept on line C is the PI of the bitumen

Fig. 7.12 Nomograph for penetration index (SP/pen)

The degree to which their behaviour is viscous and elastic is a function of both temperature and period of loading (usually referred to as 'loading time'). At high temperatures or long times of loading, they behave as viscous liquids whereas at very low temperatures or short times of loading they behave as elastic (brittle) solids. The intermediate range of temperature and loading times, more typical of conditions in service, results in visco-elastic behaviour.

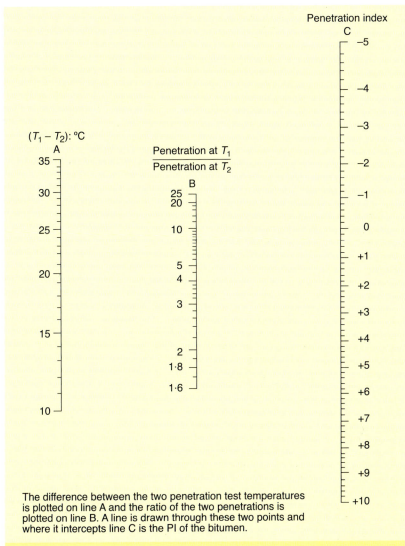

The difference between the two penetration test temperatures is plotted on line A and the ratio of the two penetrations is plotted on line B. A line is drawn through these two points and where it intercepts line C is the PI of the bitumen.

Fig. 7.13 Nomograph for penetration index (pen/pen)

In order to define the visco-elastic properties, Van der Poel[174] introduced, in 1954, the concept of stiffness modulus as a fundamental parameter to describe the mechanical properties of bitumens by analogy with the elastic modulus of solids. Thus, if a tensile stress, σ, is applied at loading time $t = 0$, a strain, ε, is instantly attained that does not increase with loading time. The elastic modulus, E, of the material is

174 VAN DER POEL C. *A general system describing the visco-elastic properties of bitumen and its relation to routine test data, J App Chem, vol 4, pp 221–236, 1954.*

expressed by Hooke's law:

$$E = \frac{\sigma}{\varepsilon}$$

In the case of visco-elastic materials such as bitumen, a tensile stress, σ, applied at loading time $t = 0$, causes a strain, ε, that increases, but not proportionately, with loading time. The stiffness modulus, S_t, at loading time t, is defined as the ratio between the applied stress and the resulting strain at loading time t:

$$S_t = \frac{\sigma}{\varepsilon_t}$$

It follows from the above that the value of the stiffness modulus is dependent on the loading time that is due to the special nature of bitumen. Similarly, the stiffness modulus will also depend on temperature, T. Consequently, it is necessary to state both temperature and time of loading of any stiffness measurement:

$$S_{t,T} = \frac{\sigma}{\varepsilon_{t,T}}$$

The effect of changes in temperature and loading time on the stiffness modulus of three different bitumens are shown in Figs. 7.14, 7.15 and 7.16. Figure 7.14 shows a bitumen of low PI, -2.3. At very short loading

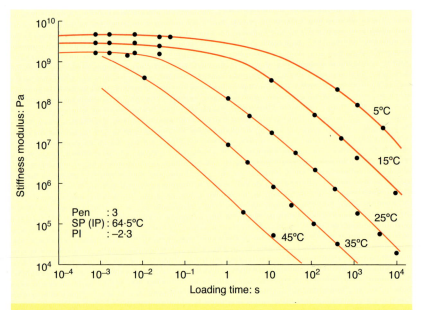

Fig. 7.14 The effect of temperature and loading time on the stiffness of a low PI bitumen

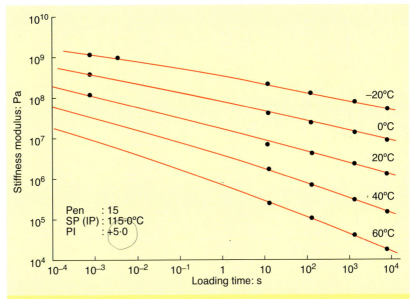

Fig. 7.15 The effect of temperature and loading time on the stiffness of 115/15 bitumen

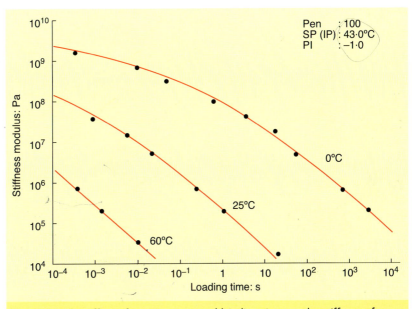

Fig. 7.16 The effect of temperature and loading time on the stiffness of a 100 pen bitumen

times, the stiffness modulus is virtually constant, asymptotic towards 2·5 to 3·0 × 10^9 Pa, and is, in this region, largely independent of temperature and loading time, i.e. $S = E$. The effect of PI is clearly illustrated by comparing Figs. 7.14 and 7.15. The bitumen with the higher PI (+5) (Fig. 7.15) is considerably stiffer at higher temperatures and longer loading times, i.e. it is less temperature susceptible.

Figure 7.16 shows the relationship for a 100 pen bitumen. At a loading time of 0·02 s (which equates to a vehicle speed of around 50 km/h) the stiffness modulus is approximately 10^7 Pa at 25°C but falls to approximately 5 × 10^4 Pa at 60°C. At low temperatures, the stiffness modulus is high and therefore permanent deformation does not occur. However, at higher temperature or longer loading times (stationary traffic), the stiffness modulus is substantially reduced and, under these conditions, permanent deformation of the road surface is much more likely to occur.

In order to fully appreciate the significance of the stiffness modulus and its measurement, it is necessary to consider the deformation under stress of simple solids and liquids. The deformation behaviour of visco-elastic materials can then be derived.

7.6.2 Determination of the stiffness modulus of bitumen

The methods used to measure the stiffness modulus of bitumen are based on shear deformation, as shown in Fig. 7.17. The resistance to shear is

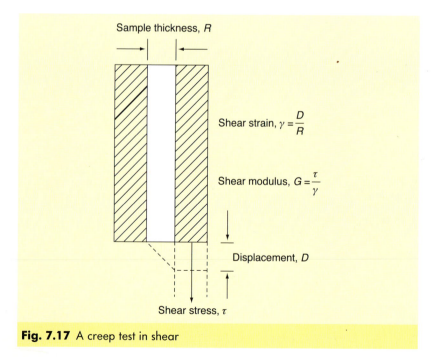

Sample thickness, R

Shear strain, $\gamma = \dfrac{D}{R}$

Shear modulus, $G = \dfrac{\tau}{\gamma}$

Displacement, D

Shear stress, τ

Fig. 7.17 A creep test in shear

expressed in terms of the shear modulus, G, which is defined as:

$$G = \frac{\tau}{\gamma} = \frac{\text{shear stress}}{\text{shear strain}}$$

The elastic modulus and shear modulus are related by the equation:

$$E = 2(1 + \mu)G$$

where μ = Poisson's ratio. The value of μ depends on the compressibility of the material and may be assumed to be 0·5 for almost incompressible pure bitumens while values of <0·5 have to be considered for asphalt mixtures. Thus:

$$E \approx 3G$$

Shear stress can be determined statically in a creep test or dynamically by application of a sinusoidal load. In a creep test, the shear stress is applied from the beginning of the test. The deformation at loading times from about 1 to 10^5 s or longer can be measured. In dynamic tests, the shear stress is usually applied as a sinusoidally varying stress of constant amplitude and fixed frequency. The deformation of the material under test also varies sinusoidally with the same frequency as the applied stress. This is illustrated in Fig. 7.18.

The shear modulus, G_f, at frequency f, is given by the ratio of the amplitudes of shear stress, τ, and shear strain, γ, according to:

$$G_f = \left(\frac{\tau}{\gamma} \right)_f$$

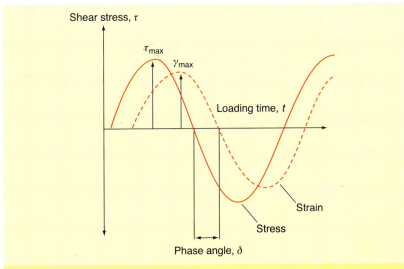

Fig. 7.18 A dynamic test in shear

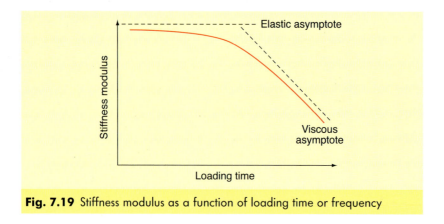

Fig. 7.19 Stiffness modulus as a function of loading time or frequency

It follows that the stiffness modulus under dynamic conditions is

$$S_f = 3G_f$$

Figure 7.18 shows the phase angle, δ, between shear stress and shear strain. This phase angle is a measure of the degree of elasticity of the bitumen under the test conditions. A purely elastic material would not show any phase difference between the stress and strain compared with a phase angle of 90° or one quarter cycle for a purely viscous material. With a visco-elastic material such as bitumen, the phase angle between stress and strain is between 0° and 90°, depending on the type and grade of bitumen, temperature and frequency. Small phase angles are found at low temperature and high frequency and vice versa, indicating that, under these conditions, the bitumen approximates respectively to elastic and viscous behaviour.

By combining creep tests with dynamic tests, a considerable range of stiffness moduli and loading times can be covered. The stiffness modulus as a function of loading time is, therefore, often represented in a graph with logarithmic scales. This is illustrated in Fig. 7.19 in which the asymptotes represent the approximation of elastic and viscous response at short and long loading times respectively.

Measurements of the stiffness as a function of loading time at various temperatures result in a graph of the type shown in Fig. 7.20. It appears that the stiffness–loading time curves obtained at different temperatures on one grade of bitumen all have the same shape and, if shifted along the loading time axis, would coincide. In this case, the bitumen is said to be 'thermorheologically simple'. Most road and industrial grade bitumens belong to this category.

Measurement of stiffness modulus: dynamic shear rheometry
This type of test applies an oscillatory shear force to a bitumen sample sandwiched between two parallel plates. The typical arrangement of

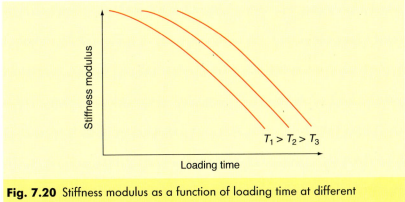

Fig. 7.20 Stiffness modulus as a function of loading time at different temperatures

dynamic shear rheometers (DSRs) is of a fixed lower plate and an oscillating upper plate through which the shear force is applied to the specimen as shown in Fig. 7.21 (the apparatus is shown in Fig. 7.22). The centre line of the upper plate, described by point A in Fig. 7.21, moves to point B then passes through its original position to point C and then returns to point A, representing one cycle. This movement is then repeated continuously throughout the duration of the test.

Two types of DSR exist, controlled stress and controlled strain. In a controlled stress arrangement, a fixed torque is applied to the upper plate to generate the oscillatory motion between points B and C. Because the applied stress level is fixed, the distance the plate moves on its oscillatory path may vary between cycles. In controlled strain testing, the upper plate moves precisely between the extremities of amplitude at the specified frequency and the torque necessary to maintain the oscillation is measured.

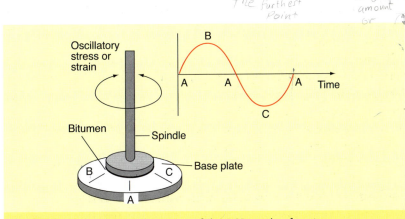

Fig. 7.21 Schematic representation of the DSR mode of testing

145

Fig. 7.22 Dynamic shear rheometer

DSRs can be used to characterise both viscous and elastic behaviour of bitumens by measuring the complex shear modulus, G^*, and the phase angle, δ, from a single test run. For visco-elastic materials like bitumen, the shear modulus is composed of a loss modulus (viscous component, G'') and a storage modulus (elastic component, G'), the relative magnitude of which dictates how the material responds to applied loads. The two components are linked by the phase angle to the complex modulus which can be simply shown by drawing vector arrows as in Fig. 7.23. Hence, the different components can be described by the following equation:

$$G^* = \sqrt{(G'')^2 + (G')^2}$$

The values of G^* and δ for bitumens are dependent on temperature and frequency of loading, so very accurate temperature control of the specimen must be maintained throughout the duration of the test. This is normally achieved by a circulating fluid bath that prevents any temperature gradient from developing within the sample.

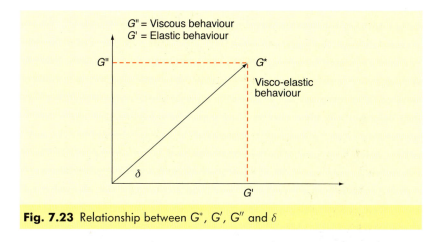

Fig. 7.23 Relationship between G^*, G', G'' and δ

Under normal pavement temperatures and traffic loading, bitumens will exhibit both viscous and elastic characteristics, the relative response of which will vary as conditions change. By identifying the linear visco-elastic range of the bitumen, frequency sweeps can be performed at a series of different temperatures to build up a 'master curve' for the bitumen under test. The master curve showing G^* and δ gives a full rheological profile of the bitumen describing the behaviour of bitumen over a wide range of conditions.

DSR testing provides fundamental data on the properties of bitumens and such data can be used as guidelines to the performance of the bitumens under test. In Section 4.7, the SHRP philosophy for the specification of binders was outlined where binders are categorised according to their fundamental properties when tested at different temperatures with dynamic shear rheometry being used for all high service temperature testing.

Prediction of stiffness modulus

If direct measurement of stiffness modulus is not feasible, it can be predicted using the Van der Poel nomograph[175]. Van der Poel showed that two bitumens of the same PI at the same time of loading have equal stiffnesses at temperatures that differ from their respective softening points by the same amount. Over forty bitumens were tested whose PIs varied from +6·3 to −2·3 at many temperatures and frequencies using both creep and dynamic tests. From the test data, Van der Poel produced a nomograph from which, using only penetration and softening

175 VAN DER POEL C. *A general system describing the visco-elastic properties of bitumen and its relation to routine test data, J App Chem, vol 4, pp 221–236, 1954.*

point[176], it is possible to predict, within a factor of 2, the stiffness modulus of a bitumen for any conditions of temperature and time of loading. Figure 7.24 shows a Van der Poel nomograph with the stiffness modulus determined for a 40/60 pen bitumen at a loading time of 0·02 s and a test temperature of 5°C.

7.6.3 Tensile strength

Another important engineering property of bitumen is its breaking stress and/or tensile strain at break. The tensile properties of bitumen are, like stiffness modulus, dependent on both temperature and loading time. Assuming that the breaking properties of bitumen are dependent on its rheology at the instant of failure, its behaviour can be expressed in terms of the amounts of elastic, delayed elastic and viscous strain as depicted in Fig. 7.25. This shows, accepting the above assumption, that at stiffnesses greater than 5×10^6 Pa, strain at break is a function of stiffness modulus. Only at low stiffness, where viscous deformation is large, is PI a significant factor. At high stiffness, $S = E$. As the strain at break is defined as the ratio between change in length and the instantaneous length, it will be equal to the elongation, i.e. increase in length divided by the initial length, as long as the increase in length remains low.

In practice, break takes place under conditions of large stress. Large stresses normally occur at low temperature, i.e. at high stiffness. Strain at break is then at its minimum. At higher temperatures, i.e. at low stiffness moduli, there will normally be no break at the same load but there will be deformation.

Strain at break can be predicted using the Heukelom nomograph[177], shown in Fig. 7.26, that has been derived from the Van der Poel nomograph. In the example, strain at break of a bitumen with a softening point (IP) of 53·5°C and a PI of zero is given at 12°C. At a loading time of 0·02 s, strain at break is 0·075 (7·5%) and at 10 s is 1·00 (100%).

7.6.4 Fatigue strength

Like many other materials, the strength of bitumen can be reduced by repeated loading, i.e. it fatigues. This is illustrated in Fig. 7.27 where

---

176 AMERICAN SOCIETY for TESTING and MATERIALS. *Standard test method for softening point of bitumen (Ring and ball apparatus), D36-95, 1 Nov 1995.* ASTM, Philadelphia.

177 HEUKELOM W. *Observations on the rheology and fracture of bitumens and asphalt mixes, Proc Assoc Asph Pav Tech, vol 35, pp 358–399, 1966.* Association of Asphalt Paving Technologists, Seattle.

178 VAN DER POEL C. *A general system describing the visco-elastic properties of bitumen and its relation to routine test data, J App Chem, vol 4, pp 221–236, 1954.*

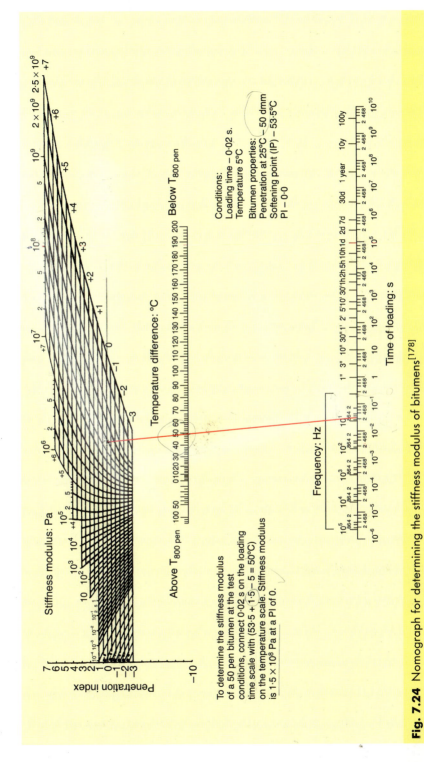

Fig. 7.24 Nomograph for determining the stiffness modulus of bitumens[178]

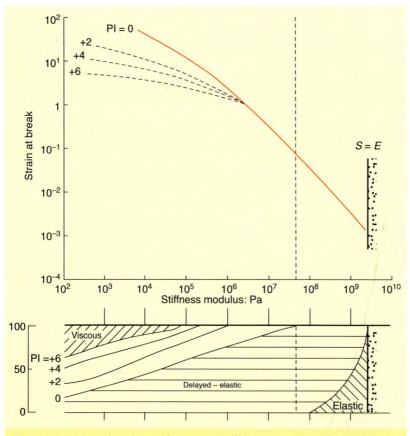

Fig. 7.25 Strain at break as a function of stiffness modulus and penetration index[179]

the curves for fatigue strength and breaking strength are shown as a function of stiffness modulus. All the tests have been carried out by bending at constant stress amplitude. The fatigue strength is the stress that causes failure after 10^3, 10^4, 10^5, and 10^6 load cycles while the breaking strength corresponds to one cycle. The strength is reduced as the number of loading cycles increases. This figure shows that, at high stiffness moduli, the effect of load repetitions is significantly reduced and, at the maximum stiffness modulus ($S = E = 2.7 \times 10^9$ Pa), fatigue strength is independent of the number of load repetitions. For a single

179 HEUKELOM W and P W O WIJGA. *Bitumen testing, 1973.* Koninklijke/Shell-Laboratorium, Amsterdam.

180 HEUKELOM W. *Observations on the rheology and fracture of bitumens and asphalt mixes, Proc Assoc Asph Pav Tech, vol 35, pp 358–399, 1966.* Association of Asphalt Paving Technologists, Seattle.

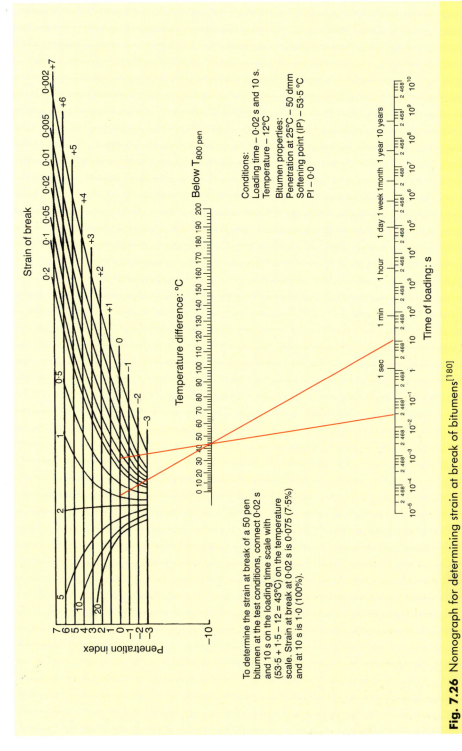

Fig. 7.26 Nomograph for determining strain at break of bitumens[180]

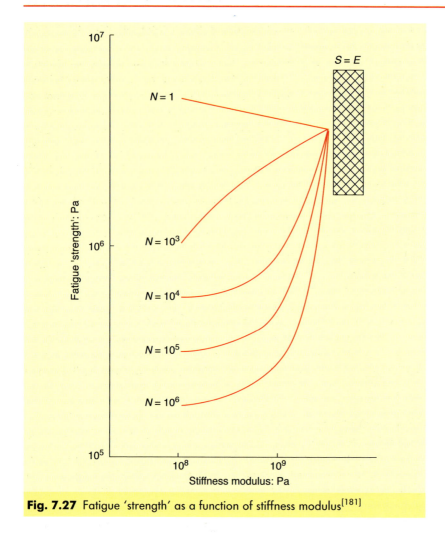

Fig. 7.27 Fatigue 'strength' as a function of stiffness modulus[181]

bend, the fatigue strength is virtually constant, around 4×10^6 Pa for stiffness moduli greater than 5×10^7 Pa.

7.7 Other bitumen tests
7.7.1 Force ductility test
In Section 4.5.1, the traditional ductility test[182] was described. It was explained that the cohesive strength of bitumen is characterised by its

181 HEUKELOM W. *Observations on the rheology and fracture of bitumens and asphalt mixes, Proc Assoc Asph Pav Tech, vol 35, pp 358–399, 1966.* Association of Asphalt Paving Technologists, Seattle.
182 AMERICAN SOCIETY for TESTING and MATERIALS. *Standard test method for ductility of bituminous materials, D113-99, 1 Nov 1999.* ASTM, Philadelphia.

ductility at low temperature. A European version has been developed and is called the force ductility test[183].

The procedure is very similar to that adopted in the traditional method. Samples of bitumen under test are formed into 'dumb-bells' and immersed in a water bath and stretched at a constant rate of 50 mm per minute until fracture occurs. The distance the specimen is stretched before failure is reported as the ductility. The test temperature is adjusted depending on the penetration of the bitumen under test. In these conditions, the test has been found to discriminate between bitumens of different cohesive strengths.

However, some standard bitumens form a very thin thread at large values of strain. This produces misleading test results since the cohesive strength is essentially zero. It has been shown that the modification of bitumen with elastomers produces much thicker threads in the force ductility test, indicating a greater cohesion in such binders compared to that found in unmodified bitumens.

To quantify this effect, the force required to stretch the bitumen sample is recorded during the measurement phase. The area below the resulting force–distance function can be calculated and represents the cohesive energy. Penetration grade bitumen and polymer modified bitumen produce significantly different force–ductility curves and this is illustrated in Figs. 7.28(a) and (b).

7.7.2 Toughness and tenacity test

Polymer modified bitumen emulsions for surface dressing in the UK began to enjoy significant usage in the 1980s and the early part of the 1990s. Throughout that period, there was little by way of laboratory testing available to assist in predicting the likely performance of these binders on site.

The development of the toughness and tenacity test enabled the industry to quantitatively differentiate not only between both modified and unmodified bitumens but also between levels and the nature of modification. It was therefore one of the first 'specialist' tests to be included in tender documents for proprietary surface dressing programmes.

Developed from an American test[184], the procedure involves the pulling of a tension head assembly test piece with a semi-submerged spherical head from within a pre-prepared sample of bitumen (either recovered from the emulsion or sampled pre-emulsification) at constant speed and

183 EUROPEAN COMMITTEE FOR STANDARDIZATION. *Bitumen and bituminous binders, Determination of the tensile properties of modified bitumen by the force ductility method, prEN 13589.* CEN, Brussels.

184 BENSON J R. *New concepts for rubberised asphalts, Roads and Streets, vol 98, no 4,* pp 138–142, Apr 1955.

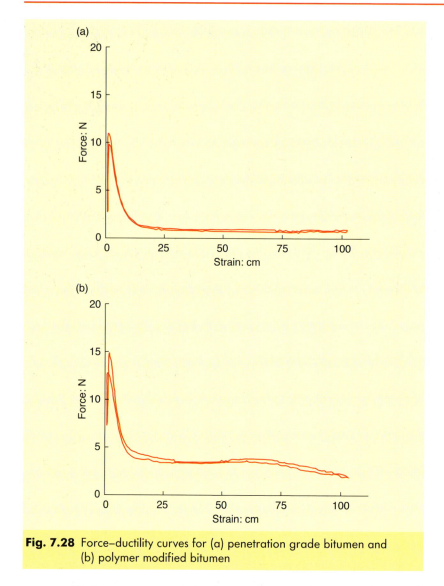

Fig. 7.28 Force–ductility curves for (a) penetration grade bitumen and (b) polymer modified bitumen

at a specified temperature. A load–extension graph is recorded, from which areas under the curve (i.e. work done) can be calculated.

Essentially, the curve is split into two parts, defining two modes of behaviour of the bitumen. An initial peak describes the yielding of the bitumen under tensile force. This part of the curve is primarily due to the behaviour of the base bitumen and results in a low area under the curve. The second part of the graph reflects the elastomeric behaviour of the bitumen and is primarily dependent on the level and type of polymer modification. This portion of the curve may show a secondary increase in force over a prolonged extension, significantly increasing

Fig. 7.29 The Vialit pendulum test

the extension prior to break, therefore substantially increasing the area under the curve.

Toughness is defined as the area under the total curve, i.e. the total work done in removing the test piece from the sample, whereas the tenacity is represented by the area under the curve after the bitumen contribution has been removed.

In 1994, the method was formalised and entered into the Institute of Petroleum 'Standard methods for analysing and testing of petroleum and related products' as a proposed method PM/BQ94, at which level it remained. The method lost favour within the wider industry in the mid to late 1990s for two key reasons; firstly, the correlation between predicted and actual performance for surface dressing binders was not proven and secondly, significant difficulty was encountered in achieving an acceptable level of reproducibility between laboratories. Other tests, such as the Vialit cohesion and DSR began to find favour. As a result, the toughness and tenacity test has been dropped from tender documents since the latter part of the 1990s.

7.7.3 Vialit cohesion

The Vialit pendulum test[185] assesses the degree of adhesion between an aggregate and a bitumen when subjected to a sudden impact. The

---

185 EUROPEAN COMMITTEE FOR STANDARDIZATION. *Bitumen and bituminous binders, Determination of cohesion of bituminous binders for surface dressings, prEN 13588, 1999.* CEN, Brussels.

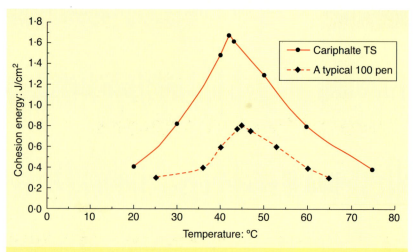

Fig. 7.30 Comparison of cohesion curves for a penetration grade bitumen and a polymer modified bitumen

apparatus used for carrying out the test is shown in Fig. 7.29. The procedure involves placing a thin film of binder between two cubes and measuring the energy required to remove the upper block. It is of greatest significance in situations where aggregate is placed in direct contact with traffic stresses, for example in surface dressings and the chippings in hot rolled asphalt surface courses. The maximum impact energy is usually significantly increased by polymer modification as is the overall energy across the entire temperature range (see Fig. 7.30). Its significance for other materials has yet to be fully evaluated[186].

In addition, this test can be used to assess the effect on the degree of cohesion of variables such as the amount and type of dust, moisture, the type of binder, adhesion agents, etc.

186 LANCASTER I M. (R N HUNTER, ed). *Asphalts in Road Construction, pp 45–106, 2000*. Thomas Telford, London.

Chapter 8

Durability of bitumens

Long-term studies have shown that if an asphalt surfacing is to achieve its design life, it is important that the bitumen is not excessively hardened during hot storage, the asphalt manufacturing process or in service on the road. Bitumen, in common with many organic substances, is affected by the presence of oxygen, ultraviolet radiation and by changes in temperature. In bitumen, these external influences cause it to harden, resulting in a decrease in penetration, an increase in softening point and, usually, an increase in penetration index (PI). In recent years, the phenomenon of the hardening of bitumen and, hence, the hardening of an asphalt mixture has been viewed as being beneficial in the pavement layers as it increases the stiffness of the material and, therefore, the load spreading capabilities of the structure. This hardening is known as 'curing' and is believed to extend the life of a pavement[187].

In the surface course, where the material is exposed to the environment, hardening of the bitumen can have a detrimental effect on its performance and can lead to fretting and/or cracking. This effect is still referred to as 'hardening', the term implying that the change in bitumen properties is detrimental to the service life of the surface course.

8.1 Bitumen hardening

The tendency for bitumen to harden under the influence of the atmosphere has been known and studied for many years. As many as fifteen different factors that influence bitumen ageing have been identified and these are detailed in Table 8.1.

The most important of these fifteen mechanisms of bitumen hardening are:

---

187 NUNN M E, A BROWN, D WESTON and J C NICHOLLS. *Design of long-life flexible pavements for heavy traffic, Report 250, 1997*. TRL, Crowthorne.

Table 8.1 Mechanisms of bitumen ageing[188]

Factors that influence bitumen ageing:	Influenced by:					Occurring:	
	Time	Heat	Oxygen	Sunlight	Beta and gamma rays	At the surface	In the mixture
Oxidation (in dark)	✓	✓	✓			✓	
Photo-oxidation (direct light)	✓	✓	✓	✓		✓	
Volatilisation	✓	✓				✓	✓
Photo-oxidation (reflected light)	✓	✓	✓	✓		✓	
Photo-chemical (direct light)	✓	✓		✓		✓	
Photo-chemical (reflected light)	✓	✓		✓		✓	✓
Polymerisation	✓	✓				✓	✓
Steric or physical	✓					✓	✓
Exudation of oils	✓	✓				✓	
Changes by nuclear energy	✓	✓			✓	✓	✓
Action by water	✓	✓	✓	✓		✓	
Absorption by solid	✓	✓				✓	✓
Absorption of components at a solid surface	✓	✓				✓	
Chemical reactions	✓	✓				✓	✓
Microbiological deterioration	✓	✓	✓			✓	✓

- oxidation
- volatilisation
- steric or physical factors
- exudation of oils.

8.1.1 Oxidation

Like many organic substances, bitumen is slowly oxidised when in contact with atmospheric oxygen. Polar groups containing oxygen are formed and these tend to associate into micelles of higher micellar weight thereby increasing the viscosity of the bitumen. Polar hydroxyl, carbonyl and carboxylic groups are formed, resulting in larger and more complex molecules that make the bitumen harder and less flexible. The degree of oxidation is highly dependent on the temperature, time and the thickness of the bitumen film. The rate of oxidation doubles for each 10°C increase in temperature above 100°C. Hardening due to oxidation has long been held to be the main cause of ageing to the extent that other factors have been given scant consideration. However, it has been shown that although other factors are generally less important than oxidation, they are measurable.

---

188 TRAXLER R N. *Durability of asphalt cements, Proc Assoc Asph Pav Tech, vol 32, pp 44–63, 1963.* Association of Asphalt Paving Technologists, Seattle.

8.1.2 Loss of volatiles

The evaporation of volatile components depends mainly upon temperature and the exposure conditions. Penetration grade bitumens are relatively involatile and therefore the amount of hardening resulting from loss of volatiles is usually fairly small.

8.1.3 Steric or physical hardening

Physical hardening occurs when the bitumen is at ambient temperature and is usually attributed to reorientation of bitumen molecules and the slow crystallisation of waxes. Physical hardening is reversible in that, upon reheating, the original viscosity of the bitumen is obtained.

This phenomenon can be easily reproduced in the laboratory by measuring the penetration of a sample of bitumen (without reheating) over a period of time. As time passes, the penetration of the sample will fall.

8.1.4 Exudative hardening[189]

Exudative hardening results from the movement of oily components that exude from the bitumen into the mineral aggregate. It is a function of both the exudation tendency of the bitumen and the porosity of the aggregate.

8.2 Hardening of bitumen during storage, mixing and in service

The circumstances under which hardening occurs vary considerably. During storage, the bitumen is in bulk at a high temperature for a period of days or weeks. During mixing, hot storage, transport and application, the bitumen is a thin film at high temperature for a relatively short period. In service, the bitumen is a thin film at a low or moderate temperature for a very long time. The degree of exposure to the air of asphalts in service is important and dependent upon the void content of the mixture. In dense, well compacted mixtures, the amount of hardening is relatively small whilst asphalts that have a more open constitution, such as porous asphalt, will undergo significant hardening.

8.2.1 Hardening of bitumen in bulk storage

Very little hardening occurs when bitumen is stored in bulk at high temperature. This is because the surface area of the bitumen that is exposed to oxygen is very small in relation to the volume. However, if the bitumen is being circulated and is falling from the pipe entry at the top

189 VAN GOOSWILLIGEN G, F TH DE BATS and T HARRISON. *Quality of paving grade bitumen – a practical approach in terms of functional tests, Proc 4th Eurobitume Symp, pp 290–297, Madrid, Oct 1989.*

of the tank to the surface of the bitumen, significant hardening may occur. This arises because the surface area of the bitumen will be relatively large as it falls from the entry pipe exposing it to the action of the oxygen. This effect can be minimised using the storage tank layout shown in Figure 2.4.

8.2.2 *Hardening of bitumen during mixing with aggregate*

This is described as 'short-term ageing', a term that is also applied to hardening which occurs during laying. During the mixing process, all the aggregate and filler is coated with a thin film of bitumen usually between 5 and 15 µm thick. If the bitumen from 1 tonne of dense coated macadam was spread at 10 µm thick, it would occupy an area of around $10\,000\,m^2$, the equivalent of over one and a half average-sized football pitches. Thus, when bitumen is mixed with hot aggregate and spread into thin films in an asphalt pugmill, conditions are ideal for the occurrence of oxidation and the loss of volatile fractions within the bitumen. Hardening of bitumen during this process is well known and is taken into account when selecting the grade of bitumen to be used. As a very rough approximation, during mixing with hot aggregate in a convention pugmill, the penetration of a paving grade bitumen falls by about 30%. However, the amount of hardening depends on a number of factors such as temperature, duration of mixing, bitumen film thickness, etc. The minimisation of hardening during mixing depends on careful control of all these factors. Control of the temperature and the bitumen content are particularly critical. Figure 10.2 clearly shows increasing bitumen hardening, measured by higher values of softening point, as mixing temperatures are raised. Similarly, Fig. 8.1 shows that reducing the thickness of the bitumen film significantly increases the viscosity of the bitumen. The latter is measured by the ageing index which is defined as the ratio of the viscosity of the aged bitumen, η_a, to the viscosity of the virgin bitumen, η_o.

The type of mixer used also affects the amount of hardening during mixing. It has been recognised for some time that the amount of hardening in a drum mixer is often less than that which occurs in a conventional batch mixer[190]. This is probably the result of shorter bitumen/aggregate mixing times and the presence of steam in the drum, which limits the availability of oxygen. However, the multiplicity of different designs of drum mixers means that variation in the amount of hardening between different designs of plant is almost inevitable. Notwithstanding, a study carried out by Shell Bitumen on two different drum mixers showed

190 HAAS S. *Drum-dryer mixes in North Dakota, Proc Assoc Asph Pav Tech, vol 34, p 417, 1974.* Association of Asphalt Paving Technologists, Seattle.

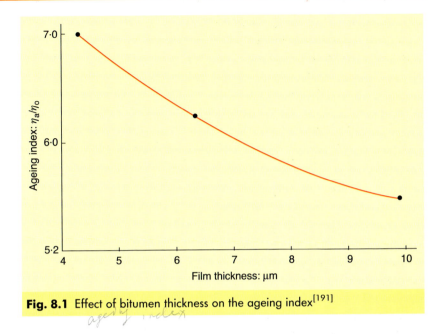

Fig. 8.1 Effect of bitumen thickness on the ageing index[191]

that, for equivalent mixing temperatures, the overall reduction in penetration and increase in softening point can be less than half of that which occurs in a conventional batch mixer.

8.2.3 *Hardening of bitumen in asphalts during hot storage, transport and laying*

Drum mixers, which produce large volumes of asphalts, require silo storage for the mixed material to account for peaks and troughs in demand. In such circumstances, mixtures are stored in hot silos as well as in the delivery vehicle during transportation.

Some hardening of the bitumen will take place during hot storage whether it is in a silo or in a truck. It was stated above that the amount of hardening will depend principally upon the duration of exposure to oxygen, the thickness of the bitumen film and the temperature of the mixture. When a mixture is discharged into a storage silo, air enters with the mixture and some is trapped in the voids of the material. During the storage period, some of the oxygen in this entrained air will react with the bitumen. If no additional air enters the silo, oxidation of the bitumen will cease.

191 GRIFFIN R L, T K MILES and C J PENTHER. *Microfilm durability test for asphalt, Proc Assoc Asph Pav Tech, vol 24, p 31, 1955.* Association of Asphalt Paving Technologists, Seattle.

It is important that the entry and discharge gates are airtight and that there are no other openings where air can enter the silo. If the discharge gate is not airtight, the silo may behave like a chimney drawing air in at the discharge gate (which exits the loading gate) resulting in oxidation and cooling of the stored material. In addition, the silo should be as full as practicable in order to minimise the amount of free air at the top of the silo. Air remaining at the top of the silo will react with the top surface of the material. This reaction forms carbon dioxide that, because it is heavier than air, tends to blanket the surface of the mixture protecting it from further oxidation. In the USA, some silos have the facility to be pressurised with exhaust gases, containing no oxygen, from a burner. These exhaust gases purge the silo of entrained air and provide a slight positive pressure preventing more air entering the silo.

Studies carried out in the USA[192] suggest that if oxidation in the silo is limited to that induced by entrained air then little or no additional oxidation will occur during transportation and laying. It is hypothesised that this is because no significant quantity of fresh air is entrained in the mixture during discharge into the truck. Thus, little or no additional air is available for oxidation. In fact, it was observed that if the mixture was discharged directly from the pugmill into the delivery vehicle, the amount of hardening during transportation was very similar to that which occurs during silo storage.

If materials are being laid at low ambient temperatures or if the mixture has to be retained for a period in hot storage, there is a temptation to increase mixing temperatures to offset these two factors. However, increasing the mixing temperature will considerably accelerate the rate of bitumen oxidation, resulting in an increase in bitumen viscosity. Thus, a significant proportion of the reduction in viscosity achieved by increasing the mixing temperature will be lost because of additional oxidation of the bitumen which may adversely affect the long-term performance of the material.

8.2.4 *Hardening of bitumen on the road*

As explained above, a significant amount of bitumen hardening occurs during mixing and, to a lesser extent, during hot storage and transportation. However, hardening of the binder will continue on the road until some limiting value is reached. This behaviour is described as 'long-term ageing' and is illustrated in Fig. 8.2 which shows the ageing index of the bitumen after mixing, storage, transport, paving and subsequent service.

---

192 BROCK J D. *Oxidation of asphalt, Astec Industries Technical Bulletin T-103, 1986.*

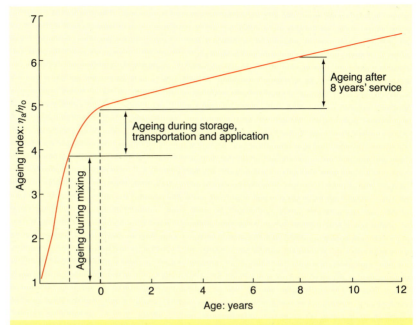

Fig. 8.2 Ageing of bitumen during mixing, subsequently during storage, transportation and application and in service.
Note that ageing index is not a fundamentally defined parameter – it is usually a ratio of two values (e.g. viscosity, stiffness, penetration) measured at different times

Bitumen hardening in asphalt surface courses

The main factor that influences bitumen hardening on the road is the void content of the mixture. Table 8.2 shows the properties of bitumens recovered from three asphaltic concrete mixtures after 15 years' service. The bitumen recovered from the mixture with the lowest void content had hardened very little. However, where the void content was high, allowing constant ingress of air, substantial hardening had occurred. The PI of the material with the highest void content had increased appreciably. This limits the amount of stress relaxation that can occur and may result in cracking of the compacted material. Similarly, Fig. 8.3 shows the in situ bitumen properties of five year old asphaltic concrete with void contents ranging from 3 to 12%[193]. At void contents less than 5%, very little hardening occurred in service. However, at void contents greater than 9%, the in situ penetration fell from 70 to less than 25.

193 LUBBERS H E. *Bitumen in de weg- en waterbouw, Apr 1985*. Nederlands Advies-bureau voor Bitumentoepassingen, Gouda.

Table 8.2 Hardening of bitumen in service[194]

Road	A	B	C
Voids in mix, %	4	5	7
Properties after mixing and laying			
Softening point (IP), °C	64	63	66
Penetration at 25°C, dmm	33	33	30
Penetration index (Pen, SP)	+0·7	+0·7	+0·9
Stiffness (S_0), Pa (calc)			
10^4 s, 25°C	$1·4 \times 10^3$	$1·4 \times 10^3$	$2·5 \times 10^3$
10^4 s, 0°C	$5·0 \times 10^5$	$5·0 \times 10^5$	$7·0 \times 10^5$
10^{-2} s, 25°C	$2·5 \times 10^7$	$3·0 \times 10^7$	$3·0 \times 10^7$
10^{-2} s, 0°C	$3·0 \times 10^8$	$3·0 \times 10^8$	$3·0 \times 10^8$
Properties after 15 years' service:			
Softening point (IP), °C	68	76	88
Penetration at 25°C, dmm	24	15	11
Penetration index (Pen, SP)	+0·8	+1·1	+2·1
Stiffness (S_{15}), Pa (calc)			
10^4 s, 25°C	4×10^3	20×10^3	150×10^3
10^4 s, 0°C	13×10^5	40×10^5	80×10^5
10^{-2} s, 25°C	4×10^7	7×10^7	8×10^7
10^{-2} s, 0°C	4×10^8	6×10^8	6×10^8
Ageing index (S_{15}/S_0)			
10^4 s, 25°C	2·8	14	60
10^4 s, 0°C	2·6	8	11
10^{-2} s, 25°C	1·6	2·3	2·7
10^{-2} s, 0°C	1·3	2·0	2·0

Note: Ageing index is not a fundamentally defined parameter. It is usually a ratio of two values (e.g. viscosity, stiffness, penetration) measured at different times.

The bitumen at the surface of the road hardens much more quickly than the bitumen in the bulk of the pavement. There are three reasons for this:

- the existence of a constant supply of fresh oxygen;
- the occasional incidence of high temperatures at the road surface; and
- the occurrence of photo-oxidation of the bitumen by ultraviolet radiation.

Photo-oxidation causes a skin, 4 to 5 µm thick, to be rapidly formed on the surface of the bitumen film. This is induced by natural ultraviolet radiation that, it is believed, is absorbed into the upper 10 µm of a film of bitumen. It has been shown that the skin formation can retard oxygen absorption and loss of volatiles. However, the oxidised material

194 EDWARDS J M. *Dense bituminous mixes, Paper presented to the Conf on road engineering in Asia and Australasia, Kuala Lumpur, Jun 1973.*

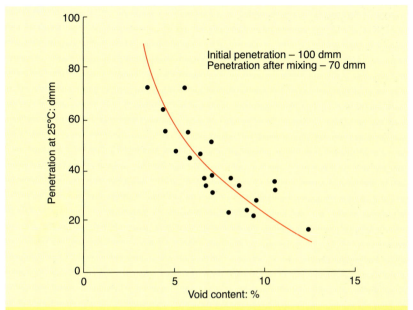

Fig. 8.3 The effect of void content on the hardening of bitumen on the road[195]

is soluble in rainwater and can be abraded away resulting in the exposure of fresh bitumen[196].

Although both continuously-graded and gap-graded surface course mixtures are regarded as dense, gap-graded materials are considered to be more durable. One factor that affects the durability is the permeability of the mixture to air. It is considered that gap-graded mixtures are generally less permeable to air than continuously-graded mixtures for similar air void contents[197]. This is because the voids in gap-graded mixtures are generally discrete and not interconnecting as may be the case with continuously-graded mixtures.

The speed of surface oxidation is illustrated in Fig. 8.4 in which penetration, softening point and PI of bitumens recovered from the surface and bulk of an asphaltic concrete mix over a number of years are plotted[195]. After seven years, the penetration at the surface is 25 whereas the bitumen in the bulk of the mixture has a penetration of 45. Similarly,

195 LUBBERS H E. *Bitumen in de weg- en waterbouw, Apr 1985*. Nederlands Adviesbureau voor Bitumentoepassingen, Gouda.

196 DICKINSON E J, J H NICHOLAS and S BOAS-TRAUBE. *Physical factors affecting the absorption of oxygen by thin films of bituminous binders, J App Chem, vol 8, p 673, 1958.*

197 MARAIS, C P. *Gap-graded asphalt surfacings – the South African scene, Proc 2nd Conf on asphalt paving for Southern Africa, vol 3, 1974.*

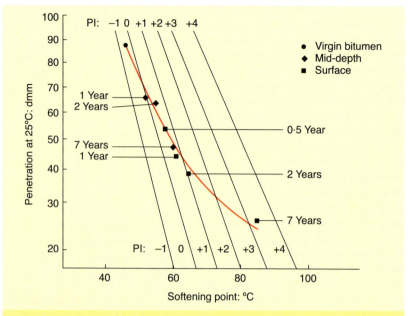

Fig. 8.4 Hardening of bitumen at the surface and at mid-depth of an asphalt concrete surface course[198]

there is a substantial difference in the PI. In the bulk of the mixture, the PI has only marginally changed whereas the PI of the bitumen at the surface has increased to above 4. Surface oxidation is, in fact, desirable because it enables bitumens to be eroded away relatively quickly exposing new aggregate surfaces and thereby improving the skid resistance of the road surface.

Bitumen content (bitumen film thickness) also plays a very important role in the speed of bitumen hardening on the road. Laboratory work has shown[199] that the oxidation of a film of bitumen at high ambient temperatures, i.e. between 40 and 60°C, is limited to a depth of about 4 μm as shown in Fig. 8.5. The significance of this will depend on the average bitumen film thickness that, as stated earlier, can range from 5 to 15 μm. It has been suggested[200] that a minimum bitumen film

198 LUBBERS H E. *Bitumen in de weg- en waterbouw, Apr 1985.* Nederlands Adviesbureau voor Bitumentoepassingen, Gouda.

199 BLOKKER P C and H VAN HOORN. *Durability of bitumen in theory and practice, Proc 5th World Petroleum Congress, paper 27, p 417, 1959.*

200 CAMPEN W H, J R SMITH, L G ERICKSON and L R MERTZ. *The relationship between voids, surface area, film thickness and stability in bituminous paving mixtures, Proc Assoc Asph Pav Tech, vol 28, p 149, 1959.* Association of Asphalt Paving Technologists, Seattle.

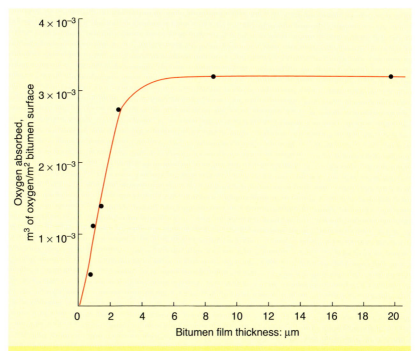

Fig. 8.5 Relationship between oxygen absorbed at 50°C and film thickness[201]

thickness of between 6 and 8 μm is required for satisfactory performance of both continuously-graded and gap-graded mixtures. Thus, in general, the lower the bitumen content, the thinner the bitumen film and the faster hardening will occur.

The bitumen film thickness can be determined using the methods given in Appendix 4. Typical bitumen film thicknesses are as follows:

- dense coated macadam   >5 μm
- hot rolled asphalt   >7 μm
- porous asphalt   >12 μm

This method of calculating bitumen film thickness determines surface area factors for the aggregate down to 75 μm. Clearly, some of the material passing this sieve will be substantially finer than 75 μm, particularly for materials where limestone filler is added to the mixture. Therefore, for high filler content materials such as hot rolled asphalt, the bitumen film thickness will be lower than the value given above. However, if it is assumed that the bitumen and filler coat the coarse

201 MARAIS, C P. *Gap-graded asphalt surfacings – the South African scene, Proc 2nd Conf on asphalt paving for Southern Africa, vol 3, 1974.*

and fine aggregate then the above values underestimate the binder film thickness.

Bitumen curing in asphalt bases

Contrary to the effects of bitumen hardening in the surface course, a gradual hardening of the main structural layers of the pavement appears to be beneficial and is described as 'curing'. As with all hot mixed asphalts, the penetration of the bitumen in a base will harden by approximately 30% during the mixing and laying process and, despite asphalt bases being locked inside the pavement construction and shielded from exposure to the environment, the penetration of the bitumen will continue to exhibit varying rates of hardening.

The Transport Research Laboratory (TRL) has carried out an investigation into the curing of bases using small scale untrafficked test pavements comprising a granular foundation over which was laid a 100 pen dense bitumen macadam (DBM) base overlaid by a surfacing[202]. Over a two year period, cores were taken from the trial section and properties of both the macadam and recovered bitumen were measured. Stiffness values of the cores and recovered binder penetrations were obtained to monitor the effect of curing on the materials.

Although the observed curing behaviour was variable, the test pavements showed that the stiffness modulus of the DBM base could double over the first 12 months of service – a feature largely attributable to binder hardening. Information from the trial sections was complemented with data obtained by the TRL from a large number of live pavements that showed that after 15 years' service, the bitumen penetration had fallen from an initial value of 100 to lie in a range between 20 and 50. Corresponding increases in the stiffness values of cores from the aged pavements illustrated that after 20 years' service, stiffnesses ranged from 3 to 15 GPa compared to initial values of 1·25 to 3 GPa.

Clearly, the increase in stiffness of asphaltic bases over time will improve the structural competence of a road pavement and has implications with respect to the structural design of flexible roads (see Chapter 18).

8.3 Bitumen ageing tests

It is clearly desirable that there should be laboratory tests that quantitatively determine the resistance of bitumens to hardening at the various stages during the production process. A number of tests already exist to measure the effect of heat and air on bitumen. The main aim of

202 NUNN M E, A BROWN, D WESTON and J C NICHOLLS. *Design of long-life flexible pavements for heavy traffic, Report 250, 1997.* TRL, Crowthorne.

these tests is to identify bitumens that are too volatile or are too susceptible to oxidation to perform well in service.

The thin-film oven test (TFOT)[203] simulates practical conditions. In
this test, the bitumen is stored at 163°C for five hours in a layer 3·2 mm
thick. It is claimed that in this test, the amount of hardening that takes
place is about the same as that which occurs in practice. However, diffusion in the bitumen film is also limited and it is not possible to obtain
homogeneous hardening or ageing. Accordingly, the test is far from
ideal. This test was adopted initially by the American Society for Testing
Materials (ASTM) in 1969 as method ASTM D1754 and has been
modified since that time to include improvements[204].

In 1963, the State of California Department of Public Works, Division
of Highways, developed a test that more accurately simulates what
happens to a bitumen during mixing. It is called the rolling thin-film
oven test (RTFOT)[205,206]. In this test, eight cylindrical glass containers
each containing 35 g of bitumen are fixed in a vertically rotating shelf.
During the test, the bitumen flows continuously around the inner surface
of each container in a relatively thin film with preheated air blown
periodically into each glass jar. The test temperature is normally 163°C
for a period of 75 minutes. The method ensures that all the bitumen is
exposed to heat and air and continuous movement ensures that no skin
develops to protect the bitumen. A homogeneously aged material, similar
to that which is produced during full-scale mixing, is obtained. Clearly,
the conditions in the test are not identical to those found in practice
but experience has shown that the amount of hardening in the RTFOT
correlates reasonably well with that observed in a conventional batch
mixer.

The RTFOT was accepted in 1970 by the ASTM as method ASTM
D2872 and was included as part of the European specification for
paving grade bitumens in BS EN 12591[207].

203 LEWIS R H and J Y WELBORN. *Report on the properties of the residues of 50-60 and
85-100 penetration asphalts from oven tests and exposure, Proc Assoc Asph Pav Tech, vol
11, pp 86–157, 1940.* Association of Asphalt Paving Technologists, Seattle.

204 AMERICAN SOCIETY for TESTING and MATERIALS. *Standard test method
for effect of heat and air on asphaltic materials, D1754-97, 1 Jan 1997.* ASTM, Philadelphia.

205 HVEEM F N, E ZUBE and J SKOG. *Proposed new tests and specifications for paving
grade asphalts, Proc Assoc Asph Pav Tech, 1963.* Association of Asphalt Paving
Technologists, Seattle.

206 SHELL BITUMEN. *The rolling thin-film oven test, Shell Bitumen Review 42, p 18,
1973.* Shell International Petroleum Company Ltd, London.

207 BRITISH STANDARDS INSTITUTION. *Bitumen and bituminous binders, Specifications for paving grade bitumens, BS EN 12591: 2000.* BSI, London.

Over the years there have been a number of attempts to simulate the long-term ageing of bitumen in situ. This has proved to be extremely difficult because of the number of variables that affect binder ageing – void content, mixture type, aggregate type, etc. The US Superpave specification[208] uses the RTFOT to simulate initial ageing followed by ageing over 20 hours at elevated temperature (90, 100 or 110°C) and pressure 2070 kPa in a pressure ageing vessel (PAV)[209,210]. After this ageing procedure, the residue is used for dynamic shear rheometry, bending beam rheometry and direct tension testing. The PAV using modified conditions is currently being considered in Europe as a method for ageing bitumens in the laboratory. The artificial ageing of binders in the PAV to simulate ageing in situ has still to be fully validated.

208 ASPHALT INSTITUTE. *Superpave performance graded asphalt binder specification and testing, Superpave Series No 1, pp 15–19, 1997.* AI, Kentucky.
209 ANDERSON D A, D W CHRISTIANSEN, H U BAHIA, R DONGRE, M G SHARMA, C E ANTLE and J BUTTON. *Binder characterization, vol 3: Physical characterization, SHRP-A-369, Strategic Highways Research Program, 1994.* National Research Council, Washington DC,
210 HARRIGAN E T, R B LEAHY and J S YOUTCHEFF. *The SUPERPAVE mix design system manual of specifications, Test methods and practices. SHRP-A-379, Strategic Highways Research Program, 1994.* National Research Council, Washington DC,

Chapter 9

Adhesion of bitumens

The primary function of bitumen is to act as an adhesive. It is required either to bind aggregate particles together or to provide a bond between particles and an existing surface. Although the incidence of premature failure attributed to adhesion is relatively rare, failures when they occur may involve considerable expense. The need to ensure adhesion between the aggregate and the bitumen is very important. In the UK, the traditional dense, high bitumen content mixtures, such as hot rolled asphalt, have now been replaced by high stone content materials that contain less bitumen. These materials tend to be laid in thinner layers where in situ stressing places more demand on the adhesive properties of the bitumen in the mixture.

The adhesion of bitumen to most types of dry and clean aggregate presents few problems. However, aggregate is easily wetted by water, the presence of which can result in unexpected problems. These may occur at any time during the life of an asphalt from the initial coating of the aggregate during the mixing process to maintaining an adequate bond between the bitumen and aggregate under trafficking conditions. The aim of this chapter is to draw attention to aggregate/bitumen adhesion and how it may be possible to limit the likelihood of associated premature failures in service.

9.1 The principal factors affecting bitumen/aggregate adhesion

Table 9.1 summarises the main factors that influence bitumen/aggregate adhesion. It is considered that approximately 80% of these factors are controllable during production and construction. The importance of aggregate mineralogical composition has been recognised for many

171

Table 9.1 Material properties and external factors that can affect the bitumen/aggregate bond

Aggregate properties	Bitumen properties	Mixture properties	External factors
Mineralogy	Rheology	Void content	Rainfall
Surface texture	Electrical polarity	Permeability	Humidity
Porosity	Constitution	Bitumen content	Water pH
Dust		Bitumen film thickness	Presence of salts
Durability		Filler type	Temperature
Surface area		Aggregate grading	Temperature cycling
Absorption		Type of mixture	Traffic
Moisture content			Design
Shape			Workmanship
Weathering			Drainage

years[211,212]. The physico-chemical character of the aggregate is important with adhesion related to chemical composition, shape, structure, residual valency and surface area. Generalisations about the influence of mineralogy are difficult because of the effects of grain size, shape and texture.

The majority of adhesive failures have been associated with siliceous aggregates such as granites, rhyolites, quartzites, cherts etc. The fact that satisfactory performance is achieved with these same aggregates and that failures occur using aggregates that have good resistance to stripping, e.g. limestone, emphasises the complexity of bitumen/aggregate adhesion and the possibility that some other factors may play a role in the failure.

Failure of the aggregate/bitumen bond is commonly referred to as 'stripping'. One of the main factors is the type of aggregate. This has a considerable influence on bitumen adhesion due to differences in the degree of affinity for bitumen. The vast majority of aggregates are classified as 'hydrophilic' (water loving) or 'oleophobic' (oil hating). Aggregates with a high silicon oxide content, e.g. quartz and granite (i.e. acidic rocks) are generally more difficult to coat with bitumen than basic rocks such as basalt and limestone.

The residual valence or surface charge of an aggregate has been shown to be important[213,214]. Aggregates with unbalanced surface charges

211 SAAL R N J. *Adhesion of bitumen and tar to solid road building materials, Bitumen, vol 3, p 101, 1933*.

212 WINTERKORN H F. *Surface chemical aspects of the bond formation between bituminous materials and mineral surfaces, Proc Assoc Asph Pav Tech, vol 17, pp 79–85, 1936.* Association of Asphalt Paving Technologists, Seattle.

213 LEE A R and J H NICHOLAS. *The properties of asphaltic bitumen in relation to its use in road construction, J Inst Pet, vol 43, pp 235–246, 1957.* Institute of Petroleum, London.

214 ORCHARD D F. *Properties and testing of aggregates, Concrete Technology, vol 3, 3rd ed, p 281, 1976.* Applied Science Publications, London.

possess a surface energy. If the aggregate surface is coated with a liquid of opposite polarity, the surface energy demands can be satisfied and an adhesive bond will result. Where two liquid phases are present, e.g. bitumen and water, the liquid that can best satisfy the energy requirement will adhere most tenaciously to the aggregate. The phenomenon of stripping of the bitumen in the presence of water can therefore be related to the surface charges. Where water equilibrates the surface charge better than bitumen, the bitumen may separate from the aggregate surface in favour of the water.

Physico-mechanical adsorption of bitumen into the aggregate depends on several factors including the total volume of permeable pore space, the size of the pore openings[215], the viscosity and surface tension of the bitumen. The presence of a fine microstructure of pores, voids and micro-cracks can bring about an enormous increase in the adsorptive surface available to the bitumen. It has also been shown[216,217] that fractions of the bitumen are strongly adsorbed in the aggregate surface to a depth of approximately $180 \text{ Å} (18 \times 10^{-9} \text{ m})$.

Other factors affecting the initial adhesion and subsequent bond are the surface texture of the aggregate, the presence of dust on the aggregate and, to a lesser extent, the pH of the water in contact with the interface. It is generally agreed that rougher aggregate surfaces have better adhesion characteristics. However, a balance is required between wetting of the aggregate (smooth surfaces being more easily wetted) and rougher surfaces which hold the bitumen more tenaciously once wetting has been achieved.

It has been suggested that the good mechanical bond achieved on a rough aggregate can be more important than the aggregate mineralogy in maintaining bitumen/aggregate adhesion[218]. As shown in Table 9.1, the properties of the bitumen are also important in the acquisition and subsequent retention of the bitumen/aggregate bond. In particular, the viscosity of the bitumen during coating and in service, polarity and

215 THELAN E. *Surface energy and adhesion properties in asphalt/aggregate systems, Highway Research Board Bulletin, vol 192, pp 63–74, 1958.* Transportation Research Board, Washington.

216 PLANCHER H, S M DORRENE and J C PETERSEN. *Identification of chemical types strongly absorbed at the asphalt-aggregate interface and their displacement by water, Proc Assoc Asph Pav Tech, vol 46, pp 151–175, 1977.* Association of Asphalt Paving Technologists, Seattle.

217 SCOTT J A N. *Adhesion and disbonding mechanisms of asphalt used in highway construction and maintenance, Proc Assoc Asph Pav Tech, vol 47, pp 19–48, 1978.* Association of Asphalt Paving Technologists, Seattle.

218 LEE A R and J H NICHOLAS. *The properties of asphaltic bitumen in relation to its use in road construction, J Inst Pet, vol 43, pp 235–246, 1957.* Institute of Petroleum, London.

constitution all influence adhesion characteristics. However, it is the nature of the aggregate that is, by far, the most dominant factor influencing bitumen/aggregate adhesion.

9.2 The main disbonding mechanisms[219]

Many studies have been carried out to determine the mechanism of bitumen disbonding in asphalts[220,221]. There are two main methods by which the bitumen/aggregate system may fail, i.e. adhesive and cohesive mechanisms. If the aggregate is clean and dry and the mixture is effectively impermeable, the mode of failure will be cohesive. However, in the presence of water, the failure mode will almost certainly be due to a loss of adhesion caused by stripping of the bitumen from the aggregate surface. Several mechanisms of disbonding are possible and these are discussed below.

9.2.1 Displacement

The displacement theory[222] relates to the thermodynamic equilibrium of the three-phase bitumen/aggregate/water system. If water is introduced at a bitumen/aggregate interface then consideration of the surface energies that are involved shows that the bitumen will retract along the surface of the aggregate. Figure 9.1 shows an aggregate particle embedded in a bitumen film with point A representing the equilibrium contact position when the system is dry. When in contact with water, the equilibrium point shifts and the new interface moves or retracts over the surface to point B. This new equilibrium position has a contact angle that will depend on the type and viscosity of the bitumen used.

9.2.2 Detachment

Detachment occurs when a thin film of water or dust separates the bitumen and aggregate with no obvious break in the surface of the bitumen film being apparent. Although the bitumen film completely encapsulates the aggregate particle, no adhesive bond exists and the bitumen can easily be peeled from the aggregate surface. This process may be reversible, i.e.

219 ASPHALT INSTITUTE. *Cause and prevention of stripping in asphaltic pavements, Educational Series No 10, 1981.* AI, Kentucky.

220 HUGHES R I, D R LAMB and O PORDES. *Adhesion of bitumen macadam, J App Chem, vol 180, pp 433–440, 1960.*

221 TAYLOR M A and N P KHOSLA. *Stripping of asphalt pavements: State of the art, Transportation Research Record 911, pp 150–157, 1983.* Transportation Research Board, Washington.

222 BLOTT J F T, D R LAMB and O PORDES. *Weathering and adhesion in relation to the surface dressing of roads with bituminous binder, Adhesion and adhesives: Fundamentals and practice, 1954.* Society of Chemical Industry, London.

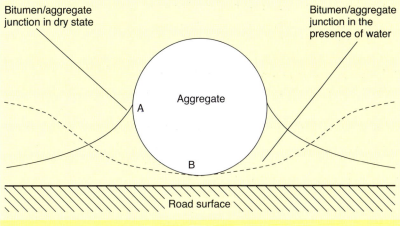

Fig. 9.1 Retraction of the bitumen/water interface in the presence of water[223]

if the water is removed the bitumen may re-adhere to the aggregate. A second mechanism of disbonding must occur to allow the ingress of water between the bitumen and the aggregate.

9.2.3 *Film rupture*

Film rupture may occur despite the fact that the bitumen fully coats the aggregate. At sharp edges or asperities on the aggregate surface where the bitumen film is thinnest, it has been shown[224] that water can penetrate through the film to reach the surface of the aggregate. This movement of water to the aggregate surface may occur with the water in either a vapour or liquid form. Once this process has started, it is possible for the water to spread between the bitumen and aggregate surface to produce a detached film of bitumen.

The speed with which the water can penetrate and detach the bitumen film will depend on the viscosity of the bitumen, the nature of the aggregate surface, the thickness of the bitumen film and the presence of filler and other components such as surface active agents. Once significant detachment of the bitumen film from the aggregate has occurred, stresses imposed by traffic will readily rupture the film and the bitumen will retract, exposing water-covered aggregate.

223 ASPHALT INSTITUTE. *Cause and prevention of stripping in asphaltic pavements, Educational Series No 10, 1981.* AI, Kentucky.

224 HUGHES R I, D R LAMB and O PORDES. *Adhesion of bitumen macadam, J App Chem, vol 180, pp 433–440, 1960.*

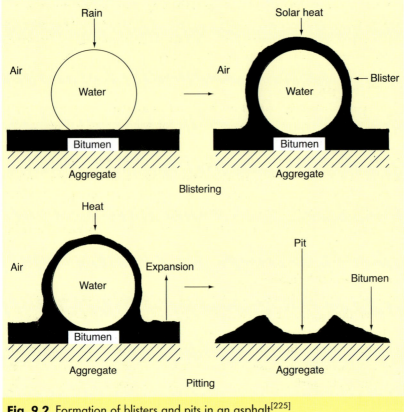

Fig. 9.2 Formation of blisters and pits in an asphalt[225]

9.2.4 *Blistering and pitting*[225,226]

If the temperature of the bitumen in a pavement increases, the viscosity of the bitumen will reduce. If this is associated with a recent rainfall, the bitumen may creep up the edges of water droplets to form a blister, as shown in Fig. 9.2. If the temperature is increased, the blister will expand, leaving a hollow or a pit which may allow water to access the surface of the aggregate.

9.2.5 *Hydraulic scouring*

Hydraulic scouring or pumping occurs in the surface course and is caused by the action of vehicle tyres on a saturated pavement surface, i.e. water is forced into surface voids in front of the vehicle tyre. On passing, the

225 THELAN E. *Surface energy and adhesion properties in asphalt/aggregate systems, Highway Research Board Bulletin, vol 192, pp 63–74, 1958.* Transportation Research Board, Washington.

226 HUGHES R I, D R LAMB and O PORDES. *Adhesion of bitumen macadam, J App Chem, vol 180, pp 433–440, 1960.*

action of the tyre sucks up this water, thereby inducing a compression–tension cycle in these surface voids, which may result in disbonding of the bitumen from the aggregate. Suspended dust and silt in the water can act as an abrasive and can accelerate disbonding.

9.2.6 Pore pressure

This type of disbonding mechanism is most important in open or poorly compacted mixtures where it is possible for water to be trapped as the material is compacted by traffic. Once the material becomes effectively impermeable, subsequent trafficking induces a pore water pressure[227]. This creates channels around the bitumen/aggregate interface leading to loss of bond. Higher temperatures acting on the entrapped water result in expansive stresses accelerating water migration and disbonding. Low temperatures may also lead to the formation of ice, which is equally destructive.

9.2.7 Chemical disbonding

Diffusion of water through a bitumen film can lead to double layers of water at the aggregate surface. The presence of the water causes the aggregate surface to exhibit a negative surface charge against a slightly negatively charged bitumen. This results in two negatively charged surfaces in contact and repulsion is the result. As more water is attracted to the aggregate surface, disbonding of the bitumen film will finally occur[228].

9.3 Methods of measuring and assessing adhesion

Given the potential for premature failure due to adhesion-related problems, the need for a predictive laboratory test is self-evident. A number of different types of test have been developed to compare combinations of aggregate, bitumen and water. However, the problem with many of these methods is a lack of information relating the laboratory prediction to performance in practice.

Typically, aggregate is coated with bitumen, immersed in water under controlled conditions and the effect of stripping determined after a period of time. The various methods differ in the type of specimen used, the conditions under which the sample is immersed in water and the method that is used to assess the degree of stripping.

227 LEE A R and J H NICHOLAS. *The properties of asphaltic bitumen in relation to its use in road construction, J Inst Pet, vol 43, pp 235–246, 1957.* Institute of Petroleum, London.

228 GUSFELT K H and J A N SCOTT. *Factors governing adhesion in bituminous mixes. Proc 2nd Int Symp devoted to tests on bitumens and bituminous materials, vol 1, pp 45–65, Budapest, 1975.*

In the majority of these tests, the coated aggregate is immersed in water. However, salt-based or de-icing solutions and those containing fuel oils may also be used. Certain types of test may tend to coat an aggregate in the presence of water. In such cases, the degree of coating obtained is used as an index of adhesivity.

Adhesion tests fall into a number of categories. There may be several different tests within each type but, in most cases, the individual tests of one type differ in detail rather than in principle. The categories are:

- static immersion tests
- dynamic immersion tests
- chemical immersion tests
- immersion mechanical tests
- immersion trafficking tests
- coating tests
- adsorption tests
- impact tests
- pull-off tests.

Each is now discussed in more detail.

9.3.1 *Static immersion tests*

This is the simplest type of test and consists of aggregate being coated with bitumen that is then immersed in water. The degree of stripping is estimated by a visual inspection after a period of time. For example, in the total water immersion test (a test carried out by Shell in-house) 14 mm single-size chippings are coated with a known quantity of bitumen. The coated aggregate is then immersed in distilled water at 25°C for 48 hours. The percentage of bitumen stripped off the aggregate is assessed visually.

The fundamental problem with this method is its subjective nature, resulting in poor reproducibility. However, the experienced operator may be capable of ranking the aggregates in relation to their performance in situ. It must be recognised that in some cases, an aggregate with good laboratory performance may perform poorly occasionally on the road and those with poor static immersion test results may perform satisfactorily in practice.

9.3.2 *Dynamic immersion tests*

This type of test is very similar to the static immersion test but the sample is agitated mechanically by shaking or kneading. Again, the degree of stripping is estimated visually together with a subjective judgement of whether the mixture remains cohesive or separates into individual particles of aggregate. The reproducibility of this type of test is also very poor.

9.3.3 *Chemical immersion tests*

In this type of test, aggregate coated in bitumen is boiled in solutions containing various concentrations of sodium carbonate. The strength of sodium carbonate solution in which stripping is first observed is used as a measure of adhesivity. However, the artificial conditions of the test are unlikely to predict the likely performance on the road.

9.3.4 *Immersion mechanical tests*

Immersion mechanical tests involve the measurement of a change in a mechanical property of a compacted asphalt after it has been immersed in water. Thus, the ratio of the property after immersion divided by the initial property is an indirect measure of stripping (usually expressed as a percentage).

Different types of mechanical properties can be measured and these include shear strength, flexural strength and compressive strength. The two most common are known as the retained Marshall stability test and the retained stiffness test. Given the increasing availability of asphalt stiffness test apparatus, the retained stiffness test is now the most popular method in the UK.

Retained Marshall stability test

There are a number of different versions of each type of test. In Shell's retained Marshall test, at least eight Marshall specimens are manufactured using a prescribed aggregate type, aggregate gradation, bitumen content and void content. Four specimens are then tested using the standard Marshall method to give a standard stability value. The four remaining specimens are vacuum treated under water at a temperature of between 0 and 1°C to saturate the pore volume of the mixture with water. They are then stored in a water bath at 60°C for 48 hours after which their Marshall stability is determined. The ratio of the Marshall stability of the soaked specimens to the standard Marshall stability is termed the 'retained Marshall stability'. A value >75% retained Marshall stability is usually regarded as acceptable.

Retained stiffness test

The principle of the retained stiffness test is similar to that of the retained Marshall test. The major difference is that the test specimens are assessed using the indirect tensile stiffness modulus (ITSM) test[229] carried out using a Nottingham asphalt tester (NAT) or similar. This has the additional benefit in that the method is non-destructive allowing the same

---

229 BRITISH STANDARDS INSTITUTION. *Method for determination of the indirect tensile stiffness modulus of bituminous mixtures, DD 213, 1993.* BSI, London.

test specimens to be used after soaking. A number of versions of the method exist where the soaking period and water temperature may be varied.

Retained Cantabro test

The Cantabro test was originally developed to measure particle loss for porous asphalts. It can also be used for testing water sensitivity[230]. A minimum of five moulds of 101 mm diameter × 63·5 mm height are abraded, one at a time in a Los Angeles machine. The loss in mass expressed as a percentage after 300 turns is termed the 'dry particle loss' (Pl_d). When measuring water sensitivity, a further five moulds are soaked in water for 68 hours followed by 24 hours in an oven at 25°C. Again, these are subjected to 300 revolutions and the loss in mass expressed as a percentage termed the 'wet particle loss' (Pl_w). This gives the 'particle loss index' (PLI) from the equation:

$$PLI = 100 \times \frac{Pl_w}{Pl_d}$$

9.3.5 Immersion trafficking tests

A major problem with most types of adhesion tests is that they do not consider the effect of trafficking in causing stripping. A method that considers this is the immersion wheel-tracking test[231] as shown in Fig. 9.3. In this test, a specimen is immersed in a water bath and traversed by a loaded reciprocating solid rubber tyre. In the standard method, three specimens that have been compacted in moulds are tracked using a 20 kg load at 25 cycles per minute at a water temperature of 40°C until failure occurs.

The development of the rut that forms is measured until stripping starts to occur. As shown in Fig. 9.4, this is typically marked by a steep increase in the rut depth and surface ravelling of the test specimen. Good correlation has been shown to exist for stripping failures of heavily trafficked roads. It has been found that factors such as aggregate shape, aggregate interlock, bitumen viscosity and sample preparation affect failure times.

Dry wheel-track testing using air to control temperature superseded the original method of wet wheel-track testing. This dry method does not assess stripping but rather considers permanent deformation.

230 EUROPEAN COMMITTEE FOR STANDARDIZATION *Testing bituminous materials, Test: Particle loss from porous asphalt, CEN TC 227/WG1, TG2 reference number 1.15, TC 227. CEN, Brussels.*

231 MATHEWS D H and D M COLWILL. *The immersion wheel-tracking test, J App Chem, pp 505–509, 1962.*

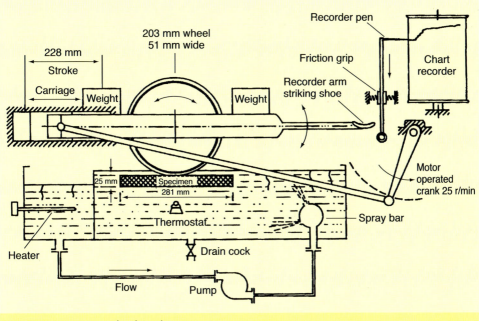

Fig. 9.3 Immersion wheel-track test equipment

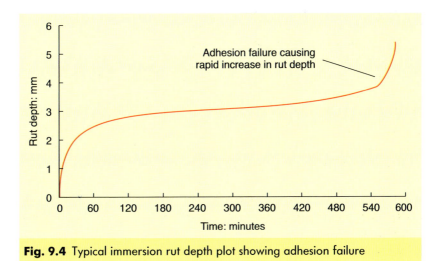

Fig. 9.4 Typical immersion rut depth plot showing adhesion failure

However, in more open and thinner mixtures with a high coarse aggregate content, the possibility for stripping-related failure rather than rutting failure is more prevalent.

The wet wheel-track test is now being reconsidered as a means of assessing asphalts such as porous asphalts and high stone content thin surfacings. The method is easily adaptable where temperature, immersion

181

medium and loading may be varied. The test may be performed for a specific time or until failure occurs.

A locked-wheel method was developed to assess the likelihood of ravelling due to the presence of moisture[232]. The only alteration to the equipment involved the provision of a locking ratchet that allowed the wheel to travel freely in one direction but also to be locked in position as it is dragged back across the specimen surface on the remainder of the cycle. Surface ravelling has been related to grading, compaction, bitumen and aggregate characteristics. This method is, however, very aggressive and may be most suitable for highly stressed trafficking locations.

9.3.6 Coating tests

This type of testing attempts to assess adhesion between aggregate and bitumen when water is also present. For example, in the immersion tray test[233], aggregate chippings are applied to a tray of bitumen covered with a layer of water. By careful examination of the chippings, it may be possible to determine whether surface active agents improve adhesion under wet conditions.

9.3.7 Adsorption tests

The Strategic Highways Research Program (SHRP) in the USA investigated the effect of moisture damage as one of six major distress areas requiring further investigation. The result was the net adsorption test method[234]. This test combined a fundamental measurement of aggregate/bitumen adhesion with a measure of moisture sensitivity. The method was able to match an aggregate and bitumen to give optimum performance conditions.

In the test, a stock solution of bitumen is first prepared by adding 1 g of bitumen to 1000 ml of toluene. The aggregate is crushed to produce 50 g test samples of 5 mm to dust grading, then 140 ml of the stock solution is added to a conical flask containing the graded aggregate placed on a rotating table to agitate the mix. After 6 hours, a 4 ml sample is removed and diluted to 25 ml using toluene. The absorbance is measured using a spectrophotometer at 410 nm.

232 MCKIBBEN D C. *A study of the factors affecting the performance of dense bitumen macadam wearing courses in Northern Ireland, PhD Thesis, University of Ulster, 1987.*

233 NICHOLLS J C et al. *Design guide for road surface dressing, Appendix D, Road Note 39, 5th ed, 2002.* TRL Ltd, Crowthorne.

234 CURTIS C W, K ENSLEY and J EPPS. *Fundamental properties of asphalt-aggregate interactions including adhesion and adsorption, Strategic Highway Research Program, Report SHRP-A-341, 1993.* National Research Council, Washington DC.

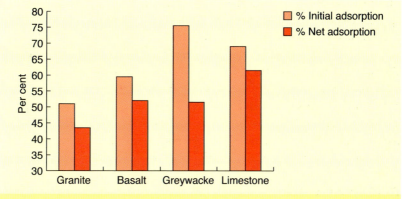

Fig. 9.5 Net adsorption data for four aggregates and a single 160/220 pen bitumen

The reading is compared to that obtained from the stock solution and the percentage initial adsorption calculated. The flask then has 2 ml of distilled water added and is agitated overnight. Another 4 ml is removed and assessed as before. The value obtained at this stage is the percentage net adsorption and should be lower than the percentage initial adsorption value as some of the bitumen will have been removed from the aggregate due to the introduction of moisture into the system.

Figure 9.5 shows net adsorption data for four aggregates and a single 160/220 pen bitumen. The effect of aggregate type on both initial adsorption and the reduction due to addition of moisture is apparent.

9.3.8 Impact tests

There are basically two impact tests that may be used to assess the adhesive qualities of bitumen – the Vialit pendulum test and the Vialit plate test. Both methods are readily adaptable to predict a wide range of in situ conditions and are of the greatest significance in situations where the aggregate is in direct contact with trafficking stresses such as surface dressings.

The Vialit pendulum test was discussed in Section 7.7.3.

In the Vialit plate test, aggregate particles are pressed onto a tray of bitumen. This is turned upside down and a steel ball is dropped onto the reverse side. The impact of the ball may cause detachment of the aggregate particles depending on test conditions. The number of detached aggregate particles together with an increasing number of impacts may be used as an indication of performance. Visual inspection of the detached aggregate can usually determine the type of failure.

A wide range of variables may be assessed using this simple equipment. For example, Fig. 9.6 shows aggregate chipping loss for a dry clean and a

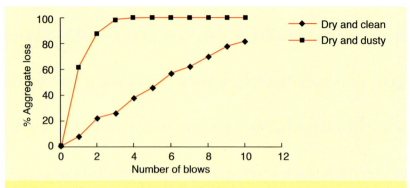

Fig. 9.6 Simple example showing Vialit plate data for a dry, clean and a dry, dusty dolerite aggregate

dry dusty dolerite aggregate[235]. Different binder types and application rates may be applied to the steel plate. The test specimens may be subjected to a wide range of conditioning procedures such as high/low temperatures, ageing or hardening, etc.

9.3.9 Pull-off tests

Bitumen adhesion may be assessed using different types of pull-off tests. The Instron pull-off test[235] uses an Instron tensile apparatus to extract aggregate test specimens from containers of bitumen under controlled laboratory conditions. Test variables such as rock type, dust coatings, test temperature, rate of loading, bitumen type, etc. have been shown to alter the results. An example of the data obtained for four different rock types with increasing dust contents is shown in Fig. 9.7 and demonstrates how the maximum stress during testing varied for the different aggregates.

The limpet pull-off test[235] was developed to measure, quantitatively, the bond strength between the aggregate of a surface dressing and the underlying surface. This uses a limpet apparatus that was originally developed for measuring the tensile strength of concrete. A 50 mm diameter metal plate is fixed to the road surface and the maximum load to achieve pull-off determined. The method may be used both in the laboratory and on the actual road surface.

Figure 9.8 shows data obtained for both dusty and clean chippings with both a K1-70 and 100 sec bitumen. The effect of both dust content and bitumen type on the adhesion bond can be seen. It was found that the dust content of an aggregate could adversely affect adhesion between

---

235 CRAIG C. *A study of the characteristics and role of aggregate dust on the performance of bituminous materials. PhD Thesis, University of Ulster, 1991.*

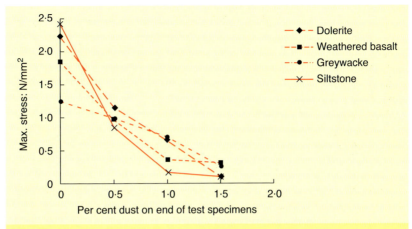

Fig. 9.7 Instron pull-off data showing the effect of dust content for four different aggregates

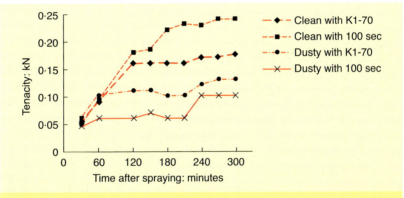

Fig. 9.8 Vialit plate test using K1-70 and 100 sec binders and clean and dusty chippings

aggregate and bitumen. Flaky chippings had better adhesion than cubic chippings due to the lower value of average least dimension. Washed aggregate or the use of adhesion agents was shown to improve adhesion of a surface dressing and, as expected, the poorest adhesion results occurred when dusty chippings and low bitumen spray rates were used together.

9.4 Improving bitumen/aggregate adhesion

Typically, the adhesion of bitumen to aggregate is not a problem. However, in the presence of water, unexpected adhesion-related problems may occur. There are a number of traditional methods used to reduce the likelihood of this happening, i.e. by using higher viscosity bitumen,

185

hydrated lime or surface active agents that improve the bond between the bitumen and the aggregate. Whilst modifying bitumen viscosity may be easily achieved, this may result in workability and compaction related problems particularly for high stone content thin surfacing mixtures.

The use of 1 to 3% hydrated lime as part of the filler content has traditionally been used as an anti-stripping agent. It has been suggested[236] that the hydrated lime reacts with the carboxyl acids present in the bitumen and allows other carbonyl groups such as ketones to attach themselves to the aggregate surface. These ketones are not as easily removed by water as the acids and so the mixture is less susceptible to stripping.

It has also been suggested[237] that a hydrated lime solution will result if water is present at the bitumen/aggregate interface. The calcium ions in this solution cause the surface of the aggregate to become basic. The electro-chemical balance forces the water away from the aggregate and into an emulsion in the bitumen. The balance will then force attachment to the hydrophobic surface of the aggregate. Bitumen/aggregate adhesion may also be improved by the addition of chemical additives. These act in two main ways:

- they may change the interfacial conditions between the aggregate and the bitumen so that the bitumen preferentially wets the aggregate which improves adhesion; or
- they may improve the adhesive bond between the aggregate and the bitumen, thus increasing the long-term resistance to bitumen detachment due to water.

Typically, 0·1 to 1·0% fatty amines is the main type of additive that is used to improve adhesion. It is believed that the amine groups are attracted to the surface of an aggregate whilst the fatty groups remain in the bitumen. The result is an ionically bonded cross-link between aggregate and bitumen. However, these additives may be relatively unstable at bitumen storage temperatures and can become deactivated. It is also possible that a given additive will not improve the adhesion of all aggregate types, i.e. they may be rock type specific. There is also the issue that whilst they may improve initial adhesion they may have no or limited long-term effect.

It is recommended that prior to their use, laboratory testing should be carried out to optimise the type and amount of additive used for a given

236 PLANCHER H, E GREEN and J C PATERSON. *Reduction of oxidative hardening of asphalts by treatment with hydrated lime – a mechanistic study, Proc Assoc Asph Pav Tech, vol 45, pp 11–24, 1976.* Association of Asphalt Paving Technologists, Seattle.

237 ISHAI I and J CRAUS. *Effect of the filler on aggregate-bitumen adhesion properties in bituminous mixtures, Proc Assoc Pav Tech, vol 46, pp 228–257, 1977.* Association of Asphalt Paving Technologists, Seattle.

Table 9.2 Net and initial adsorption results with different binders

Aggregate type	Initial adsorption, %		Net adsorption, %	
	100 pen bitumen	100 pen bitumen with 0·5% adhesion agent	100 pen bitumen	100 pen bitumen with 0·5% adhesion agent
Quartz dolerite	52·2	63·1	46·4	49·6
Greywacke	64·6	67·6	54·3	50·4

aggregate/bitumen combination. This may be undertaken using the wide range of the test methods already mentioned in this chapter. For example, Table 9.2 shows data obtained using the net adsorption test for two different aggregates and a 100 pen bitumen with and without 0·5% adhesion agent. It can be seen that while the adhesion agent increased initial adsorption by 10·9% for the quartz dolerite, the adsorption only improved 3% for the greywacke. In terms of improving moisture sensitivity, there was a small improvement for the quartz dolerite and no improvement for the greywacke. This example shows how aggregate properties influence attempts at improving adhesion. For the greywacke, the addition of adhesion agents essentially did little to improve performance.

Chapter 10

Influence of bitumen properties on the performance of asphalts

Despite the enormous range of applications to which asphalts are put and the substantial variations in weather and loading to which they are subjected, the vast majority of asphalt pavements perform well for many years. However, failures do occur. In some cases, these arise as a result of some fault in the construction process. An understanding of the mechanisms that cause asphalt pavements to fail is important if designers, contractors and producers are to employ specifications, manufacturing techniques, equipment and methods that will minimise the possibility of defects occurring.

The performance of asphalts in service is significantly influenced by the rheological (or mechanical) properties and, to a lesser extent, the chemical constitution of the bitumen. The latter is particularly important at the road surface because the constitution of the bitumen influences the rate of oxidation and, thereby, how rapidly the bitumen is eroded by traffic. These factors are, in turn, influenced by changes due to the effect of air, temperature and water on the bitumen. There are, of course, many other factors influencing behaviour, including the nature of the aggregate, mixture composition, bitumen content (i.e. bitumen film thickness), degree of compaction, etc. – all of which influence long-term durability (see Chapter 8).

Bitumens are visco-elastic materials and their behaviour varies from purely viscous to wholly elastic depending upon loading time and temperature. During the mixing and compaction of asphalts and at high service temperatures, the properties can be considered in terms of viscosity but for most service conditions, bitumens behave visco-elastically and their properties can be considered in relation to their stiffness modulus.

The rheological requirements for bitumen during mixing, compaction and in service are illustrated in Fig. 10.1 and the critical requirements are summarised in Table 10.1. This chapter considers a number of failure mechanisms with particular emphasis on the role that is played by the bitumen in avoiding their development.

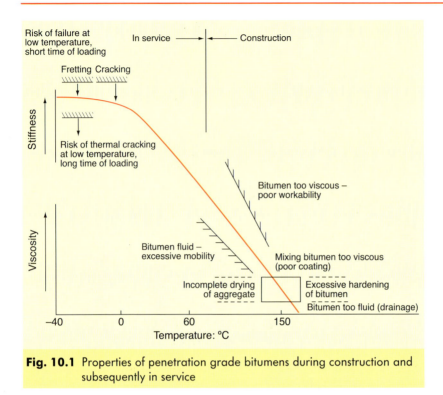

Fig. 10.1 Properties of penetration grade bitumens during construction and subsequently in service

Table 10.1 Engineering requirements for bitumens during application and in service[238]

Behaviour during:	Condition		Significant property of the bitumen in the mix
	Temperature: °C	Time of loading: s	
Application			
Mixing	High (>100°C)	–	Viscosity, approximately 0·2 Pa s
Laying	High	–	Viscosity
Compaction	High	–	Viscosity, minimum 5 Pa s, maximum 30 Pa s
In service			
Permanent deformation	High road temperature (30–60°C)	Long >10⁻²	Minimum viscosity determined by penetration index and the softening point of the bitumen
Fatting up	High road temperature (30–60°C)	Long >10⁻²	
Cracking			
– Traffic stresses	Low road temperature	Short (10⁻²)	Maximum bitumen stiffness
– Thermal stresses	Low road temperature	Long	
Fretting	Low road temperature	Short (10⁻³)	

238 DORMON G M. *Some observations on the properties of bitumen and their relation to performance in practice and specifications, Shell Construction Service Report, 1969.* Shell International Petroleum Company Ltd, London.

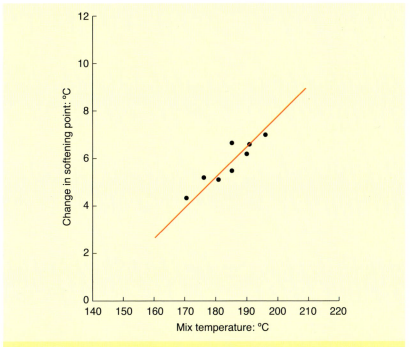

Fig. 10.2 Relationship between the temperature of the mixture and change in softening point[239]

10.1 The influence of bitumen properties during construction
10.1.1 Mixing and transport

During mixing, the dried hot aggregate has to be coated by the hot bitumen in a relatively short mixing time (typically 30 to 90 s). Whilst the mixing temperature must be sufficiently high to allow rapid distribution of the bitumen on the aggregate, the use of the minimum mixing time at the lowest temperature possible is advocated. The higher the mixing temperature, the greater the tendency of the bitumen exposed in thin films on the surface of the aggregate to oxidise. This is illustrated in Fig. 10.2 where an increase of 5·5°C in the mixing temperature, for a standard mixing time of 30 s, results in an increase of 1°C in the softening point of the bitumen[239]. There are, therefore, upper and lower limits of mixing temperature. If macadams are mixed too hot, drainage of bitumen from the aggregate may occur during hot storage or transport to site, leading to variations in bitumen content. In such circumstances, the filler serves an important function. As it is generally added to the mixture

239 WHITEOAK C D and D FORDYCE. *Asphalt workability, its measurement, and how it can be modified, Shell Bitumen Review 64, pp 14–17, Sep 1989.* Shell International Petroleum Company Ltd, London.

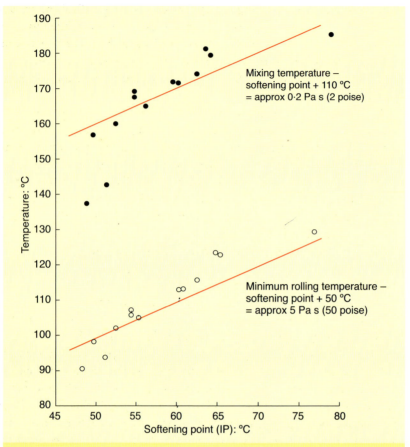

Fig. 10.3 Relationship between the softening point of the bitumen and equi-viscous temperatures for mixing and rolling of hot rolled asphalt surface course[240]

after coating of the aggregate has been completed, it 'stabilises' or increases the apparent viscosity of the bitumen, reducing drainage. In thin surfacings, stone mastic asphalts and porous asphalts, binder drainage can be prevented by the addition of fibres to the mixture or by using a polymer modified bitumen.

These different considerations combine to give an optimal bitumen viscosity of 0·2 Pa s (2 poise) at the mixing temperature. It has been shown[240] that the temperature required to achieve a viscosity of 0·2 Pa s can be crudely estimated by simply adding 110°C to the softening

240 JACOBS F A. *Hot rolled asphalt: Effect of binder properties on resistance to deformation, Laboratory Report 1003, 1981.* Transport and Road Research Laboratory, Crowthorne.

point of the bitumen (see Fig. 10.3). The disadvantage of this method is that it does not take account of the penetration index (PI) of the bitumen. A more precise estimate can be determined using both the penetration and softening point of the bitumen on the bitumen test data chart (see Fig. 7.8). The various conditions described above are summarised in Fig. 10.1.

When materials are being laid at low ambient temperatures, or if haulage over long distances is necessary, mixing temperatures are often increased to offset these factors. However, increasing the mixing temperature will tend to accelerate the rate at which the bitumen oxidises and this increases the viscosity of the bitumen. Thus, a significant proportion of the reduction in viscosity achieved by increasing the mixing temperature may be lost because of additional oxidation of the bitumen. If asphalts are transported in properly sheeted, well insulated vehicles, the loss in temperature is very low, about $2°C$ per hour.

10.1.2 Laying and compaction

In discharging the mixture into a paver, a loss of temperature of up to $20°C$ will be experienced; spreading the material into a mat will further reduce the temperature. The temperature loss will be dependent on a number of factors, including the thickness of the layer, ambient temperature, wind speed and the temperature of the substrate upon which the new material is being placed. The two most crucial factors influencing the cooling of the layer are wind speed and layer thickness[241]. Once the mat has been spread, it must be sufficiently workable to enable the material to be satisfactorily compacted with the available equipment. For effective compaction to take place, the viscosity of the bitumen should be between 5 Pa s (50 poise) and 30 Pa s (300 poise). At viscosities lower than 5 Pa s, the material will probably be too mobile to compact and at viscosities greater than 30 Pa s, the material will be too stiff to allow further compaction. The minimum rolling temperature can be estimated by adding $50°C$ to the softening point of the bitumen[242] (see Fig. 10.3).

The slope of the viscosity/temperature relation or PI in this temperature region is particularly important because it determines the temperature range over which the viscosity of the bitumen remains at a suitable level for compaction. Figure 10.4 clearly shows that bitumens

241 DAINES M E. *Cooling of bituminous layers and time available for their compaction, Research Report 4, 1985.* Transport and Road Research Laboratory, Crowthorne.

242 JACOBS F A. *Hot rolled asphalt: Effect of binder properties on resistance to deformation, Laboratory Report 1003, 1981.* Transport and Road Research Laboratory, Crowthorne.

193

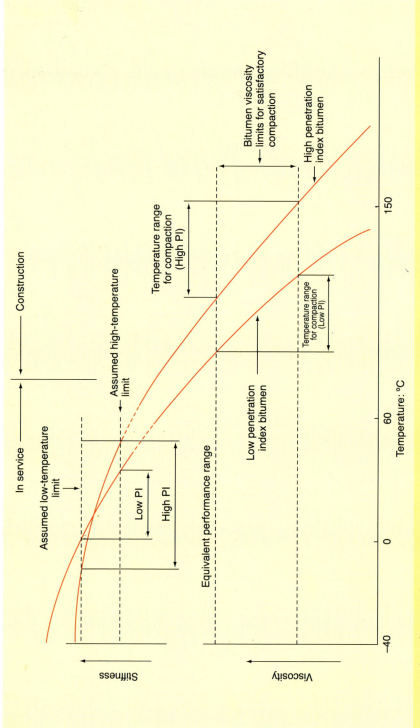

Fig. 10.4 The influence of penetration index on behaviour during construction and subsequent performance in service

with a high temperature susceptibility, i.e. a low PI, have a much narrower temperature 'window' within which satisfactory compaction of the material can be achieved.

10.2 The influence of bitumen properties on the performance of asphalts in service

After the mixture has been manufactured, laid and compacted, its behaviour in service can be considered, in particular by considering the circumstances surrounding the occurrence of specific types of failures.

10.2.1 Cracking

Pavement cracking is a complex phenomenon that can be caused by several factors. It is associated with stresses induced in the asphalt layers by wheel loads, temperature changes or a combination of the two. Furthermore, the volume of bitumen in the mixture and its rheological behaviour has a major bearing on the susceptibility of the asphalt to cracking. An asphalt mixture, by virtue of the bitumen it contains, displays visco-elastic behaviour. If an asphalt test specimen is strained to a predetermined point, and held constant, a stress will be induced. Depending on temperature, this stress will dissipate more or less quickly. This process is called 'relaxation'. At high temperatures, the viscous component dominates and total stress relaxation may take a few minutes. At very low temperatures, relaxation can take many hours or even days.

Cracking occurs when the tensile stress and related strain induced by traffic and/or temperature changes exceeds the breaking strength of the mixture. At elevated temperatures, stress relaxation will prevent these stresses reaching a level that can cause cracking. On the other hand, at low temperatures, the tensile condition will persist and, therefore, pavement cracking is more likely.

It is also recognised that bitumen in a mixture ages during its service life. This results in a progressive increase in the stiffness modulus of the asphalt, together with a reduction in its stress relaxation capability. This will further increase the likelihood of pavement cracking.

During mixing and laying, the penetration of bitumen generally reduces to about 70% of the pre-mixed value. This is known as 'short-term ageing'. This ageing continues during its service life, albeit at a lower rate, and is known as 'long-term ageing'. Several processes contribute to long-term ageing and these are described in more detail in Chapter 8. The most dominant of these mechanisms is oxidative hardening and this is influenced by the thickness of the binder film, the air voids content in the mixture and the temperatures experienced.

195

Pavement cracking can take several forms. The most frequent are:

- longitudinal cracking, occurring generally in the wheel track;
- transverse cracking, which can be in any area and it is not necessarily associated with the wheel path;
- alligator or crocodile cracking, where longitudinal and transverse cracks link up forming a network of cracks; and
- reflective cracking, resulting from an underlying defect (typically joints in concrete layers).

Traditionally, pavement design has only considered load-associated cracking in which cracks are initiated in the underside of the asphalt base caused by repeated pavement flexure under traffic. This has led to the design of pavements of greater thickness to withstand the higher flow rates predicted for future commercial traffic. A programme of research carried out at the Transport Research Laboratory[243] to investigate how these thicker asphalt pavements were performing revealed that cracks in thick asphalt pavements invariably initiate at the surface and propagate downwards. Similar observations have been made by other organisations[244,245].

Thermal cracking

Cracking that results from extreme cold is generally referred to as low-temperature cracking whereas cracking that develops from thermal cycling is normally referred to as thermal fatigue cracking. Thermal cracking will occur when the bitumen becomes too stiff to withstand the thermally induced stress and it is related to the coefficient of thermal expansion and the relaxation characteristics of the mixture. Both these properties are related to the nature of the bitumen and the risk of thermal cracking increases with the age of the pavement as the binder hardens as a result of oxidation or time-dependent physical hardening.

The coefficient of thermal expansion of bitumen is an order of magnitude higher than that of the aggregate in the mixture. A rough estimate for the thermal expansion coefficient of the mixture can be determined

243 NUNN M E, A BROWN, D WESTON and J C NICHOLLS. *Design of long-life flexible pavements for heavy traffic, Report 250, 1997.* TRL, Crowthorne.

244 SCHORAK N and A VAN DOMMELEN. *Analysis of the structural behaviour of asphalt concrete pavements in SHRP-NL test sections, International Conf: SHRP and Traffic Safety, Prague, 1995.*

245 UHLMEYER J F, L M PIERCE, K WILLOUGHBY and J P MAHONEY. *Top-down cracking in Washington State asphalt concrete wearing courses, Transportation Research Board 79th annual meeting, Washington, 2000.* TRB, Washington.

using the equation[246]:

$$a_{m} = \frac{a_{b}v_{b} + a_{a}v_{a}}{v_{b} + v_{a}}$$

where a_{m} = the coefficient of expansion of the mixture, v_{b} = proportion of the bitumen by volume, v_{a} = proportion of the aggregate by volume, a_{b} = coefficient of thermal expansion of the bitumen and a_{a} = coefficient of thermal expansion of the aggregate. The coefficient of volumetric expansion of bitumen is approximately $6 \times 10^{-4}/°C$. Typically, the linear coefficient of thermal expansion of an asphalt mixture is between 2 and $3 \times 10^{-5}/°C$.

Two different thermal cracking mechanisms can occur. At low pavement temperatures, transverse cracks that run the full depth of the pavement can suddenly appear. Pavement temperatures generally have to fall below about $-30°C$ to induce this form of cracking. Accordingly, documented cases of 'low-temperature cracking' in the UK are extremely rare.

For milder conditions, cracks may develop at a slower rate, taking several seasons to propagate through the asphalt layers. This form of cracking initiates at the surface and propagates relatively slowly with each thermal cycle. This is generally described as 'thermal fatigue cracking'.

The general mechanism responsible for these two forms of cracking is considered to be similar. The main differences are:

- low-temperature cracking is a single-event phenomenon that is the result of the full depth of asphalt being put into thermal tension under conditions where stress relaxation cannot occur;
- thermal fatigue cracking is more dependent on the properties of the surface course material, and cracks first have to initiate at the surface and propagate through the surface course before they affect the lower asphalt layers.

Low-temperature cracking. The mechanism of low-temperature cracking is illustrated in Fig. 10.5. The asphalt layer is subjected to a tensile stress distribution with depth. These stresses are caused by the contraction of the asphalt as it cools and they are a function of the temperature change and the relaxation characteristics, stiffness modulus and coefficient of expansion of the asphalt. They are not necessarily uniform because pavement temperature can vary with depth. These stresses can, potentially, cause a crack to propagate down from the surface.

---

246 EUROPEAN COST ACTION 333. *Development of new bituminous pavement design method, Final report of the action, 1999.*

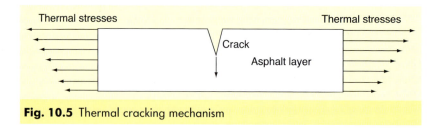

Fig. 10.5 Thermal cracking mechanism

As asphalt becomes colder, its tensile strength (β_z) increases initially and then begins to decrease as a result of micro-cracking in the binder matrix. This is caused by the differential contraction that results from the large difference between the coefficients of thermal expansion of the aggregate and bitumen. These fractures can be detected as acoustic events using a sensitive microphone to record their increasing occurrence with falling temperature[247].

At the same time, thermal stress (σ) builds up as the material loses its relaxation ability and, at some point, the thermal stress will exceed the strength of the material. This defines the probable low-temperature fracture temperature (spontaneous cracking). The difference between the tensile strength and the low-temperature stress is known as the tensile strength reserve $(\Delta\beta_z)$ and it is this reserve that is available to accommodate additional superimposed stresses (for example, traffic-induced stresses). This is illustrated in Fig. 10.6.

The curves shown in Fig. 10.6 can be derived using the thermal stress restrained specimen test (TSRST) and the isothermal direct tensile test over a range of low temperatures. These tests were developed within the US Strategic Highways Research Program (SHRP)[248]. Although expensive and time consuming, these tests are considered[249] to offer a reliable means of predicting the temperature at which a pavement will crack due to excessive thermal stresses. The results are reasonably reproducible and the relative ranking of materials is consistent with field performance. Various studies of low-temperature cracking have generally concluded that, in order to reduce cracking, the binder stiffness

247 VALKERING C P and D J JONGENEEL. *Acoustic emission for evaluating the relative performance of asphalt mixes under thermal loading conditions, Proc Assoc Asph Pav Tech, vol 60, 1991.* Association of Asphalt Paving Technologists, Seattle.

248 JUNG D H and T S VINSON. *Final report on test selection – low temperature cracking, SHRP, 1992.* National Research Council, Washington DC.

249 KING G N, H W KING, O HARDERS, W ARAND and P-P PLANCHE. *Influence of asphalt grade and polymer concentration on the low temperature performance of polymer modified asphalt, Proc Assoc Asph Pav Tech, vol 62, 1993.* Association of Asphalt Paving Technologists, Seattle.

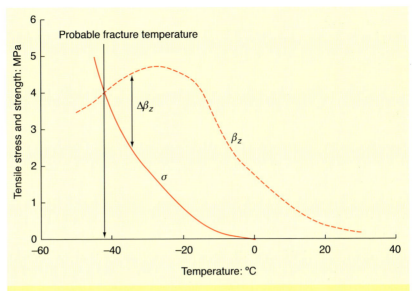

Fig. 10.6 Low-temperature stress (σ) and tensile strength of asphalt (β_z) as a function of temperature

must not exceed some defined limit at the coldest pavement temperature[250,251].

In SHRP, the bending beam rheometer (BBR) is used to produce a master curve for the bitumen stiffness as a function of loading time and temperature. This is used to predict the temperature at which the bitumen has a stiffness of 200 MPa at a loading time of 60 s. This was the SHRP specification limit for binders in cold climates[252]. However, after the SHRP programme was completed, initial experience suggested that this specification was restrictive and the specification limit was raised to 300 MPa[253].

250 RUTH B E, L A K BLOY and A A AVITAL. *Prediction of pavement cracking at low temperatures, Proc Assoc Asph Pav Tech, vol 51, 1982.* Association of Asphalt Paving Technologists, Seattle.

251 BAHIA H U. *Low temperature physical hardening of asphalt cements, Research Report, Pennsylvania State University, 1991.*

252 ANDERSON D, D CHRISTENSEN, H BAHIA, R DONGRE, M SHARMA, C ANTLE and J BUTTON. *Binder characterization and evaluation, SHRP A-369, vol 3, Physical characterization and evaluation, SHRP, 1994.* National Research Council, Washington DC.

253 ANDERSON D, L CHAMPION-LAPALU, M MARASTEANU, Y LE HIR, J-P PLANCHE and M DIDIER. *Low temperature thermal cracking of asphalt binders as ranked by strength and fracture properties, Transport Research Board 80th Annual Meeting, Washington, 2001.* TRB, Washington.

Jung and Vinson[254] concluded that the TSRST provided an excellent indicator of low-temperature cracking. Furthermore, King *et al.*[255] demonstrated that results using the TSRST correlated very well with the temperature prediction using the BBR ($R^2 = 0.96$) and that the BBR is certainly a better performance indicator. However, recent work[256] suggests that fracture toughness should be included in the criteria to account for the performance of polymer modified bitumen.

Low-temperature cracking is of particular concern in Canada, North America, Northern and Eastern Europe and Northern Asia where temperatures can fall as low as $-40°C$. To counter this problem, softer binders are often used. It is recognised that the risk of low-temperature cracking is related primarily to the properties of the bitumen, with viscosity and temperature susceptibility of the binder being the most important properties. The risk increases as the hardness of the bitumen rises. Variations in the grading and type of aggregate have very little effect and increasing the binder content reduces the mixture susceptibility to thermal cracking only slightly.

Thermal fatigue and load associated surface cracking. Cracking in the surface of asphalt pavements is relatively common. These cracks can be transverse or longitudinal and are usually located in the wheel tracks. Longitudinal wheel-track cracking has often been regarded as evidence of conventional fatigue in which cracks are assumed to have initiated at the bottom of the base and then propagated to the surface. However, investigation by coring strong asphalt pavements has invariably found that either the cracks partially penetrate the thickness of asphalt or if the crack is full depth, the propagation is downwards rather than upwards. A typical example of this type of cracking in a major UK motorway is shown in Fig. 10.7. This crack extended for several hundred metres. Coring showed that only at the most seriously cracked locations had cracks progressed further than the top 100 mm of asphalt. None of the cracks had penetrated the full thickness of asphalt.

254 JUNG D H and T S VINSON. *Final report on test selection – low temperature cracking, SHRP, National Research Council, Washington DC, Oct 1992.*

255 KING G N, H W KING, O HARDERS, W ARAND and P-P PLANCHE. *Influence of asphalt grade and polymer concentration on the low temperature performance of polymer modified asphalt, Proc Assoc Asph Pav Tech, vol 62, 1993.* Association of Asphalt Paving Technologists, Seattle.

256 ANDERSON D, L CHAMPION-LAPALU, M MARASTEANU, Y LE HIR, J-P PLANCHE and M DIDIER. *Low temperature thermal cracking of asphalt binders as ranked by strength and fracture properties, Transport Research Board 80th Annual Meeting, Washington, 2001.* TRB, Washington.

Fig. 10.7 Longitudinal crack in the wheel track

Investigations of a number of relatively thick UK Motorways showed that this behaviour was typical. However, crack investigations by coring in the USA and Holland have shown that in fully flexible pavements with less than 160 mm of asphalt, the cracking is likely to be full depth[257,258]. Above this thickness, an increasing proportion of the pavements have cracks that are confined to the asphalt surfacing, i.e. the binder course and the surface course. In the Dutch study, all pavements less than 160 mm thick had full-depth cracking and all pavements with 300 mm

257 SCHORAK N and A VAN DOMMELEN. *Analysis of the structural behaviour of asphalt concrete pavements in SHRP-NL test sections. International Conf: SHRP and Traffic Safety, Prague, 1995.*

258 UHLMEYER J F, L M PIERCE, K WILLOUGHBY and J P MAHONEY. *Top-down cracking in Washington State asphalt concrete wearing courses, Transportation Research Board 79th Annual Meeting, Washington, 2000.* TRB, Washington.

or more of asphalt had cracks that were confined to the top 100 mm of asphalt.

Penetration tests on the binder recovered from the hot rolled asphalt surface course in the UK study showed somewhat lower values in cracked areas compared with uncracked lengths. However, binder penetration is not always a good indicator of the susceptibility of the surface course to surface cracking and it is evident that more detailed studies will be required to understand the loading, material and environmental factors that determine the initiation and propagation of cracks.

Surface cracking is not always longitudinal; transverse or block cracking can occur. This is not confined to wheel tracks and it can occur at any location across the carriageway. Figure 10.8 shows an example of transverse cracks in a UK motorway that had carried traffic totalling some 30 msa (million standard axles, discussed in Section 18.3.1) since it was laid 24 years previously. During this time, it had not been resurfaced or overlaid. The penetration of recovered binder from the surfacing was 15 dmm and it was concluded that the cracking was not traffic related but due to brittle surfacing resulting from aged binder. As with longitudinal cracking, transverse cracks generally only penetrated to a depth of about 100 mm.

The phenomenon of surface cracking has received relatively little attention. However, there are a considerable number of observations of surface cracking of flexible pavements that span all climatic regions, from cold to tropical regions. This suggests that higher temperatures associated with the warmer season cause binder to harden with age and this reduces its capacity to withstand the thermal stresses generated during the cooler nights, particularly during the coldest season.

With all types of surface cracking, age hardening of the binder in the surface course will play a part. As stated above, this hardening over time progressively reduces the ability of the surface course to withstand the thermal and traffic generated stresses at the surface. Recent binder rheological studies[259] have shown that the binder in the top few millimetres of the surface course becomes particularly hard with age. Figure 10.9 gives an example of the results obtained using a DSR on testing binder extracted from the top 10 mm of the surface course and from between 20 and 30 mm below the surface.

Figure 10.9 shows that the binder close to the surface is substantially harder than binder that is deeper in the layer. Investigations of four

259 NUNN M E, A BROWN, D WESTON and J C NICHOLLS. *Design of long-life flexible pavements for heavy traffic, Report 250, 1997*. TRL, Crowthorne.

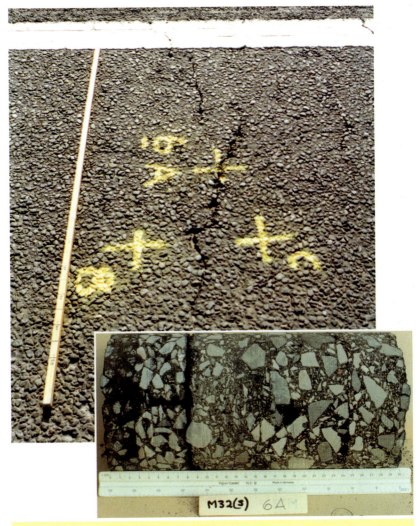

Fig. 10.8 Transverse cracks in the M32

sites that were 18 to 24 years old indicated that the penetration of the binder recovered from the top 10 mm of surface course was approximately 50% of that obtained from the lower layer. This aged, and hence hard, upper skin of the surface course is considered to be a major factor in the initiation of surface cracks.

Modelling surface cracking

The mechanism of surface cracking is complex and, to date, there is no completely satisfactory explanation of this phenomenon. It is now being recognised that tyre stresses can induce a tensile condition at the

203

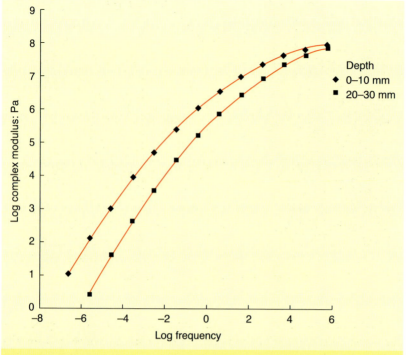

Fig. 10.9 DSR master curves at 25°C, illustrating increased binder ageing near surface

pavement surface and this can initiate a crack. De Beer *et al.*[260] have shown that the tyre can transmit non-uniform horizontal and vertical stresses to the pavement, and pavement stresses in the locality of the tyre need to be taken into account in modelling crack behaviour. Finite element modelling predicts that a tensile condition can be created in the surface course close to the edge of the tyre as shown in Fig. 10.10. Similar conditions are also predicted near the edge of tyre treads[261].

These near-tyre tensile stresses, shown in red, extend only approximately 10 mm into the asphalt. Consequently, they may initiate a longitudinal crack in the wheel track but another mechanism is required to account for crack propagation to any depth. There is increasing conviction that these surface cracks are propagated by thermal stresses.

260 DE BEER M, C FISHER and F JOOSTE. *Determination of pneumatic tyre/pavement interface contact stresses under moving loads and some effects on pavements with thin asphalt surfacing layers, Proc 8th Int Conf on asphalt pavements, Seattle, Washington, 1997.*

261 MYERS A L and R ROQUE. *Evaluation of top-down cracking in thick asphalt pavements and implications for pavement design, TRB/NRC, Transport Research Circular No 503, Washington 2001.* Transportation Research Board, Washington.

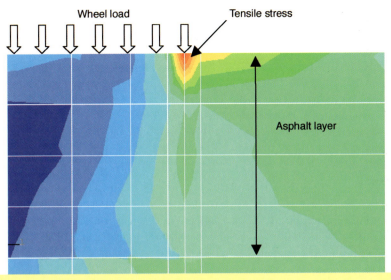

Fig. 10.10 Pavement stresses induced by tyre loading

Thermal fatigue is the mechanism responsible for the development of transverse surface cracks with crack propagation resulting from the cyclic, diurnal thermal stresses. However, transverse rolling cracks, sometimes seen at the time of compaction, are likely to play a major part in the initiation of transverse cracks.

The models predict[262] that, for a thick asphalt pavement, a thermally propagated crack will stabilise at some point and will not propagate through the full thickness of asphalt. On the other hand, the prediction for a thin pavement is that the rate of crack propagation will accelerate as the remaining thickness of asphalt is subjected to greater thermal and traffic loading.

Fatigue cracking

Fatigue cracking has received more attention from the research community than any other deterioration mechanism. Fatigue is the phenomenon of cracking under the repeated application of a stress that is less than the tensile strength of the material.

When a wheel load passes over a point in an asphalt pavement, the pavement flexes and a tensile strain is induced at the underside of the base layer. Continuous flexure and relaxation over many years produces

---

262 NESNAS N and D M MERRILL. *Development of advanced models for the understanding of deterioration in long-life pavements, 6th Bearing Conf on roads and airfields, Workshop on modelling materials and structures, 2002.*

the possibility of fatigue cracks initiating at the underside of the asphalt base and propagating upwards.

Fatigue is a major component of all modern analytical pavement design methods (see Chapter 18). These methods use a simple shift factor, or calibration factor, to relate the laboratory determined fatigue life to the performance of the material in the road. The conditions and the behaviour of material in service are much more complex[263] than those used in laboratory studies. For example, the nature of the stress system and long-term physico-chemical changes in the bitumen are not considered.

The development of the long-life pavement concept[264] has demonstrated that long-term changes in the bitumen properties can result in an increased fatigue life of the pavement. Also, in the USA where this concept is also being developed, it has been suggested that, at low strain amplitudes, asphalt has a fatigue endurance limit[265,266].

The fatigue resistance of an asphalt mixture is especially sensitive to binder volume. The simplest means of increasing the predicted fatigue life of the pavement is to construct the pavement using a binder-rich lower asphalt layer[267].

Reflective cracking

A composite pavement consists of a continuously laid cementitious base under asphalt surfacing. A regular pattern of thermally induced transverse cracks can appear in the cementitious base soon after laying and these begin to appear as reflection cracks in the asphalt surfacing several years later, as shown in Fig. 10.11.

Until recently, these cracks were assumed to be caused either by the crack in the cementitious layer opening and closing as the result of thermal expansion and contraction or by a flexing and shearing action caused by wheel loads passing over the crack. These mechanisms will

---

263 THROWER E N. *A parametric study of a fatigue prediction model for bituminous pavements, Laboratory Report 892, 1979.* Transport Research Laboratory, Crowthorne.

264 NUNN M E, A BROWN, D WESTON and J C NICHOLLS. *Design of long-life flexible pavements for heavy traffic, Report 250, 1997.* TRL, Crowthorne.

265 NEWCOMBE D E, I J HUDDLESTON and M BRUNCHER. *US perspective on design and construction of perpetual asphalt pavements, Proc 9th Int Conf on asphalt pavements, Copenhagen, 2002.*

266 NISHIZAWA T, S SHIMENO and M SEKIGUCHI. *Fatigue analysis of asphalt pavements with thick asphalt mixture layer, Proc 8th Int Conf on asphalt pavements, Seattle, 1997.*

267 HARM E and D LIPPERT. *Developing Illinois' extended life hot-mix asphalt pavement specifications, Transportation Research Board 81st Annual Meeting, 2002.* TRB, Washington.

Fig. 10.11 Reflection cracks on a Motorway with a composite pavement

induce a stress concentration in the asphalt immediately above the crack and this causes a crack in the asphalt to initiate and propagate towards the surface. However, extensive coring of in-service roads in the UK has shown that reflection cracks often start at the surface of the road and propagate downwards to meet an existing crack or joint in the underlying concrete layer (Fig. 10.12). Furthermore, this study[268] has shown that environmental effects rather than traffic loading cause reflection cracks in as-laid composite pavements. When the pavement is new, the surface course is ductile enough to withstand thermally induced stresses but as the pavement ages it will progressively lose this capability. The study showed that the occurrence of reflection cracking in as-laid pavements correlated with the strain that the surface course could accommodate before cracking. This reduced with age and it is related to the type and volume of binder used.

---

268 NUNN M E. *An investigation of reflection cracking in composite pavements, Proc RILEM Conf on reflective cracking, Liege, 1989.*

Fig. 10.12 Core of reflection crack initiating from surface[269]

   In their early stages of development, the cracks are not considered to present a structural problem. Once they propagate through the asphalt layer, water infiltration and the pumping action of the traffic will weaken the foundation layers. At the same time, the load transfer across the slab will deteriorate, and under these conditions, the cementitious slabs will move under heavy traffic and further cracking, spalling and general deterioration will result. The reflection cracking of a strengthening overlay over a crack of this nature will be dependent on traffic-induced forces and the severity of the crack.

   Modelling reflection cracking in as-laid composite pavements has generally treated the asphalt as a passive layer that has to respond to movements in the concrete layer. These models either assume that the thermal opening and closing of the crack in the cement bound layer or shearing caused by traffic will induce a high stress concentration in the

269 NUNN M E. *An investigation of reflection cracking in composite pavements, Proc RILEM Conf on reflective cracking, Liege, 1989.*

asphalt immediately above the crack causing it to initiate and propagate upwards. These models have not considered that:

- the thermal coefficient of expansion of the asphalt surfacing is several times greater than that of concrete;
- larger diurnal temperature changes occur close to the surface;
- age hardening of the bitumen is more severe close to the surface.

The latest development in modelling[270] takes into consideration all these factors and uses the finite element technique to model the behaviour of the asphalt and cement bound layers as a single system as illustrated in Fig. 10.13(a). Figure 10.13(b) shows the thermally induced stress contours predicted by this model. It predicts that the highest thermal stresses will occur at the surface immediately above the crack in the cement bound base. Furthermore, it predicts that a compressive stress condition will exist at the underside of the asphalt layer adjacent to the crack in the cement bound layer. This implies that it is not possible for a crack to propagate upwards. This model provides a qualitative explanation for the observed manner in which cracks initiate at the surface and propagate downwards.

10.2.2 Deformation

Deformation (often called 'rutting') may be restricted to one or more of the asphalt layers or it may extend throughout the entire pavement and into the subgrade. In practice, however, the majority of deformation defects result from plastic deformation of the surface course or the surfacing. An example is shown in Fig. 5.8(a). This can occur under moving or stationary traffic and particularly under the high shearing stresses imposed by braking, accelerating or turning traffic. The primary factor influencing plastic deformation is the composition of the mixture. Plastic deformation is greatest at high service temperatures, for which 60°C may be taken as a maximum in situ temperature. At such temperatures, the cumulative effect of repeated loadings of short duration will be determined by bitumen viscosity. It has been estimated that during the long hot summer experienced in the UK during 1976, deformation in the wheel tracks of hot rolled asphalt surface courses was between two and four times the rate of an average UK summer.

The significance of the PI of the bitumen was confirmed by an extensive full-scale road trial carried out on the Colnbrook bypass. Bitumens of widely differing rheological characteristics were employed in hot

---

270 BENSALEM A. *Private communication, 2001*. TRL Ltd, Crowthorne.

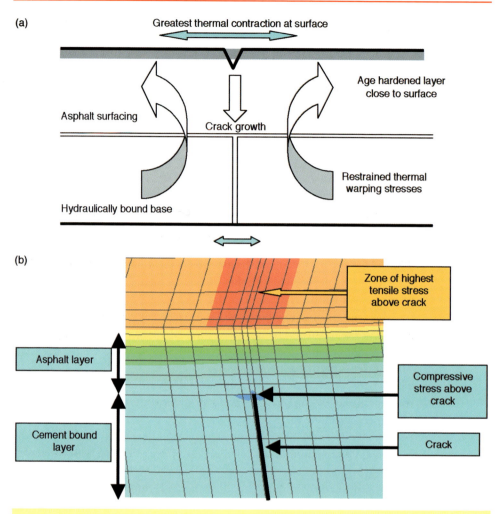

(a)

Greatest thermal contraction at surface

Age hardened layer close to surface

Asphalt surfacing

Crack growth

Restrained thermal warping stresses

Hydraulically bound base

(b)

Zone of highest tensile stress above crack

Asphalt layer

Compressive stress above crack

Cement bound layer

Crack

Fig. 10.13 (a) Schematic model of surface-initiated reflection cracking. (b) Predicted thermally induced stress contours

rolled asphalt surface course mixtures used in this project. Figure 10.14 shows the relationship between rut depth on the road after 8 years and the PI of the bitumen. The advantage of using higher PI bitumens is evident.

Work in several laboratories has shown that deformation occurring in a wheel-tracking test can be predicted using the uniaxial unconfined creep compression test. Creep curves of mixture stiffness (S_{mix}) plotted against bitumen stiffness (S_{bit}) can be used to compare different mixtures and to predict their relative deformations if subjected to the same temperature and loading conditions.

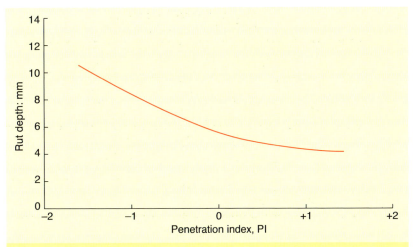

Fig. 10.14 Rut depth as a function of in situ PI measured on the Colnbrook bypass after eight years in service[271]

The shape of a creep curve for a typical hot rolled asphalt surface course is such that it can be approximated for most of its length by a straight line having a slope of 0·25, thus:

$$S_{mix} = k(S_{bit})^{0·25}$$

For two mixtures having S_{bit} values in a known ratio:

$$S_{mix} \text{ ratio} = (S_{bit} \text{ ratio})^{0·25}$$

For asphalts in similar road structures subjected to the same traffic:

$$\text{Deformation} = k\frac{1}{S_{mix}}$$

For two mixtures having S_{mix} values in a known ratio:

$$\text{Deformation ratio} = \frac{1}{S_{mix}} \text{ ratio}$$

The above relationships can be used to compare the effect of changing the penetration and/or PI of the bitumen.

Calculations made on this basis show that if the deformation of a 50 pen bitumen is regarded as unity, a mixture manufactured using a 30 pen bitumen subjected to the same temperature and loading regime will deform by 0·75 units. This assumes that the loading time for

271 DORMON G M. *Some observations on the properties of bitumen and their relation to performance in practice and specifications, Shell Construction Service Report, 1969.* Shell International Petroleum Company Ltd, London.

Table 10.2(a) Relative deformation of mixtures at 40°C, effect of bitumen penetration

Bitumen penetration	PI	Viscosity at 40°C: Pa s	Relative S_{bit}	Relative S_{mix}	Relative deformation
100	−0·8	6×10^3	0·2	0·67	1·5
60	−0·6	18×10^3	0·6	0·88	1·1
50	−0·5	30×10^3	1	1	1
40	−0·5	40×10^3	1·3	1·07	0·94
30	−0·4	90×10^3	3·0	1·32	0·75

Table 10.2(b) Relative deformation of mixtures at 40°C, effect of bitumen penetration index

Bitumen penetration	PI	Viscosity at 40°C, Pa s	Relative S_{bit}	Relative S_{mix}	Relative deformation
40	−0·5	4×10^4	1	1	1
40	+0·5	$1·5 \times 10^5$	3·8	1·41	0·71
40	+2·0	6×10^5	15	1·97	0·50

determination of viscosity and stiffness are the same. Conversely, if a 100 pen bitumen is used, 1·5 units of deformation will occur (see Table 10.2(a)). Clearly little improvement in deformation resistance is achieved purely by using a harder bitumen.

Increasing the PI of the bitumen significantly improves resistance to deformation. For example, at 40°C a bitumen with a penetration of 40 and a PI of −0·5 has a viscosity of 4×10^4 Pa s, whereas a bitumen with the same penetration but having a PI of +2·0 has a viscosity of 6×10^5 Pa s at 40°C. This gives a factor of 15 in the viscosity and, hence, in the stiffness of the two bitumens. Thus, applying the 0·25 power formula stated above gives an increase of approximately 2 in mixture stiffness. Accordingly, deformation using a mixture made with a bitumen having a PI of +2·0 would be half of one made with a standard bitumen (see Table 10.2(b)). Figure 10.15 shows the above theoretical relationship for relative deformation plotted as a function of PI. This clearly shows that the theoretical relationship is supported by both simulative laboratory wheel-tracking tests and full-scale road trials.

Correlation of bitumen properties with both Marshall and wheel-tracking tests[272] shows that the relationship between bitumen penetration and Marshall stability is poor, whereas softening point relates very

---

272 JACOBS F A. *Hot rolled asphalt: Effect of binder properties on resistance to deformation, Laboratory Report 1003, 1981.* Transport and Road Research Laboratory, Crowthorne.

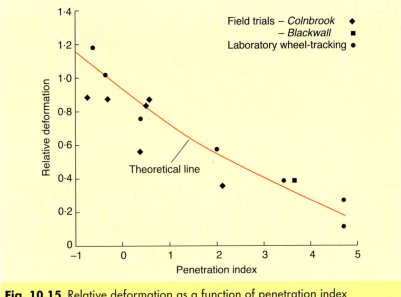

Fig. 10.15 Relative deformation as a function of penetration index

well to both Marshall stability and deformation in the wheel-tracking test (see Fig. 10.16). For every 5°C increase in softening point, the Marshall stability increased by over 1·3 kN (Fig. 10.16(b)) and the wheel-tracking rate almost halved (Fig. 10.16(c)).

The zero shear viscosity concept

In terms of permanent deformation, it is thought that the binder contribution to this mode of distress is primarily associated with viscosity. However, identifying the relevant viscosity in terms of shear, shear rate and stress is a complex problem and a single quantifiable regime must be established for assessment purposes. Pavement design is based on the assumption that the material response to traffic loading is within the linear regime and to remain consistent with this premise, the corresponding linear viscosity of the binder should be examined. This is the zero shear viscosity, η_0, sometimes termed 'Newtonian viscosity' as stress is proportional to shear. In this regime, stress and strain are linearly proportional to each other so the stiffness modulus is independent of stress or strain level and the resultant viscosity, η_0, is independent of the shear rate.

By measuring binder properties within the linear regime, conclusions can be made as to the likelihood of a bitumen contributing to or mitigating the rutting process. Zero shear viscosities can be measured using a dynamic shear rheometer, which also permits the elastic and viscous

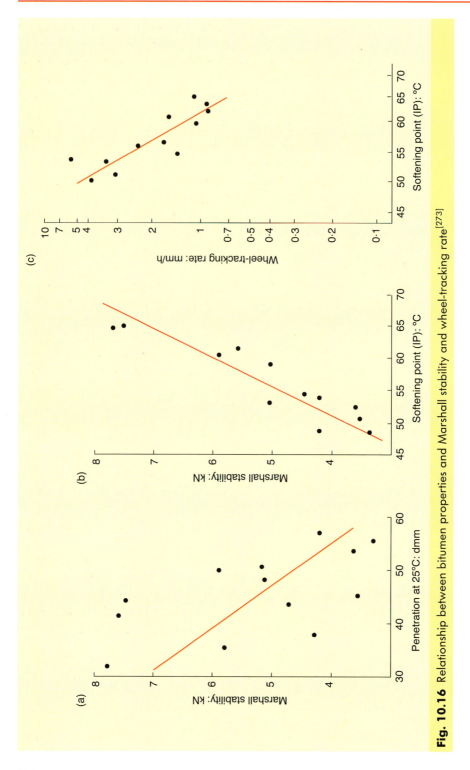

Fig. 10.16 Relationship between bitumen properties and Marshall stability and wheel-tracking rate[273]

components to be identified. The higher the value of zero shear viscosity, the lower the influence of the bitumen in relation to permanent deformation. Research has shown that the useful upper limit on zero shear viscosity of unaged bitumen is 10^5 Pa s, above which further resistance to deformation gives diminishing returns[274]. For bitumens that exhibit zero shear viscosities below this value, the relative value of the zero shear viscosity coupled with the elastic component of the binder can be used to interpret the deformation behaviour of the bitumen.

The SHRP deliberations arrived at a similar parameter in their methodology for providing a guideline on a 'rutting factor' designated by the value of $G^*/\sin\delta$, which represents a measure of the high-temperature stiffness (for more on the SHRP specification, see Section 4.7).

10.2.3 Fatting-up

One reason why fatting-up occurs is as a result of secondary compaction of the aggregate in the mixture by traffic. The void content is reduced, eventually squeezing bitumen from within the structure to the surface. This will be exacerbated if the bitumen content is too high or if the void content, after full compaction, is too low. Migration of bitumen to the surface results in a smooth, shiny surface that has poor resistance to skidding in wet weather. Fatting-up is most likely to occur at high service temperatures. Accordingly, reducing the softening point or viscosity of the bitumen at 60°C will limit this failure mechanism.

10.2.4 Fretting

Fretting is the progressive loss of interstitial fines from the road surface. It occurs when traffic stresses exceed the breaking strength of the asphalt itself or the asphalt mortar depending on the nature of the mixture. Fretting is more likely to occur at low temperatures and at short loading times when the stiffness of the bitumen is high.

The major factors influencing fretting are the bitumen content of the mixture and the degree of compaction. Loss of aggregate can be due either to the loss of adhesion between the aggregate and the bitumen or to brittle fracture of the bitumen film connecting particles of aggregate. The first condition should not arise if suitable aggregates/bitumen are

273 JACOBS F A. *Hot rolled asphalt: Effect of binder properties on resistance to deformation, Laboratory Report 1003, 1981.* Transport and Road Research Laboratory, Crowthorne.

274 PHILLIPS M and C ROBERTUS. *Binder rheology and asphalt pavement permanent deformation: the zero shear viscosity concept, Proc Eurasphalt–Eurobitume Congress, Session 5 paper 134, 1996.*

selected. In the majority of cases, fretting is associated with a low degree of compaction or inadequate bitumen content. The choice of bitumen grade may control the degree of compaction that can be achieved. This is especially true if operations are being conducted in adverse weather conditions or if compaction is effected at inappropriate temperatures. The higher the penetration of the bitumen or the lower the stiffness at low temperatures, the greater will be the resistance to fretting.

Fretting of macadam surface course

For coated macadam mixtures, the following three different forms of fretting are identifiable.

- *Superficial fretting.* This involves the loss of interstitial fines from the road surface. The material subsequently closes up under traffic without further deterioration or detrimental effect on performance.
- *Severe fretting.* If superficial fretting does not close up under traffic, the loss of interstitial fines results in a reduction of the level of mechanical interlock that is likely to lead to the loss of coarse aggregate on the surface and a substantial reduction of internal cohesion due to the ingress of water. The presence of water breaks down the adhesive bond between the bitumen and the aggregate resulting in stripping of the bitumen. In addition, the uncoated fines and water form a slurry that is pumped through the material, as a result of pore pressures induced by moving traffic, abrading the coarse aggregate and exacerbating the problem. This will eventually lead to collapse of the material and a lack of internal stability, often visible on the road as slight rutting or as potholes. Bitumen/aggregate adhesion and the effect of prolonged contact with water is discussed in depth in Chapter 9.
- *Ravelling.* Ravelling differs from the two modes described above in that it involves plucking out of surface aggregate by traffic without any loss of cohesion of internal fines. It occurs when individual aggregate particles move under the action of traffic. If the tensile stress (induced in the bitumen as a result of the movement) exceeds the breaking strength of the bitumen, cohesive fracture of the bitumen will occur and the aggregate particle will be detached from the road surface. Thus, ravelling is most likely to occur at low temperatures and at short loading times when the stiffness of the bitumen is high.

Fretting of hot rolled asphalt surface course

In 1984, a survey was carried out by the British Aggregate Construction Materials Industries (BACMI – now the Quarry Products Association,

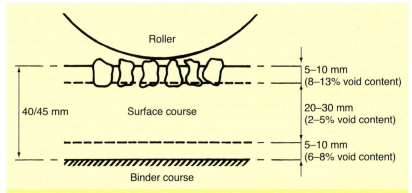

Fig. 10.17 Compaction of hot rolled asphalt surface course under adverse weather conditions

QPA) as part of a wider survey into chipping loss by the Institution of Highways and Transportation[275,276]. The BACMI survey showed[277] that over half of the locations where chipping loss had occurred were high-stress low-speed sites such as roundabouts and junctions where a minimum 1·5 mm texture depth had been needlessly specified.

Failure to achieve adequate chipping embedment usually occurs as a result of one or more of the following factors:

- the asphalt is unworkable, i.e. too stiff;
- the temperature of the asphalt is too low;
- the asphalt cools prematurely due to poor weather conditions during laying;
- excessive application of pre-coated chippings; and/or
- inadequate compaction of the material, see Fig. 10.17.

One method of treating patches where the chippings are inadequately embedded is to reheat the surface using an infrared heater and, when the asphalt is sufficiently hot, embed the chippings further into the asphalt. Used judiciously, this technique should not significantly harden the bitumen except at the surface. This surface material will be

275 JOHNSON W M and G F SALT. *Loss of chippings from rolled asphalt wearing courses, Paper presented at the Institution of Highways and Transportation National Conf, Keele University, Sep 1985*. IHT, London.

276 JOHNSON W M and G F SALT. *Loss of chippings from rolled asphalt wearing courses, Highways and Transportation, vol 32, pp 22–27, Jul 1985*. Institution of Highways and Transportation, London.

277 WHITE M J. *Flexible pavements and bituminous materials, Residential course at Newcastle University, Section G, Bituminous materials – special requirements, problems and processes, pp G1–G16, Sep 1989*.

eroded by traffic over time to expose the aggregate in the asphalt, a situation that will not threaten the long-term integrity of the surface.

Some chipping loss may occur as a result of a failure to achieve or maintain bond between the pre-coated chippings and the asphalt. It has been found that the bitumen on pre-coated chippings can be 'coked' and made non-adhesive by storage of the chippings in large stockpiles immediately after manufacture. The adhesivity of the bitumen coating can be checked using the hot sand test[278].

When the chippings have been plucked out of the asphalt, it exposes the asphalt mortar to traffic. It is likely that this material will be poorly compacted and, as a result, it will fret rapidly.

278 BRITISH STANDARDS INSTITUTION. *Sampling and examination of bituminous mixtures for roads and other paved areas, Methods for determination of the condition of the binder on coated chippings and for measurement of the rate of spread of coated chippings, BS 598-108: 1990.* BSI, London.

Chapter 11

Aggregates in asphalts

Behavioural aspects of bitumen are dealt with throughout this book. However, bitumen is frequently mixed with aggregates to produce asphalts. Thus, an appreciation of how aggregates behave is important if the performance of asphalts is to be understood. This chapter considers the origins, types and key parameters of aggregates.

11.1 Origins and types of rock

Asphalts consist primarily of aggregates. The properties of aggregates (sometimes described as 'mineral aggregates') have a significant influence on the behaviour of asphalts. Thus, an appreciation of the origins and properties of aggregates is one of the key elements to understanding asphalt technology.

11.1.1 Natural aggregates

Rock is as old as the Earth itself, having been formed as part of the creation of the planet. The exact development of the Earth is still not completely understood but, according to current knowledge, it is believed a red hot gas ball developed in space about three billion years ago. The action of cooling down to a temperature between 1000 and 2000°C caused the initial stages of a crust to form on the surface of the Earth.

Further cooling took place, reaching the inner parts of the planet and the thickness of the solid zone increased. Large stresses developed between the hotter and the cooler zones and, at the same time, the solidified rock further compressed the inner, liquid parts of the planet. This resulted in deformation followed by rearrangement of the Earth's crust.

Clods and blocks of rock developed and shifted towards each other, sometimes remelting. This continuous change of the surface of the Earth was accelerated by weathering since further cooling led to the formation of water.

Table 11.1 Structure and texture of rock types

Category	Structure	Texture
Igneous	Glassy, fine to coarse, crystalline, porphyric	Dense, smooth
Sedimentary	Varies according to type and size of the components	Less dense, rough texture
Metamorphic	Crushed minerals	Variable according to origin

The thickness of the Earth's crust is estimated to be around 15 km. The melted rock in this zone is called magma. It is subdivided into three classifications according to how each was formed:

- igneous rock resulting from solidification;
- sedimentary rock resulting from sedimentation;
- metamorphic rock resulting from transformation.

'Minerals' means, in literal terms, any material that is mined but in the context of this book refers to natural aggregates. Minerals consist of chemical elements or combinations of chemical elements. Some minerals have a uniform composition and structure whilst others consist of mixtures of different minerals and have heterogeneous structures.

Aggregates are classified according to their chemical composition, their mineral contents and the structure of the minerals. 'Structure' denotes the mineral composition and 'texture' means the arrangement of the minerals. The characteristics of the three main rock types in terms of these properties are illustrated in Table 11.1.

'Porphyric' indicates a special kind of structure of igneous rock. Porphyric rocks are formed when aggregates are pushed up suddenly from the deeper parts of the Earth's crust to the surface. Slow cooling is then followed by very rapid cooling which leads to a finely grained structure. Traditionally, porphyry was a hard rock quarried in ancient Egypt composed of crystals of white or red feldspar in a red matrix. However, it now denotes an igneous rock with large crystals scattered in a matrix of much smaller crystals.

Sedimentary rocks cannot be classified easily using the system illustrated in Table 11.1. If the constituents are angular and cemented together by finer material, the term 'breccia' is used for the structure. If the material is composed of coarse but rounded grains that are cemented together, it is called 'conglomerate'. Sandstones are composed of finer grains.

Igneous rocks

Igneous rocks were formed by volcanic or magmatic action. ('Magma' is a fluid or semi-fluid material under the Earth's crust from which lava and other igneous rocks are formed.) Igneous rocks are divided into three subgroups:

- plutonic
- dike
- volcanic.

Plutonic rocks cooled down slowly in deeper parts of the Earth's crust, below other, older minerals. They were named after the Greek god of the underworld. Examples are granite, syenite, diorite, and gabbro.

Dike rocks were probably formed when liquid magma poured into cracks and crevices in cooled rock. These cracks and crevices arise because of temperature stress. In the cracks, rapid cooling took place. This category comprises all stones with porphyric structure such as granite porphyry, syenite porphyry and diabase porphyry.

Volcanic rock, in contrast to the formation of plutonic rock, emerged from the Earth's crust and cooled on the surface of the Earth. This still happens nowadays at active volcanoes. Very often, porphyric structures are formed in this process. This group includes diabase, basalt, phono-lithe and, in part, porphyry.

Sedimentary rocks

Sedimentary rocks are subdivided into mechanical sedimentary (gravel, breccia, conglomerate), chemical sedimentary (dolomite), and biogenetic sedimentary (coal). Only rocks that evolved from the first two of those processes are used in road construction.

Mechanically formed sedimentary rocks can originate either from magmatic rock or from chemical sediments like moraine exposed to weathering after their formation. Rainwater and melted snow transported loosened particles into river beds, seas, and oceans where they were deposited. Chemically formed sedimentary rocks are residual or precipitated stones. Their suitability for road construction depends on their hardness.

The most remarkable feature of this type of structure is the layering. It is for this reason that sedimentary rocks are sometimes described as bedded rock.

Metamorphic rocks

Metamorphic rocks originated as igneous rocks because the first phase of their formation was volcanic. However, rearrangement of the Earth's crust shifted the materials back to deeper layers where higher temperatures existed. Remelting and mixing formed a variety of structures with a slate-like appearance. Typical examples of metamorphic rock are gneiss and amphibolite. There are numerous types of metamorphic rock and the suitability of any single variety for road construction should be checked carefully.

11.1.2 Synthetic aggregates

Industrial by-products like blast furnace slag or recycled building materials are utilised in the manufacture of asphalts. The properties of these synthetic aggregates vary substantially according to their origin. The technical suitability and environmental properties have to be tested in every case before they can be considered as suitable for use in an asphalt.

11.2 Aggregate extraction

Natural sand and gravel can be obtained with relatively simple, inexpensive equipment. Since single grains are unbound, they can be separated with ease. For dry extraction, an excavator is adequate. Where material extraction is undertaken below water, a floating excavator is needed. The gravel is separated into granular classes or broken down to produce crushed gravel sand.

More effort is required to crush bedrock. In modern quarries, complete walls are blasted at different locations at the same time. The resultant large boulders are crushed and screened to the required granular sizes.

Blast furnace slags are formed from the hot liquid in large tubs and are subsequently crushed and screened.

11.3 The European aggregate Standard

In January 2004 the UK quarrying industry made one of the biggest changes to the way aggregates were called and categorised since metrication was introduced. The changes were introduced as a direct result of EU Directives to remove barriers to trade, and form part of the national standards of all countries within the EU community. The changes impacted on all aggregates that used size as a means of categorising them and aggregates that complied with a European Standard. It is important to remember that the changes were designed to only impact on the way that aggregates are described or categorised and that the aggregates should require only slight changes to processing to meet the new Standard.

For this Standard to work across European countries it has been written in a 'menu' style, providing sieve sizes, categories, properties and test methods for all countries to select appropriate limiting values. BS EN 13043 *Aggregates for bituminous mixtures and surface treatments for roads, airfields and other trafficked areas*[279] is not an easy document to use. To aid the understanding of the European Standard, the UK adopted the

---

279 BRITISH STANDARDS INSTITUTION. *Aggregates for bituminous mixtures and surface treatments for roads, airfields and other trafficked areas, BS EN 13043: 2002.* BSI, London.

use of guidance documents, known as the PD 6682 series (PD stands for published document) and these are published by BSI. The PD 6682 series can only recommend categories and limiting values in line with the relevant European Standard and does not replace it. The guidance document for BS EN 13043 is PD 6682-2: 2003,[280] and the two documents should be read together.

PD 6682-2: 2003 recommends the properties of aggregates and filler obtained by processing natural, manufactured or recycled materials for use in bituminous mixtures and surface treatments. PD 6682-2: 2003 contains recommended grading tables, previously found in the superseded British Standards specifications BS 63 and BS 1047. These grading tables show the recommended new aggregate size, grading category and grading limits.

PD 6682-2: 2003 also recommends the use of new test methods, detailing the key differences between previous British Standards such as the BS 812 series with the new European Standards. For a clear step by step approach to the changes in test methods, PD 6682-9: 2003[281] provides guidance on all European test method Standards.

11.3.1 Quality control and CE marking

The European aggregate Standard also introduces quality control, requiring the producer to demonstrate Factory Production Control (FPC) to a third party, leading ultimately to certification from a notified body.

This is one of the biggest implications of the European Standard for the aggregate producer requiring CE Marking. CE Marking is a legal declaration by the producer indicating the properties of their aggregates, enforced by Trading Standards and constituting a criminal offence if falsely declared. Previously, suppliers of aggregates would have provided data sheets to customers showing the most recent suite of physical, mechanical and chemical test results undertaken on the aggregates sourced from a quarry. CE Marking requires that this is undertaken for each size of aggregate at each source and declared as per the requirements of Annex ZA of the relevant European Standard.

11.3.2 Size and grading category

Size was originally described using one dimension, such as 10 mm single size. However since the introduction of the European Standard, size now

280 BRITISH STANDARDS INSTITUTION. *Aggregates for bituminous mixtures and surface treatments for roads, airfields and other trafficked areas – Guidance on the use of BS EN 13043, PD 6682-2: 2003.* BSI, London.

281 BRITISH STANDARDS INSTITUTION. *Guidance on the use of European test method standards, PD 6682-9: 2003.* BSI, London.

Table 11.2 Recommended BS EN 13043 grading categories and corresponding grading limits for coarse aggregates used in the manufacture of asphalt.[282]

Recommended BS EN 13043 aggregate (mm)	20/40	20/31·5	10/20	6·3/14	4/10	2/6·3	1/4
Equivalent BS 63 single sized aggregate (mm)	40	28	20	14	10	6	3
Recommended BS EN 13043 grading category	G_C85/20	G_C85/35	G_C85/20	G_C85/20	G_C85/20	G_C80/20	G_C90/10
Sieve size (mm)	Percentage by mass passing ISO 565 sieve						
80	100	–	–	–	–	–	–
63	98 to 100	100	–	–	–	–	–
40	85 to 99	98 to 100	100	–	–	–	–
31·5	–	85 to 99	98 to 100	100	–	–	–
20	0 to 20	0 to 35	85 to 99	98 to 100	100	–	–
14	–	–	–	85 to 99	98 to 100	100	–
10	0 to 5	0 to 5	0 to 20	–	85 to 99	98 to 100	–
8	–	–	–	–	–	–	100
6·3	–	–	–	0 to 20	–	80 to 99	98 to 100
4	–	–	0 to 5	–	0 to 20	–	90 to 99
2·8	–	–	–	0 to 5	–	–	–
2	–	–	–	–	0 to 5	0 to 20	–
1	–	–	–	–	–	0 to 5	0 to 10
0·500	–	–	–	–	–	–	0 to 2

follows the convention of using two dimensions, small '*d*' and large '*D*', such as 4/10 mm single size. Also introduced with the European Standard is the way aggregates are categorised with the use of Grading Categories, expressed as G_C for coarse aggregate, G_F for fine aggregate and G_A for all-in aggregate. A grading category gives more information about the grading of an aggregate. Grading categories work by indicating the oversize and undersize of the two sieve sizes '*d/D*' used to describe the aggregate size. Table 11.2, shows how the BS 63 size is translated into the BS EN 13043 size.

Other changes brought about from the implementation of the European aggregate Standard can be seen in Table 11.2. Firstly, the 37·5 mm sieve is replaced by the 40 mm sieve, the 28 mm sieve is replaced by the 31·5 mm sieve. The UK has taken a pragmatic approach and the 31·5 mm size may be described as 32 mm.

All fine sieves are replaced by a new set. Asphalts will use the 2 mm sieve as the controlling sieve for fine aggregate, and the 75 mm sieve is replaced by a fines category, expressing the fines content as f_x, where x

282 BRITISH STANDARDS INSTITUTION. *Aggregates for bituminous mixtures and surface treatments for roads, airfields and other trafficked areas – Guidance on the use of BS EN 13043, Table 3, PD 6682-2: 2003.* BSI, London.

Table 11.3 BS EN test method groups

BS EN test method	Group
BS EN 932 (all parts)	General properties of aggregates
BS EN 933 (all parts)	Tests for geometrical properties of aggregates
BS EN 1097 (all parts)	Tests for mechanical and physical properties of aggregates
BS EN 1367 (all parts)	Tests for thermal and weathering properties of aggregates
BS EN 1744 (all parts)	Tests for chemical properties of aggregates

is the maximum percentage passing the 0·063 mm sieve. Finally the use of microns (micrometres) has gone and mm used for all sizes.

11.3.3 *Key properties of aggregates*

Properties of aggregates have, in the past, been expressed as actual measured values. European aggregate standards require the use of categories and classes to define an aggregate property.

As mentioned above, BS 812 test methods are replaced with new EN test methods listed above in Table 11.3. A key change is the measurement of aggregate strength, where the ten per cent fines value, the aggregate crushing value and the aggregate impact value have been replaced by the Los Angeles coefficient (LA). LA values are expressed as categories such as LA_{30}, setting a requirement for a maximum LA value of 30. In this way an actual measured value of 27 would be expressed as LA_{30}.

There are occasions when the actual measured value exceeds the highest category available in the European Standard. In these cases the supplier declares the found value, such as $LA_{Declared55}$. The PSV values in BS EN 13043 Table 13, *Categories for minimum values of resistance to polishing*, are the only exception to this rule, allowing intermediate values to be expressed as a supplier declared value, such as $PSV_{Declared65}$.

One other key aggregate property fundamental to the performance of asphalt is shape. Shape has, in the past, been reported as the flakiness index, and continues to be so. But, confusion comes when comparing results from the previous BS 812 Part 105.1: 1989 method with the replacement BS EN 933-3.[283] Both methods measure the percentage passing a slot width specified for that particle size fraction, the flaky particles are those which pass through the slots. This is as far as the similarity goes. The new BS EN 933-3 method categorises a particle as flaky if its minimum dimension is less than half its upper sieve size (*D*). The size fractions used in the new test method are significantly different

---

283 BRITISH STANDARDS INSTITUTION. *Tests for geometrical properties of aggregates. Determination of particle shape. Flakiness index, BS EN 933-3: 1997.* BSI, London.

from those previously used in the BS 812 method resulting in a flakiness index value lower than previously found. As required by the European standard the flakiness index is reported as a category, FI_{20}, setting a requirement for a maximum flakiness index value of 20. Previous flakiness index requirements of 25% for surface dressing aggregates are now specified as FI_{20}, indicating a drop of 5% to allow for the differences in test methods.

The behavior of individual aggregates in relation to a number of key properties is important in ensuring that a particular asphalt performs adequately in service. These key properties are as follows.

- *Shape.* Shape is an important determinant in asphalts. If individual pieces of aggregate are elongated or very flat they can form inherent weaknesses in the mixture.
- *Strength.* In strength terms, the aggregate has to be able to withstand the applied loads without crushing and should be able to resist stresses created by the sudden imposition of loads.
- *Roughness.* The surface roughness of pieces of aggregate is a major contributor to the frictional characteristics of a pavement. This roughness, or rugosity as it is sometimes called, has to be maintained if safety is not to be compromised, hence the need for aggregates to be resistant to abrasion in service.
- *Susceptibility to polishing.* It is a fact that all aggregates polish with the action of traffic. However, this polishing should take place at a rate such that the carriageway provides acceptable braking characteristics for a reasonable period of time. Thus, the degree and speed of polishing is important.
- *Soundness.* Some aggregates that were thought to have sufficient initial strength have been found to have failed whilst in the pavement. Soundness seeks to deal with such materials and may be described as a measure of the durability of aggregates in service.

All of the above properties are now expressed as categories and defined in BS EN 13043. Specifiers are required to implement the European Standard in full and need to define BS EN 13043 categories for properties that are relevant to a particular end use. Table 11.4 shows the preferred specification for aggregates for surface treatments. At all times it is recommended the PD 6682 tables be used to define categories for the properties of aggregates.

An option to use a 'No Requirement' category is also provided. For example, a resistance to abrasion from studded tyres category of $A_N NR$, NR for no requirement, is included in BS EN 13043 and it means that there is no specified requirement for this test, particularly apt in the UK where studded tyres are not permitted. When a specifier

Table 11.4 Example specification for 6/10 mm aggregate for surface treatments[284]

Property	Category
Grading	6·3/10 mm G_C85/20
Fines content	f_1
Flakiness index	FI_{20}
Resistance to fragmentation	LA_{30}
PSV	PSV_{62} (Contract Specific)
AAV	$AAV_{Declared}$ (Contract Specific)
Durability:	
Water absorption to BS EN 1097-6:2000, Clause 7	$WA_{24}2$
For WA > 2%, magnesium sulphate soundness	MS_{25}

does not want to dictate a property that is not relevant for an end use the 'No Requirement' designation should be used.

11.3.4 Implications for other Standards

BS EN 13043 supersedes BS 63-1: 1987, BS 63-2: 1987, and BS 1047: 1983 but has implications on current British Standards for asphalt, BS 594, BS 4987, BS 5273 and the *Specification for Highway Works* (SHW), all of which contain requirements for aggregates. In 2003, these asphalt Standards were updated to incorporate the changes. Key changes were the introduction of new sieves, the replacement of 3·35 mm and 2·36 mm sieves in coated macadam and hot rolled asphalt respectively with the 2 mm sieve. This triggered the amendment of binder correction tables in BS 598 used for analysis.

40 mm dense base mixtures, which used the 37·5 mm sieve to control upper size, have been replaced by 0/32 mm size mixtures, as the 37·5 mm sieve is no longer part of the sieve series used in the UK. 0/32 mm size mixtures have replaced all 28 mm base and binder course mixtures, as the 28 mm sieve is also no longer available.

The way asphalts are described by size also changed, following the same rules as aggregates, d/D. Examples are given in Table 11.5.

11.3.5 Assessment of key properties

In the UK, the major arterial highways (called 'Trunk Roads', which includes motorways) are the responsibility of national government. The executive agencies that look after the Trunks Roads are the Highways Agency in England, the National Assembly for Wales (Cynulliad

284 BRITISH STANDARDS INSTITUTION. *Aggregates for bituminous mixtures and surface treatments for roads, airfields and other trafficked areas – Guidance on the use of BS EN 13043, annex A, table A.1, PD 6682-2: 2003.* BSI, London.

Table 11.5 Examples of size designation for asphalts specified in BS 4987 and BS 594

Previous Description	Revised Description
BS 4987-1: 2003 Coated macadam	
28 mm DBM designed base	0/32 mm DBM designed base
6 mm medium graded surface course	0/6 mm medium graded surface course
20 mm porous asphalt	6/20 mm porous asphalt
BS 594-1: 2003 Hot rolled asphalt	
50/14 HRA binder course	50% 0/14 HRA binder course
35/14 HRA Type F design surface course	35% 0/14 HRA Type F design surface course
20 mm pre-coated chippings	14/20 mm pre-coated chippings

Cenedlaethol Cymru), the Department for Regional Development in Northern Ireland and, in Scotland, the Scottish Executive Development Department. Trunk Roads represent some 4% by length of the road network but they carry some 30% of the total traffic and some 50% of the heavy goods vehicles. Contracts for works of new construction or maintenance on Trunk Roads are contained within a suite of documents entitled the *Manual of Contract Documents for Highway Works*[285] (MCHW). Specification requirements for asphalts including aggregates are contained in Volumes 1 and 2 of this suite (entitled the *Specification for Highway Works*[286] and the *Notes for Guidance on the Specification for Highway Works*[287] respectively). These documents are also used by the majority of local authorities for works on the remaining 96% of the highway network. Table 11.6 lists some of the current requirements for aggregates used on Trunk Road schemes. Typical testing requirements for UK Trunk Roads are set out in Table 11.7.

285 HIGHWAYS AGENCY *et al. Manual of contract documents for highway works, vols 0–6, Various dates.* The Stationery Office, London.

286 HIGHWAYS AGENCY *et al. Manual of contract documents for highway works, Specification for highway works, vol 1, 2002.* The Stationery Office, London.

287 HIGHWAYS AGENCY *et al. Manual of contract documents for highway works, Notes for guidance on the specification for highway works, vol 2, 2002.* The Stationery Office, London.

Table 11.6 Some of the current requirements for aggregates on UK Trunk Roads[288]

Property	European Standard test	Limiting value
Size	Sieve analysis (dry or wet)	—
Shape *Flakiness*	Flakiness index (FI)	FI_{35}
Strength *Crushing*	Los Angeles coefficient (LA)	LA_{30}
Abrasion	Aggregate abrasion value (AAV)	$AAV_{Declared}$
Polishing	Polished stone value (PSV)	$PSV_{Declared}$
Soundness	Magnesium sulphate soundness value (MSSV)	MS_{25}

Table 11.7 Typical testing requirements for UK Trunk Roads[289]

SHW clause	Type of material	Property to be assessed	Test	Frequency of testing
901, 925, 937, 938	**Aggregate for bituminous materials**	Hardness	LA (N)	Monthly[†]
		Durability	MSSV (N) WA (N)	1 per source as required
		Cleanness	Sieve test (% <0·075 mm) (N)	Monthly
		Shape	FI (N)	Monthly
		Skid resistance	PSV (N)*	1 per source
		Abrasion	AAV (N)*	1 per source
915, 925	**Coated chippings**	Grading	Grading (N)	1 per stockpile
		Binder content	Binder content (N)	1 per stockpile
		Shape	FI (N)	1 per source
		Skid resistance	PSV (N)	1 per source
		Abrasion	AAV (N)	1 per source

Notes
All tests to be carried out by the contractor.
The frequency of testing is given for general guidance and is only indicative of the frequency that may be appropriate.
(N) indicates that UKAS (United Kingdom Accreditation Service) sampling and a test report or certificate is required.
WA – water abrasion.
*Applicable to coarse aggregates for surface courses only.
†To be introduced in 2004.

288 WOODWARD D (R N HUNTER, ed). *Assessing aggregates for use in asphalts, Asphalts in Road Construction, p 13, 2000.* Thomas Telford, London.
289 HIGHWAYS AGENCY *et al. Manual of contract documents for highway works, Notes for guidance on the specification for highway works, vol 2, table NG 1/1 (part only), 2002.* The Stationery Office, London.

Chapter 12

Types and applications of different asphalts

Despite the fact that asphalts are little more than a mixture of aggregate and bitumen, there are many different mixtures available and their characteristics can vary quite significantly.

For the majority of the last half of the twentieth century, standard practice on major roads in the UK was to construct the base and binder course using either hot rolled asphalt or dense coated macadam. The surface course was invariably constructed using chipped hot rolled asphalt, a material that was virtually unique to the UK. However, in the last decade of the twentieth century, a number of new materials, mainly variants of mixtures used in France and Germany, began to enjoy more widespread use. The availability of these materials, together with a vastly improved understanding of the failure mechanisms in flexible pavements, led to a move towards what are termed 'stiff' mixtures. Such mixtures produce pavements that have a very long structural life, providing distress in the form of cracking or rutting is addressed in a timely and effective manner.

Traditionally, asphalts were designated as either gap graded or continuously graded. This referred to the constitution of the asphalt in terms of the size of the individual aggregate fractions within mixtures. Size is measured by passing a sample of aggregate through a series of sieves. Thus, a continuously graded mixture was one that contained fractions of various (but not necessarily all) sizes throughout the range. An example of a continuously graded mixture is dense coated macadam base. A gap graded mixture is one where sizes are discontinuous. An example of a gap graded material is a hot rolled asphalt surface course consisting of 6/14 mm aggregate with the next size down being small enough to pass a 2 mm sieve.

Examples of the gradings of these materials are shown in Table 12.1, with two of the grading envelopes depicted in Fig. 12.1.

Table 12.1 Gradings for common asphalts

Sieve size	Percentage by weight passing each sieve			
	Coated macadams		Hot rolled asphalts	
	0/20 mm size dense binder course	0/10 mm size close graded surface course	Designation 50% 0/14 mm base, binder course and regulating course	Designation 35% 0/14 mm type F surface course
31·5 mm	100			
20 mm	95 to 100		100	100
14 mm	65 to 85	100	90 to 100	87 to 100
10 mm	52 to 72	95 to 100	65 to 100	55 to 88
6·3 mm	39 to 55	55 to 75		
2 mm	24 to 36	19 to 33[a]	30 to 55	55 to 67
1 mm		15 to 30		
0·500 mm			13 to 50	40 to 67
0·250 mm	7 to 21		6 to 31	15 to 55
0·063 mm	2 to 9	3 to 8	1 to 8	6·0 to 10·0

[a]May be increased to 38 when sand fine aggregate is used.

The traditional British Standards that specify asphalts emerged from the description of an asphalt as being gap graded or continuously graded. BS 594 deals with asphalts (gap graded materials) whilst BS 4987 applies to coated macadams (continuously graded materials). This definition was never particularly scientific, somewhat inconsistent and thus was never satisfactory. An example is fine cold asphalt which was categorised as being continuously graded. Nowadays, European harmonisation requires that the term 'asphalt' is used to describe all bituminous mixtures. Thus, grading is irrelevant, in terms of description.

Recently, the adoption of relatively hard bitumens and the emergence of new materials combined with an improved understanding of the behaviour of asphalts has resulted in a vast increase in the usage of what were traditionally called continuously graded mixtures, the coated macadams. Nowadays there are numerous asphalts available to the purchaser. Some are general purpose, such as bases or binder courses, whilst others are produced for particular applications, e.g. fuel-resisting or deferred-set asphalts.

For non-proprietary mixtures, two British Standards govern all aspects of asphalt production and laying in the UK. These are BS 4987 and BS 594 and they are concerned with coated macadams and hot rolled asphalts respectively. Each of these Standards is split into two Parts,

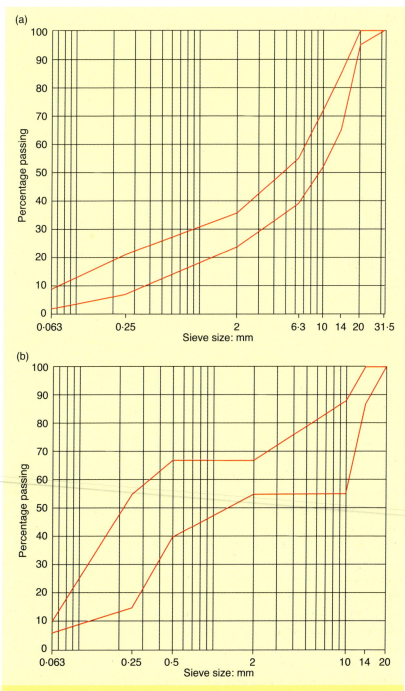

Fig. 12.1 Grading envelopes for a continuously graded and a gap graded asphalt. (a) 0/20 mm size dense binder course coated macadam. (b) Designation 35% 0/14 mm type F surface course hot rolled asphalt

one dealing with the specification and manufacture[290,291] and the other with transporting, laying and compaction[292,293].

The following three sections describe a number of these mixtures in terms of their properties and applications.

12.1 Coated macadams
12.1.1 Bases and binder courses

Coated macadams are, by far, the most common mixtures in use as binder courses and bases in road pavements in the UK. In the USA, such materials are described as asphaltic concretes or, formerly, Marshall asphalts. The key difference between these mixtures is the penetration of the bitumen although, as discussed below, that is not always the case. These materials are continuously graded and are manufactured with relatively low bitumen contents. Their strength is derived from the interlock of the coated aggregate. A number of different UK mixtures have evolved, primarily from mixtures used in other EC countries.

In the UK, high-performance base and binder course materials are required, particularly for the reconstruction of heavily trafficked roads where the available construction depth is often restricted by bridge clearance and drainage levels. In order to enhance the performance of the pavement without increasing the depth of construction, improved materials were developed. Appropriate design curves for these materials can be derived by the measurement of their properties in relation to those of the standard dense coated macadam base. One of the most important properties is that of the stiffness modulus, which strongly affects the critical strains generated in both the base and subgrade.

One of the improved dense coated macadam base and binder course materials is heavy duty macadam (HDM)[294–296]. In this material, the traditional 100/150 pen bitumen is replaced by 40/60 pen bitumen and the target filler content is increased from a range of 2 to 9% to a range of 7 to 11%. A significant increase in stiffness is obtained by using the 40/60 pen bitumen and a further, albeit smaller, rise results from the

290 BRITISH STANDARDS INSTITUTION. *Coated macadam for roads and other paved areas, Specification for constituent materials and for mixtures, BS 4987-1: 2003.* BSI, London.

291 BRITISH STANDARDS INSTITUTION. *Hot rolled asphalt for roads and other paved areas, Specification for constituent materials and asphalt mixtures, BS 594-1: 2003.* BSI, London.

292 BRITISH STANDARDS INSTITUTION. *Coated macadam for roads and other paved areas, Specification for transport, laying and compaction, BS 4987-2: 2003.* BSI, London.

293 BRITISH STANDARDS INSTITUTION. *Hot rolled asphalt for roads and other paved areas, Specification for transport, laying and compaction, BS 594-2: 2003.* BSI, London.

higher filler content. Laboratory fatigue tests have demonstrated that the harder bitumen and higher filler content had an insignificant effect on fatigue life. This increased stiffness reduces the strains induced at the bottom of the base where fatigue cracking has traditionally been assumed to originate. Tests using a pneumatic-tyred wheel-tracking machine at 30°C showed that adding 3% more filler reduced deformation by about 40% in mixtures that had been properly compacted.

During 1986, a number of proving trials were carried out on heavily trafficked roads using this material to identify any problems that might arise during manufacture or laying. The only significant difficulty experienced was that some types of mixing plant required longer mixing times than those necessary with conventional dense macadam. This is due to the higher filler content. This reduced the output and increased the cost of production. Experience with this material has shown that increasing the bitumen content slightly can do much to overcome this problem without significantly affecting the performance of the mixture.

Appropriate design thicknesses have been determined by the TRL[294]. For a given design life, the introduction of HDM in place of the conventional material leads to a reduction in design thickness, typically between 15 and 20%. This curve is included in the Highways Agency's design standard[297]. A specification for HDM is included in Cl 930 (base) and Cl 933 (binder course) of the SHW[298].

High modulus base (HMB) was developed by the TRL as a result of the evaluation of a French mixture called 'enrobé à module élevé' (EME) which was introduced as an economic measure to reduce the thickness of the main structural layer of the pavement. Tests were undertaken on a mixture with a grading that complied with BS 4987. The performance of a trial pavement confirmed that a bitumen macadam

294 NUNN M E, C J RANT and B SCHOEPE. *Improved roadbase macadams: road trials and design considerations, Research Report 132, 1987.* Transport and Road Research Laboratory, Crowthorne.

295 HIGHWAYS AGENCY *et al. Manual of contract documents for highway works, Specification for highway works, vol 1, Cls 930 & 933, 2002.* The Stationery Office, London.

296 LEECH D and M E NUNN. *Substitution of bituminous roadbase for granular sub-base, 3rd Eurobitume Symp, The Hague, pp 249–255, 1985. Also reprinted in Asphalt Technology, J Inst Asph Tech, Feb 1986.* Institute of Asphalt Technology, Dorking.

297 HIGHWAYS AGENCY *et al. Design manual for roads and bridges, Pavement design and maintenance, Pavement design, vol 7, HD 26/01, 2001.* The Stationery Office, London.

298 HIGHWAYS AGENCY *et al. Manual of contract documents for highway works, Specification for highway works, vol 1, 2002.* The Stationery Office, London.

binder course with a grading complying with BS 4987 and made with a 10/20 pen bitumen had very similar properties to EME[299].

Currently, it is more common for HMB materials to be manufactured with 30/45 pen bitumen. This mixture offers increased stiffness over more conventional materials and, as a result, can be laid thinner. It can be laid and compacted over a wide range of weather conditions with little adjustment to the mixing temperatures and compaction techniques employed with materials using a 40/60 pen bitumen.

A recent development with HMBs has been the incorporation of a high-stiffness polymer modified binder. The bitumen in this mixture has been designed to give improved flexibility, adhesion and durability whilst the mixture has a very high stiffness meaning that a thinner layer is required than would have been the case with conventional asphalts.

In the UK, base and binder course materials are being developed based on stone mastic asphalt (SMA) technology. SMA was originally developed in Germany and Scandinavia as a surfacing course in the 1960s for resisting studded tyres, and is now widely used through most of Europe[300]. SMA is a gap graded asphalt that has a coarse aggregate skeleton, similar to porous asphalt, but the voids are filled with a fine aggregate/filler/bitumen mortar.

BS 4987-1: 2003[301] also contains details of 0/20 mm size open graded binder course and 0/32 mm size single course. Open graded binder courses are used in lightly trafficked situations (technically where traffic is ≤250 commercial vehicles per lane per day). The single course mixture can be used as a base or as a binder course but it is most commonly used as a combined binder course and surface course on roads that carry light traffic. It should be sealed with bituminous grit soon after laying and before being subjected to traffic.

12.1.2 Surface courses

Generally speaking, coated macadam surface courses are used on roads that carry light traffic. Open graded mixtures were widely used in the UK, particularly in England, prior to the emergence of heavier commercial vehicles. Having an open texture means that they are accessible to

299 NUNN M E and T SMITH. *Evaluation of enrobé à module élevé (EME), A French high modulus roadbase material, Project Report 66, 1994.* Transport Research Laboratory, Crowthorne.

300 EUROPEAN ASPHALT PAVEMENT ASSOCIATION. *Heavy duty surfaces: the arguments for SMA, 1998.* EAPA, The Netherlands.

301 BRITISH STANDARDS INSTITUTION. *Coated macadam for roads and other paved areas, Specification for constituent materials and for mixtures, BS 4987-1: 2003.* BSI, London.

236

Fig. 12.2 Coated macadam – 0/6 mm size dense surface course
Coin shown = 1 Euro, 23 mm diameter

water and are likely to degrade rapidly where there are frequent freeze/thaw weather cycles. Their durability is a function of the nature of the pavement itself and the nature of the incident traffic. The use of a surface dressing may be necessary as a sealant. Close graded, dense and medium graded surface courses are less susceptible to intrusion and damage by surface water. However, they are only suitable for light traffic or applications such as footways, car parks, playgrounds, etc. Fine graded surface course was previously described as 'fine cold asphalt' and it is only suitable for footways, etc. A photograph of a 0/6 mm size dense coated macadam surface course is shown in Fig. 12.2.

One particularly important coated macadam asphalt is porous asphalt (PA). Although PA has not enjoyed widespread use in the UK, it is used in significant quantities in other EC states. PA is a gap graded material with interconnecting voids that allow water to flow freely through the material to the binder course. The water finds its way through to the binder course and then flows on top of the binder course into a channel where it is collected in a drain. The disadvantage is that the bitumen is exposed to oxygen in the atmosphere and is thus subject to oxidation and consequent embrittlement. Thicker, softer bitumen coating is, therefore, desirable and so the use of modified binders or fibres is necessary to

237

prevent the binder draining through the mixture. A useful source of information on PA is TRL Report 264[302].

PA has benefits in terms of particularly low noise and spray generation. In addition, glare in wet conditions is reduced.

PA with 100/150 pen bitumen has a life of 7 to 10 years with traffic of 6000 cv per lane per day. The tack coat is important as it adds to the waterproofing of the binder course; a polymer modified tack coat should be considered.

PA is the standard surface course used in the Netherlands. Dutch Engineers accept that the characteristics of PA outweigh the need to replace it rather more often than would be the case with other asphalts.

12.2 Hot rolled asphalt
12.2.1 Bases and binder courses

Hot rolled asphalts are gap graded mixtures that were not uncommon as base and binder courses until the last decade of the twentieth century when the stiffer coated macadams became available. They have higher bitumen contents than coated macadams and thus were comparatively unattractive from a financial standpoint. Their strength is derived from the characteristics of the bitumen/sand/filler mortar. Use of hot rolled asphalts has been largely restricted to the UK with other countries opting for coated macadams/asphalt concretes. However, nowadays, most UK contracts designate the use of coated macadams. In terms of pavement design in the UK, HRA bases are regarded as being equivalent to DBM made with 100/150 grade bitumen.

12.2.2 Surface courses

Hot rolled asphalt surface course with 14/20 mm pre-coated chippings was the standard UK surface course until the late 1990s. The mixture has some 30% of 6/14 mm aggregate but that element of the mixture plays a relatively minor role in the performance of the asphalt – again it is the bitumen/sand/filler mortar which predominates. Since the majority of the mixture is sand and filler, the finished surface is very smooth. Whilst this is ideal in some applications, it is wholly unsuitable on carriageways because of its very poor skid resistance. Accordingly, pre-coated chippings are added to impart texture to the running surface. These are placed by means of a custom-built chipping machine that follows immediately behind the paver depositing, typically, 12 kg/m² of 14/20 mm size chippings that have been coated in bitumen onto the partially compacted asphalt. The rollers then push the chippings into the surface whilst compacting the asphalt.

302 NICHOLLS J C. *Review of UK porous asphalt trials, Report 264, 1997.* TRL, Crowthorne.

It can be laid 40 mm thick but layers are generally 45 mm thick. Replacement is generally necessary because of cracks or deformation or because it has lost its skid resistance. It is generally expected to have a life of 10 to 20 years but areas where the surface is 25 or more years old are not uncommon, particularly in urban areas where texture is not as critical as it is on some high-speed roads.

12.3 Thin surfacings

Thin surfacings are now the standard materials of choice when specifying surface courses on roads of any size and capacity in the UK. It is this group of mixtures that were the reason for a revolution in the types of surface course mixtures specified on major UK highway contracts. They are proprietary mixtures that are provided against a specification which is a combination of recipe and end-product requirements.

Thin surfacings were introduced into the UK in the early 1990s using French technologies. Subsequently, German experience with stone mastic asphalts (SMAs) was adapted to meet UK conditions and specifications, in particular the requirement to exhibit adequate levels of skid resistance.

Although the early thin surfacings were well received by clients, there was a contractual difficulty since the standard UK specification did not easily recognise these asphalts as being acceptable for use on major roads. Hence the emergence of HAPAS ('Highway Authorities Product Approval Scheme'). This initiative came about following action by the English Highways Agency in conjunction with the County Surveyors Society. The HAPAS system requires all aspects of production and laying to be controlled. This includes constituent materials, equipment and the surfacing procedures. Monitoring and certification is undertaken by the British Board of Agrèment (BBA). Testing by the BBA is carried out on the laid asphalts and their performance monitored over a significant period of time. Success leads to the award of a HAPAS certificate and assessment is likely to have taken a period of two to three years. The standard UK specification requires that any thin surfacings employed are covered by a HAPAS certificate. This system gives the client confidence that materials will perform as required and has been particularly successful in the UK. The existence of HAPAS certification has allowed the client to adopt a much less rigorous specification. This specification can be found in Cl 942 of the SHW[303]. Thin surfacings are discussed further in Section 13.4.3.

---

303 HIGHWAYS AGENCY *et al. Manual of contract documents for highway works, vol 1, cl 942, Specification for highway works, 2002.* The Stationery Office, London.

Fig. 12.3 A thin surfacing – 0/14 mm size
Coin shown = 1 Euro, 23 mm diameter

Table 12.2 Flexible options for bases and binder courses on UK Trunk Roads[304]

Country	Options for binder courses and bases
England, Northern Ireland & Wales	DBM 125
	DBM 50
	HDM 50
	HMB 35
	SMA
	HRA*
Scotland	DBM 125
	DBM 50
	HDM 50
	HMB 35*
	SMA*

* On UK Trunk Roads including Motorways, approval is needed from the client before these can be used.

304 HIGHWAYS AGENCY *et al. Design manual for roads and bridges, Pavement design and maintenance, Pavement design, vol 7, HD 26/01, 2001.* The Stationery Office, London.

Table 12.3 Options for surface courses on Trunk Roads in England[305]

Road category	High-speed roads		Low-speed roads	
	New construction and major maintenance	Minor maintenance	New construction and major maintenance	Minor maintenance
Flexible				
HRA	×	R	×	R
Coated macadam	×	×	×	×
Porous asphalt	R	R	R	R
Thin surfacings	✓	✓	✓	✓
Generic SMA	×	×	×	×
Surface dressing	×	R	×	R
Slurry seal/micro surfacing	×	×	×	R
Flexible composite				
HRA	×	R	×	R
Coated macadam	×	×	×	×
Porous asphalt	R	R	R	R
Thin surfacings	✓	✓	✓	✓
Generic SMA	×	×	×	×
Surface dressing	×	R	×	R
Slurry seal/micro surfacing	×	×	×	R
Rigid				
HRA	×	R	×	R
Porous asphalt (CRCP only)	R	R	R	R
Thin surfacings	✓	✓	✓	✓
Generic SMA	×	×	×	×
Surface dressing	×	R	×	R
Slurry seal/micro surfacing	×	×	×	R
Rigid composite				
HRA	×	R	×	R
Porous asphalt	R	R	R	R
Thin surfacings	✓	✓	✓	✓
Generic SMA	×	×	×	×
Surface dressing	×	R	×	R
Slurry seal/micro surfacing	×	×	×	R

✓ For use without restriction
R Refer to overseeing organisation
× Not permitted

The striking features of thin surfacings are the relatively low levels of noise that are generated by traffic together with their superb surface regularity which gives a first-class quality of ride. In addition, where space is limited (such as in urban locations or where the road width is narrow) thin surfacings offer the opportunity of minimal road area

305 HIGHWAYS AGENCY *et al. Design manual for roads and bridges, Pavement design and maintenance, Surfacing materials for new and maintenance construction, vol 7, HD 36/99, table 2.2E, 1999.* The Stationery Office, London.

occupation since no chipping machinery and supporting plant are required. A further advantage is the speed of laying that may reduce the period during which the road needs to be closed – an important consideration in modern highway management.

A photograph of a 0/14 mm size thin surfacing is shown in Fig. 12.3. Note that a 0/14 mm stone mastic asphalt would look identical to a 0/14 mm thin surfacing.

12.4 Choice of asphalts on major carriageways

As has been demonstrated above, there are numerous applications to which asphalts are put. However, the bulk of these mixtures are used in works of highway construction and maintenance. The options for base and binder courses that are acceptable on Trunk Roads including motorways and then the acceptable surface courses are now considered. As explained above, the same design guide and specification documents are also used for the vast majority of non-Trunk Road carriageways. The importance of these options is obvious.

Table 12.2 illustrates the materials that are acceptable for use in the UK in binder courses and bases in flexible road pavements on Trunk

Fig. 12.4 Chipped hot rolled asphalt surface course
Coin shown = 1 Euro, 23 mm diameter

Fig. 12.5 Porous asphalt surface course
Coin shown = 1 Euro, 23 mm diameter

Roads. The numbers indicate the penetration of the bitumen used in the mixture.

With regard to the surface course, the options for UK Trunk Roads are set out in HD 36. There are separate tables for each of the countries in the UK. Those that apply to *England* are shown in Table 12.3. 'High-speed roads' are defined as those with an 85th percentile speed exceeding 90 km/h. The designation 'R' requires the designer to check that the proposed surface course material is acceptable to the Highways Agency or client, as appropriate.

Photographs of a chipped hot rolled asphalt surface course and a porous asphalt surface course are shown in Figs. 12.4 and 12.5 respectively.

Chapter 13

Specification, composition and design of asphalts

For most of the last century, the composition of UK asphalt mixtures was dictated by recipes that were included in various British Standard publications. However, with the passage of time, Highway Engineers have recognised that such an approach did not maximise the use of the precious limited resources in an asphalt. Furthermore, such mixtures are not appropriate for all situations, particularly those where heavy levels of traffic will act upon the carriageway. Thus, a number of methods of design have evolved to produce mixtures that are appropriate for particular situations.

In order to meet the rigours imposed on them by modern traffic conditions, asphalt pavement layers must:

- be able to resist permanent deformation;
- be able to resist fatigue cracking;
- be workable during laying, enabling the material to be satisfactorily compacted with the available equipment;
- be impermeable, to protect the lower layers of the road structure from water;
- be durable, resisting abrasion by traffic and the effects of air and water;
- contribute to the strength of the pavement structure;
- be easily maintained; and most importantly
- be cost effective.

In addition to the above, surface course materials must also:

- be skid resistant under all weather conditions;
- generate low noise levels; and
- provide a surface of acceptable riding quality.

This is a formidable list of requirements demonstrating the need for an informed approach to be made when specifying a particular mixture in order to ensure that the required performance is achieved without wasting valuable and limited resources. When it comes to heavily trafficked

245

roads, the use of traditional recipe mixtures without consideration of their properties is very unlikely to be appropriate. Consequently, highways subjected to light or moderate traffic levels use recipe-based mixtures whilst more heavily trafficked roads (including all UK Trunk Roads) employ asphalts with properties that are specifically chosen for a particular carriageway.

This chapter considers the various methods used to specify asphalt mixtures. In respect of the applicable standards and other documents, this chapter can only discuss relevant matters in outline. Accordingly, users are strongly advised to refer to source documents for full details.

13.1 Recipe specifications for bases and binder courses

A recipe specification defines a mixture in terms of a target aggregate grading and a target binder content. Recipe specifications for bases and binder courses are simple to use and are applicable to both types of mixture if the traffic level is light or moderate. They are generally based on known compositions that have performed successfully in practice and, provided there is widespread experience of different compositions, it is not difficult for the customer to specify the mixture that is considered suitable. Modern asphalt production plants are fully computerised and often operate under the control of a quality management system. Thus, customers can be confident that mixtures will be manufactured to meet the requirements of the specification.

However, recipe formulations have a number of severe limitations. The conditions of traffic, climate, etc, to which the asphalt is to be subjected will almost certainly not be the same as those existing when the recipe formulation was developed. Variations in the components of a mixture may occur that are outside previous experience. If this necessitates modification to the recipe, there is no means of assessing what these changes should be or what their effect will be.

Checking for compliance with mixture recipes is relatively simple. However, there is no means of assessing the seriousness or the practical implications of a minor failure to comply with the specification. Whilst it is important for contractors to comply with the specification, in cases where deviation occurs there is no certainty that such deviation will adversely affect the performance of the mixture in situ. There is perhaps too ready an assumption that any mixture failing to meet the specification is defective in some respect. An evaluation of the degree and nature of the deviation from the recipe supported by some laboratory and/or field assessment of the engineering properties of the mixture in question will often be appropriate. Reliance on a recipe specification alone may lead to the unnecessary and expensive removal of newly-laid material which marginally fails to comply with the specification but is entirely fit for purpose.

In addition, recipe specifications tend to be restrictive towards new developments and the use of less expensive locally available materials – a restriction that may result in unnecessary costs. The basis of recipe specifications is the aggregate grading curve and as these curves cannot, except in a very general way, take account of the aggregate properties and packing characteristics, they are not always reliable, particularly where a mixture of different aggregate sources is used.

In the UK, two Standards predominate in the specification of asphalt mixtures, BS 594-1: 2003[306] and BS 4987-1: 2003[307]. Both these Standards contain recipes for a range of asphalts and reference to design procedures.

13.1.1 Recipe specifications for coated macadam (asphalt concrete) bases and binder courses

Coated macadams, also referred to as asphalt concretes, particularly in the USA, are continuously graded mixtures and are, by far, the most common formulations for asphalts. They depend on the interlock of the aggregate particles for their strength. BS 4987-1: 2003[307] contains a range of recipe mixtures for all types of coated macadam (asphalt concrete) to be laid as:

- base   Group one
- binder course   Group two
- surface course   Group three
- porous asphalt surface course   Group four.

The recipe specification for porous asphalt surface course is rarely used in the UK. The standard is divided into sections including individual sections entitled 'Group one', 'Group two', etc. Mixtures are often described by reference to the nominal aggregate size, i.e. the nominal size of the largest aggregate fraction in the mixture, e.g. 0/32 mm base or 0/20 mm binder course or 0/14 mm surface course. With the introduction of the European Standard, the use of 'nominal size' is replaced by the use of 'grading categories'.

Generic mixtures that are included in this standard include dense coated macadam, heavy duty macadam (HDM) and high modulus base

---

306 BRITISH STANDARDS INSTITUTION. *Hot rolled asphalt for roads and other paved areas, Specification for constituent materials and asphalt mixtures, BS 594-1: 2003*. BSI, London.

307 BRITISH STANDARDS INSTITUTION. *Coated macadam (asphalt concrete) for roads and other paved areas, Specification for constituent materials and for mixtures, BS 4987-1: 2003*. BSI, London.

Table 13.1 Aggregate grading for 0/32 mm size dense base recipe mixtures[308]

Test sieve aperture size	Aggregate: crushed rock, slag or gravel, % by mass passing
40 mm	100
31·5 mm	90 to 100
20 mm	71 to 95
14 mm	58 to 82
6·3 mm	44 to 60
2 mm	24 to 36
0·250 mm	6 to 20
0·063 mm (DBM and HMB)	2 to 9
0·063 mm (HDM)	7 to 11

Table 13.2 Bitumen content and grade for 0/32 mm size dense base recipe mixtures[309]

Aggregate	Bitumen grade – 30/45 pen, 40/60 pen, 100/150 pen or 160/220 pen, % by mass of total mixture (±0·6%)
Crushed rock (including limestone)	4·0
Blast furnace slag of bulk density (in Mg/m^3):	
1·44	4·5
1·36	4·8
1·28	5·2
1·20	5·8
1·12	6·2
Steel slag	4·0
Gravel	4·5

Note: 40/60 pen is the preferred grade for machine-laid DBM.

(HMB). The Standard now requires that the bitumen penetration used in the mixture is specified, examples being:

- DBM 50 – dense coated macadam made with grade 40/60 pen bitumen;
- HMB 35 – high modulus base made with grade 30/45 pen bitumen.

For base and binder course asphalts, no 40 mm recipe mixtures are now included in BS 4987-1: 2003. This came about because of the tendency of

308 BRITISH STANDARDS INSTITUTION. *Coated macadam (asphalt concrete) for roads and other paved areas, Specification for constituent materials and for mixtures, table 3, BS 4987-1: 2003.* BSI, London.
309 BRITISH STANDARDS INSTITUTION. *Coated macadam (asphalt concrete) for roads and other paved areas, Specification for constituent materials and for mixtures, table 4, BS 4987-1: 2003.* BSI, London.

Table 13.3 Terminology for referring to recipe dense mixtures in Group one and Group two[310]

Mixture type and normal aggregate size	BS EN 12591 bitumen pen grade reference	Mixture description
Base		
0/32 mm dense	40/60	0/32 mm DBM 50 recipe base
	100/125	0/32 mm DBM 125 recipe base
	160/220	0/32 mm DBM 190 recipe base
0/32 mm heavy duty	40/60	0/32 mm HDM 50 recipe base
Binder course		
0/32 mm dense	40/60	0/32 mm DBM 50 recipe binder course
	100/125	0/32 mm DBM 125 recipe binder course
	160/220	0/32 mm DBM 190 recipe binder course
0/32 mm heavy duty	40/60	0/32 mm HDM 50 recipe binder course
0/20 mm dense	40/60	0/20 mm DBM 50 recipe binder course
	100/125	0/20 mm DBM 125 recipe binder course
	160/220	0/20 mm DBM 190 recipe binder course
0/20 mm heavy duty	40/60	0/20 mm HDM 50 recipe binder course

these materials to segregate during transportation and laying. Thus, the largest stone size of base and binder course recipe mixtures is 31·5 mm, known as 32 mm.

As an example, Table 13.1 specifies the aggregate grading and Table 13.2 specifies the bitumen content and grade for a 0/32 mm dense base coated macadam. Such recipe specifications are typical of those used in the Standard.

When referring to materials in circumstances where there is a chance of misunderstanding, e.g. when ordering deliveries, it is vital to give as much information as is necessary to avoid any possibility of confusion. Indeed, this is reflected in BS 4987-1: 2003 and the table showing terminology for recipe bases and binder courses is reproduced above as Table 13.3.

So, examples of descriptions of recipe mixtures would be 0/32 mm HDM 50 base and 0/20 mm DBM 125 binder course. When specifying fully, for example when placing an order, much more information would be necessary and this is detailed in the Standard[311].

310 BRITISH STANDARDS INSTITUTION. *Coated macadam (asphalt concrete) for roads and other paved areas, Specification for constituent materials and for mixtures, table C.C.2, BS 4987-1: 2003.* BSI, London.
311 BRITISH STANDARDS INSTITUTION. *Coated macadam (asphalt concrete) for roads and other paved areas, Specification for constituent materials and for mixtures, annex D, BS 4987-1: 2003.* BSI, London.

13.1.2 Recipe specifications for hot rolled asphalt bases and binder courses

Hot rolled asphalts (HRAs) are gap graded mixtures and their usage in asphalt pavements reduced dramatically in the last decade of the twentieth century. They depend on the characteristics of the mortar (the mixture of fine aggregate, filler and bitumen) for their strength. BS 594-1: 2003[312] contains details of a range of mixtures for hot rolled asphalt bases, binder courses, regulating courses and surface courses.

In regard to hot rolled asphalt recipe bases and binder courses, compositional details from BS 594-1: 2003 are reproduced in Table 13.4. (These data also apply to regulating courses that are used to raise the level of layers and/or smooth out any irregularities in existing surfaces.)

Over the years, a wide range of mixtures have evolved. In an attempt to limit the number and introduce a degree of rationalization, some of the asphalts are designated as a 'preferred mixture'. Users are encouraged by the Standard to use these whenever possible.

Table 13.4 gives details of aggregate grading and binder content but does not address the issue of grade of binder. Guidance on the selection and use of different grades of binder is given in Annex A of the Standard. It states that the grade of binder for bases should be:

- 40/60 pen for most applications;
- 70/100 pen where traffic loading is 'unlikely to be intense'.

In the case of binder courses, it advises that:

- 40/60 pen is generally satisfactory for most applications;
- 30/45 pen may provide greater resistance to deformation in areas of channelised traffic in southern England;
- 70/100 pen may be satisfactory for less heavily trafficked roads in Scotland and Northern Ireland.

13.2 Recipe specifications for surface courses
13.2.1 Recipe specifications for coated macadam (asphalt concrete) surface courses

UK contracts for maintenance or new construction on major roads, including Trunk Roads, now invariably require that the surface course is to be a thin surfacing. These are proprietary materials and are discussed below. For smaller schemes and where the client prefers, a mixture from BS 4987-1: 2003 may be specified. All surface course mixtures specified by

312 BRITISH STANDARDS INSTITUTION. *Hot rolled asphalt for roads and other paved areas, Specification for constituent materials and asphalt mixtures, BS 594-1: 2003.* BSI, London.

Table 13.4 Composition of base, binder course and regulating course mixtures[313]

Column number	2/1	2/2[a]	2/3	2/4	2/5[a]
Designation[b]	50% 0/10[c]	**50% 0/14[c]**	50% 0/20[c]	60% 0/20	**60% 0/32**
Nominal layer thickness (mm)	25 to 50	**35 to 65**	45 to 80	45 to 80	**60 to 120**
Percentage (m/m) of total aggregate passing test sieve:					
50 mm	–	–	–	–	**100**
31·5 mm	–	–	100	100	**90 to 100**
20 mm	–	**100**	90 to 100	90 to 100	**50 to 80**
14 mm	100	**90 to 100**	65 to 100	30 to 65[d]	**30 to 65**
10 mm	90 to 100	**65 to 100**	35 to 75	–	–
6·3 mm	–	–	–	–	–
2 mm	30 to 50	**30 to 50**	30 to 50	25 to 39	**25 to 39**
0·500 mm	13 to 50	**13 to 50**	13 to 50	8 to 39	**8 to 39**
0·250 mm	6 to 31	**6 to 31**	6 to 31	4 to 26	**4 to 26**
0·063 mm	1 to 8	**1 to 8**	1 to 8	1 to 7	**1 to 7**
Binder content percentage (m/m) of total mixture for:					
Crushed rock or steel slag	6·5	**6·5**	6·5	5·7	**5·7**
Gravel	6·3	**6·3**	6·3	5·5	**5·5**
Blast furnace slag of bulk density (in Mg/m^3):					
1·44	6·6	**6·6**	6·6	5·7	**5·7**
1·36	6·7	**6·7**	6·7	5·9	**5·9**
1·28	6·8	**6·8**	6·8	6·0	**6·0**
1·20	6·9	**6·9**	6·9	6·1	**6·1**
1·12	7·1	**7·1**	7·1	6·3	**6·3**

Note

For mixtures containing crushed rock fine aggregate, and in some instances sands or blends of sand and crushed rock fines, the binder contents given above may be reduced by up to 0·5%, where experience shows this to be advisable to avoid an over-rich mixture.

[a] Preferred mixtures in bold type.

[b] The mixture designation numbers (e.g. 50% 0/10 in column 2/1) refer to the nominal coarse aggregate content of the mixture and the size of the coarse aggregate in the mixture respectively.

[c] Suitable for regulating course.

[d] The value of 65 can be extended to 85 where evidence is available that the mixture so produced is suitable. To ensure consistency of finish of the laid mixture, supplies from any one source should be controlled within a 35% band within this permitted range.

the Standard are recipe based. Available mixtures are:

- 0/14 mm size open graded surface course
- 0/10 mm size open graded surface course
- 0/14 mm size close graded surface course (preferred mixture)
- 0/10 mm size close graded surface course (preferred mixture)

---

313 BRITISH STANDARDS INSTITUTION. *Hot rolled asphalt for roads and other paved areas, Specification for constituent materials and asphalt mixtures, table 2, BS 594-1: 2003.* BSI, London.

Table 13.5 Aggregate grading for 0/14 mm size close graded surface course[314]

Test sieve aperture size	Aggregate: crushed rock, slag or gravel % by mass passing
20 mm	100
14 mm	95 to 100
10 mm	70 to 90
6·3 mm	45 to 65
2 mm	19 to 33[a]
1 mm	15 to 30
0·063 mm	3 to 8

[a] May be increased to 38 when sand fine aggregate is used.

- 0/6 mm size dense surface course (preferred mixture)
- 0/6 mm size medium graded surface course (preferred mixture)
- 0/4 mm size fine graded surface course
- 6/20 mm size porous asphalt surface course
- 2/10 mm size porous asphalt surface course.

As in the case of hot rolled asphalts, users are encouraged by the Standard to use 'preferred mixtures' whenever possible.

The recipes for the asphalts listed above are given in tables that dictate the grading of the aggregate, the bitumen content and the grade of bitumen. Those for 0/14 mm size close graded surface course are reproduced as Tables 13.5, 13.6 and 13.7.

13.2.2 Recipe specifications for hot rolled asphalt surface courses

As explained above, thin surfacings are the default choice for surface course on most works of maintenance or new construction on UK carriageways. They are discussed in Section 13.4.3. However, for most of the second half of the twentieth century, chipped hot rolled asphalt was the default UK surface course. It was the UK's creditable preoccupation with skidding resistance and the availability of suitable coarse aggregates and fine sands that caused it to adopt this material. It has enjoyed little use in Europe and elsewhere although it was/is used in Denmark and South Africa.

BS 594-1: 2003[315] specifies surface course mixtures in three ways. These are:

---

314 BRITISH STANDARDS INSTITUTION. *Coated macadam (asphalt concrete) for roads and other paved areas, Specification for constituent materials and for mixtures, table 23, BS 4987-1: 2003.* BSI, London.

315 BRITISH STANDARDS INSTITUTION. *Hot rolled asphalt for roads and other paved areas, Specification for constituent materials and asphalt mixtures, BS 594-1: 2003.* BSI, London.

Table 13.6 Bitumen content for 0/14 mm size close graded surface course[316]

Aggregate[a]	Bitumen content, % by mass of total mixture (±0·5%)
Crushed rock (excluding limestone)	5·1
Limestone	4·9
Blast furnace slag of bulk density (in Mg/m^3):	
1·44	5·5
1·36	6·0
1·28	6·6
1·20	7·0
1·12	7·5
Steel slag	4·8
Gravel[b]	–

[a] See Annex B.

[b] The information on the bitumen contents required for these mixtures made with gravel is not sufficient for a single target value to be specified. The bitumen content to be used should be chosen within the range 5·5 to 6·5% and should be approved by the specifier. As a guide, lower values should be selected for pavements designed to carry category B traffic and higher values for footway work. The tolerance of ±0·5% should apply to the selected and approved bitumen content.

Table 13.7 Grade of bitumen for 0/14 mm size close graded surface course[317]

Aggregate	Grade of bitumen	
	Category A traffic	Category B traffic
Crushed rock, slag	100/150 pen 160/220 pen or 250/330 pen	160/220 pen or 250/330 pen –
Gravel	–	160/220 pen or 250/330 pen

Note
100/150 or 160/220 pen are the preferred grades.

- conventional recipe;
- designed mixtures where the aggregate grading is fixed but the optimum value of binder content is determined by testing the components in the laboratory; and
- performance-specified mixtures for areas of heavy traffic.

Section 13.4.2 deals with designed and performance-specified mixtures. The Standard suggests that use of the recipe surface course mixtures will generally give mixtures of satisfactory performance whilst recognising that a general specification cannot cover all suitable aggregates. In

316 BRITISH STANDARDS INSTITUTION. *Coated macadam (asphalt concrete) for roads and other paved areas, Specification for constituent materials and for mixtures, table 24, BS 4987-1: 2003.* BSI, London.

317 BRITISH STANDARDS INSTITUTION. *Coated macadam (asphalt concrete) for roads and other paved areas, Specification for constituent materials and for mixtures, table 25, BS 4987-1: 2003.* BSI, London.

Table 13.8 Composition of recipe type F surface course mixtures[318]

Column number	6/1[a]	6/2[a]	6/3	6/4	6/5
Designation[b]	0% 0/2[c]	15% 0/10[d]	30% 0/10	30% 0/14	35% 0/14
Nominal layer thickness (mm)	25	30	35	40	50
Percentage (m/m) of total aggregate passing test sieve:					
20 mm	–	–	–	100	100
14 mm	–	100	100	85 to 100	87 to 100
10 mm	–	95 to 100	85 to 100	60 to 90	55 to 88
6·3 mm	100	75 to 95	60 to 90	–	–
2 mm	90 to 100	75 to 87	59 to 71	59 to 71	55 to 67
0·500 mm	75 to 100	55 to 85	49 to 71	49 to 71	40 to 67
0·250 mm	30 to 65	25 to 75	18 to 62	18 to 62	15 to 55
0·063 mm	12·0 to 16·0	10·0 to 14·0	7·0 to 11·0	7·0 to 11·0	6·0 to 10·0
Maximum percentage of aggregate passing 2·0 mm and retained on 0·500 mm test sieves	–	20	15	15	14
Binder content percentage (m/m) of total mixture for:					
Crushed rock or steel slag					
Schedule 1A	10.3	8.9	7.8	7.8	7.4
Schedule 1B	10.8	9.4	8.3	8.3	7.9
Gravel					
Schedule 2A	10.3	8.9	7.5	7.5	7.0
Schedule 2B	10.8	9.4	8.0	8.0	7.5

[a] Preferred mixtures in bold type.
[b] The mixture designation numbers (e.g. 0% 0/2 in column 6/1) refer to the nominal coarse aggregate content of the mixture/ nominal size of the aggregate in the mixture respectively.
[c] Sand carpet.
[d] Suitable for footpaths.

terms of surface coarse mixtures, BS 594-1: 2003 mixtures are designated 'type C' or 'type F' depending on the grading of the fine aggregate. Type C mixtures are those where the grading of the fine aggregate is coarse and are generally made using a crushed rock or crushed slag fine aggregate. Type F mixtures are those where the grading of the fine aggregate is much finer and are generally made using natural sands. The fine aggregate used in type F mixtures must contain not more than 5% (m/m) of particles (m/m means proportion by mass) retained on a 2 mm sieve and not more than 9% (m/m) of particles passing the 0·063 mm sieve. Type C mixtures must contain not more than 10% (m/m) of particles

318 BRITISH STANDARDS INSTITUTION. *Hot rolled asphalt for roads and other paved areas, Specification for constituent materials and asphalt mixtures, table 6 (part only), BS 594-1: 2003*. BSI, London.

retained on a 2 mm sieve and not more than 19% (m/m) of particles passing the 0·063 mm sieve. Designed mixtures may be type C or type F but recipe mixtures can only be type F. The composition of recipe mixtures is depicted in Table 13.8.

Guidance on the selection and use of different grades of binder is given in Annex A of the Standard. It states that the grade of binder for surface courses:

- in traditional mixtures has been 40/60 pen; but
- if crushed fines or harsh sands are used, softer bitumen may be suitable (but caution should be exercised before so doing, particularly on heavily trafficked roads);
- 30/45 pen bitumen or crushed fines may be considered where traffic is 'more intense';
- 70/100 pen or 100/150 pen may be used, even with sand fines, on lightly trafficked roads and other lightly stressed areas (it is routine in some areas with 0% 0/2 mm and 15% 0/10 mm mixtures).

The use of 100/150 pen with sand fines in 0% 0/2 mm and 15% 0/10 mm mixtures for use in footways is routine in some parts of the UK.

When referring to recipe mixtures, as with all asphalts, it is vital that a full description is used. Referring to Table 13.8, an example would be Table 6, Column 6/4, 30% 0/14 mm recipe type F with 40/60 pen bitumen surface course. When specifying fully, for example when placing an order, much more information would be necessary and this is detailed in the Standard[319].

13.3 Design of bases and binder courses
13.3.1 Design of coated macadam (asphalt concrete) bases and binder courses

BS 4987-1: 2003[320] does not contain a design procedure per se. Essentially, it requires that the design shall comply with the requirements of Cl 929 of the *Specification for Highway Works*[321] (SHW). The SHW is Volume 1 of the *Manual of Contract Documents for Highway Works*[322]

319 BRITISH STANDARDS INSTITUTION. *Hot rolled asphalt for roads and other paved areas, Specification for constituent materials and asphalt mixtures, annex C, BS 594-1: 2003.* BSI, London.

320 BRITISH STANDARDS INSTITUTION. *Coated macadam (asphalt concrete) for roads and other paved areas, Specification for constituent materials and for mixtures, BS 4987-1: 2003.* BSI, London.

321 HIGHWAYS AGENCY *et al. Manual of contract documents for highway works, vol 1, cl 929, Specification for highway works, 2002.* The Stationery Office, London.

322 HIGHWAYS AGENCY *et al. Manual of contract documents for highway works, vols 0–6, Various dates.* The Stationery Office, London.

(MCHW). Whenever the SHW is consulted, users are strongly encouraged to refer to Volume 2, *Notes for Guidance on the Specification for Highway Works*[323] (NG). One of the major functions of the NG is to give guidance on the reasons for and background to the requirements of the SHW. Most of the Clauses in the SHW have an equivalent in the NG, e.g. Cl NG 929 contains guidance on the contents of Cl 929.

Use of the MCHW and, therefore, the SHW is mandatory on UK Trunk Roads. The UK Trunk Road network represents some 4% by length of the major routes in the UK. Responsibility for the remaining 96% falls to the councils within whose areas the roads lie. The vast majority of councils also use the MCHW for works of highway construction or maintenance. Thus, the contents of the SHW including Cl 929 are widely applied throughout the UK.

Cl 929 is entitled *Design, compaction assessment and compliance of base and binder course macadams*. Notwithstanding this title, Cl 929 does not contain a true design procedure, it simply sets a number of parameters with which the mixture must comply. Cl 929 is an 'end-product' specification. What it seeks to do is ensure that the void content of the placed mixture is somewhere between 1% and 7 to 8%. The reason for the former is that experience has shown that if the asphalt is overcompacted then the bitumen may act as a lubricant, forcing the coarse aggregate apart and causing the mixture to deform prematurely. The reason for the upper limit on air voids arises to ensure that the asphalt has adequate stiffness to perform as expected.

The Clause 929 procedure requires the Contractor to nominate a target aggregate grading and target binder content for each of the mixtures intended for use in the works. These target values have to fall within the limits set out in the appropriate table in BS 4987-1: 2003, but the tolerances applied to the target values may produce a control specification that lies to one side of the limits in the Standard.

The nominated design of the target values for the mixture is developed by the asphalt producer, working with the laying contractor to ensure that the mixture can be laid and compacted at the required thickness using the plant and equipment expected on site. Development of the design is a balance between conflicting needs. Increasing bitumen content aids compaction but reduces the air voids content and often the stiffness of the compacted mixture. Using a lower bitumen content will increase stiffness but may make the mixture more liable to segregation. Too

---

323 HIGHWAYS AGENCY *et al. Manual of contract documents for highway works, Notes for guidance on the specification for highway works, vol 2, cl NG 929, 2002*. The Stationery Office, London.

great a proportion of the largest aggregate fraction can also lead to segregation.

Most suppliers will start from a similar proven mixture and initially minimise the proportion of 'oversize' particles. If the air voids value is known to be low, increasing the proportion of 'middle size' aggregate can help to create more voids within the packed coarse aggregate which, in turn, can allow more fine aggregate and bitumen to be added. For some mixtures, adding a proportion of sand or similar material to modify the grading of the fine aggregate fraction can improve workability and help control bitumen content. If filler (particles passing the 0·063 mm sieve) needs to be added, the type, grading and bulk density of the chosen added filler may also influence mixture characteristics.

The design process provides an opportunity for an asphalt producer to make the best use of available resources. However, once the design of the mixture has been established, it is important that the grading and characteristics of the aggregates and any added filler used are consistent for the duration of the works.

The compaction of bases and binder courses is assessed by measuring the in situ and refusal air void contents of cores using the percentage refusal density (PRD) test[324] and in situ density using a nuclear density gauge. Nuclear density gauges assess density by means of what is called the 'backscatter method'. Photons from a nuclear source are scattered through the material under test. One or more photon detectors then assess the photons scattered back towards the gauge from differing depths. These data allow the density to be determined although proper calibration is vital for accuracy. 'Non-nuclear' electronic density gauges have been introduced in Germany, the USA and the UK in recent years with some success.

A 'job mixture approval trial' is then carried out. The purpose of this trial is to demonstrate the suitability of the designed asphalt and the effectiveness of the compaction plant and the rolling techniques. A trial area of the proposed asphalt is constructed using the compaction plant and procedures proposed for the works. In size, it is between 30 and 60 m long and of a width and depth specified in the contract. Average values of in situ air void content and air void content at refusal are obtained by procedures specified in Cl 929. The mixture, compaction plant and compaction procedures are acceptable if the following criteria are met.

---

324 BRITISH STANDARDS INSTITUTION. *Sampling and examination of bituminous mixtures for roads and other paved areas, Methods of test for the determination of density and compaction, BS 598-104: 1989.* BSI, London.

Table 13.9 Aggregate size and minimum binder volume	
Aggregate size: mm	Minimum binder volume of the total volume of the mixture: %
0/32	8·0
0/20	9·4

- The average in situ air void content of the core samples taken for the PRD test must be ≤7·0% in binder course or upper base without a binder course or ≤8·0% in other base layers.
- The average in situ air void content of a pair of core samples from each location taken for the PRD test must be ≤8·0% in binder course or upper base without a binder course or ≤9·0% in other base layers.
- The average air void content at refusal density of the core samples subjected to the PRD test must be ≥1·0%.
- The minimum binder volume at each location must be as set out in a table similar to Table 13.9.
- Compositionally, the asphalt must comply with the limits set out in the appropriate table in BS 4987-1: 2003 with adjustments to reflect the chosen target grading and binder content.
- The horizontal alignments, surface levels and surface regularity specified in Cl 702 must be met.

Having had the proposed materials, compaction plant and compaction procedure approved, a similar testing regime is applied to the materials laid in the permanent works using a nuclear density gauge and cores.

Once again, in order to avoid confusion, the Standard defines descriptions to be used for designed asphalts. The table that provides this information is reproduced here as Table 13.10. When specifying fully, for example when placing an order, much more information would be necessary and this is detailed in the Standard[325].

So, examples of descriptions of designed mixtures would be 0/32 mm DBM 50 designed base and 0/32 mm HMB 35 designed binder course.

13.3.2 Design of hot rolled asphalt bases and binder courses

In the UK, hot rolled asphalt bases and binder courses are all normally specified by means of recipe formulations. There is no doubt that a

325 BRITISH STANDARDS INSTITUTION. *Hot rolled asphalt for roads and other paved areas, Specification for constituent materials and asphalt mixtures, annex D, BS 594-1: 2003.* BSI, London.

Table 13.10 Terminology for referring to designed dense mixtures in Group one and Group two[326]

Mixture type and normal aggregate size	BS EN 12591 bitumen pen grade reference	Mixture description
Base		
0/32 mm dense	40/60	0/32 mm DBM 50 designed base
0/32 mm heavy duty	40/60	0/32 mm HDM 50 designed base
0/32 mm high modulus	30/45	0/32 mm HMB 35 designed base
Binder course		
0/32 mm dense	40/60	0/32 mm DBM 50 designed binder course
0/32 mm heavy duty	40/60	0/32 mm HDM 50 designed binder course
0/32 mm high modulus	30/45	0/32 mm HMB 35 designed binder course
0/20 mm dense	40/60	0/20 mm DBM 50 designed binder course
0/20 mm heavy duty	40/60	0/20 mm HDM 50 designed binder course
0/20 mm high modulus	30/45	0/20 mm HMB 35 designed binder course

good design method would produce mixtures that are more appropriate to particular tasks. However, the dominance of continuously graded mixtures in the UK and elsewhere makes the emergence of such a method unlikely in the foreseeable future.

13.4 Design of surface courses

13.4.1 Design of coated macadam (asphalt concrete) surface courses

The use of traditional continuously graded mixtures for surface courses is relatively rare, with most being restricted to very lightly trafficked locations. Thus, methods for designing such mixtures are no longer in common use.

13.4.2 Design of hot rolled asphalt surface courses

Traditional HRA surface courses

For the last thirty years or so of the twentieth century, the UK's hot rolled asphalt standard has referred to a laboratory method for designing the bitumen content of a chipped HRA surface course. However, the aggregate grading itself was a recipe-based formulation. The current edition of the Standard, BS 594-1: 2003, continues that approach.

Hot rolled asphalt mixtures complying with BS 594-1: 2003 may be type C or type F, as discussed above. The composition of HRA surface

---

326 BRITISH STANDARDS INSTITUTION. *Coated macadam (asphalt concrete) for roads and other paved areas, Specification for constituent materials and for mixtures, table C.C.1, BS 4987-1: 2003.* BSI, London.

Table 13.11 Composition of design type F surface course mixtures[327]

Column number	3/1	3/2a	3/3a	3/4	3/5
Designationb	0% 0/2c	**30% 0/14**	**35% 0/14**	55% 0/10	55% 0/14
Nominal layer thickness (mm)	25	**40**	**45 or 50**	40	45
Percentage (m/m) of total aggregate passing test sieve:					
20 mm	–	**100**	**100**	–	100
14 mm	–	**85 to 100**	**87 to 100**	100	90 to 100
10 mm	–	**60 to 90**	**55 to 88**	90 to 100	35 to 70
6·3 mm	100	–	–	35 to 70	–
2 mm	90 to 100	**59 to 71**	**55 to 67**	35 to 47	35 to 47
0·500 mm	75 to 100	**44 to 71**	**40 to 67**	25 to 47	25 to 47
0·250 mm	30 to 85	**18 to 62**	**15 to 55**	5 to 35	5 to 35
0·063 mm	12·0 to 16·0	**7·0 to 11·0**	**6·0 to 10·0**	4·0 to 8·0	4·0 to 8·0
Maximum percentage of aggregate passing 2 mm and retained on 0·500 mm test sieves	–	**15**	**14**	10	10
Minimum target binder content percentage (m/m) of total mixture	9·0	**6·5**	**6·4**	5·5	5·5

a Preferred mixtures in bold type.
b The mixture designation numbers (e.g. 0% 0/2 in column 3/1) refer to the nominal coarse aggregate content of the mixture/ nominal size of the aggregate in the mixture respectively.
c Suitable for regulating course.

course mixtures are to be as defined in tables given in the standard (reproduced here as Tables 13.11 and 13.12). Tables 13.11 and 13.12 contain values for minimum binder contents but the target binder content is to be assessed by means of the procedures contained in BS 598-107: 1990[328]. Determined values of stability and flow (see Section 16.3.1) are to be within specified limits.

Performance-related design mixtures
In the late twentieth century (particularly after the summers of 1976 and 1981), a significant number of traditional chipped HRA surface courses failed early in their service life because of deformation. These cases occurred primarily during the summer on uphill sections of

327 BRITISH STANDARDS INSTITUTION. *Hot rolled asphalt for roads and other paved areas, Specification for constituent materials and asphalt mixtures, table 3, BS 594-1: 2003.* BSI, London.
328 BRITISH STANDARDS INSTITUTION. *Sampling and examination of bituminous mixtures for roads and other paved areas, Methods of test for the determination of the composition of design wearing course rolled asphalt, BS 598-107: 1990.* BSI, London.

Table 13.12 Composition of design type C surface course mixtures[329]

Column number	4/1	4/2	4/3	4/4	4/5
Designation[a]	0% 0/2	30% 0/14	35% 0/14	55% 0/10	55% 0/14
Nominal layer thickness (mm)	25	40	45 or 50	40	45
Percentage (m/m) of total aggregate passing test sieve:					
20 mm	–	100	100	–	100
14 mm	–	85 to 100	87 to 100	100	90 to 100
10 mm	–	60 to 90	55 to 88	90 to 100	35 to 70
6·3 mm	100	–	–	35 to 70	–
2 mm	90 to 100	59 to 71	50 to 62	32 to 44	32 to 44
0·500 mm	30 to 65	25 to 45	20 to 45	15 to 35	15 to 35
0·250 mm	20 to 50	20 to 40	15 to 35	5 to 35	5 to 35
0·063 mm	12·0 to 16·0	7·0 to 11·0	6·0 to 10·0	4·0 to 8·0	4·0 to 8·0
Minimum target binder content percentage (m/m) of total mixture	9·0	6·5	6·4	5·5	5·5

Note
There is no preferred mixture for design type C.
[a] The mixture designation numbers (e.g. 0% 0/2 in column 4/1) refer to the nominal coarse aggregate content of the mixture/ nominal size of the aggregate in the mixture respectively.

south-facing carriageways where slow-moving traffic predominated. In order to combat such failures, a specification for performance-related design HRA mixtures was introduced. This is contained in Cl 943 of the SHW[330]. These mixtures are 35% stone content HRA surface courses with a specified maximum value of wheel-tracking rate. Most mixtures manufactured to meet this specification employ polymer modified bitumen in order to meet this wheel-tracking requirement.

The contractor is responsible for nominating the design of the mixture but the asphalt producer does the detailed work. The starting point is an assessment of the stability and flow of conventional design-type mixtures made with the aggregates available to the producer to identify those with higher stability and flow values. The constituent proportions, grading and bitumen content of the conventional design-type mixtures are generally used for the equivalent performance-related design mixture. Usually, the bitumen grade or type is the only change that is needed.

Testing is then carried out on specimens prepared in the laboratory or, if possible, road cores from material which has been laid recently. This

329 BRITISH STANDARDS INSTITUTION. *Hot rolled asphalt for roads and other paved areas, Specification for constituent materials and asphalt mixtures, table 4, BS 594-1: 2003.* BSI, London.

330 HIGHWAYS AGENCY *et al. Manual of contract documents for highway works, vol 1, cl 943, Specification for highway works, 2002.* The Stationery Office, London.

261

testing is detailed in BS 598-110: 1998[331] and assesses compliance with the specified wheel-tracking requirements. Depending upon the results, an alternative bitumen is then selected and the wheel-tracking assessment repeated.

A job mixture approval trial may then be carried out. However, Cl 943 recognises that information from a previous contract may be the source of key parameters which can be provided to the client for approval.

Despite the emergence of thin surfacings, performance-related design mixtures have a useful role in modern surfacing practice. They can be used where matching the existing materials is important but heavy slow traffic may deform a conventional mixture at an unacceptable rate. It may also be appropriate for use over a structure where the much lower permeability of rolled asphalt provides desirable additional protection to the expensive structure, working with the waterproofing system to prevent corrosion due to the ingress of water.

13.4.3 Thin surfacings

It was explained above that thin surfacings and SMAs now predominate as the surface course normally used on most trafficked roads of any size in the UK and most of Europe.

The UK national specification is contained in Cl 942 of the SHW 'Thin Surface Course Systems' (TSCS). It requires thin surfacings to have a British Board of Agrèment (BBA) HAPAS Roads and Bridges Certificate applicable to the traffic level and site classification given in the contract. In fact, the entire clause is written around the requirement that HAPAS certification is in place. This approval system for thin surface course systems has been adopted in the UK to permit the development of new surfacing asphalts whilst providing the customer with assurance that the surface course will provide the expected level of performance.

Thin surfacing asphalts are not easily controlled and specified using the conventional recipe approach. Volumetric proportions are often more important than aggregate grading and binder content. The performance of the finished layer is also related to a range of factors, including the laid thickness and the type and application rate of the tack/bond coat used to ensure adhesion to the layer below. For these reasons, the approval applies to the properties of the finished layer, not just the composition of the mixture used. Mixture composition is controlled during production by an approved quality plan, often involving a register of recipes for particular combinations of coarse and fine aggregate.

---

331 BRITISH STANDARDS INSTITUTION. *Sampling and examination of bituminous mixtures for roads and other paved areas, Method of test for the determination of wheel-tracking rate and depth, BS 598-110: 1998.* BSI, London.

Table 13.13 Grading of typical proprietary 0/14 mm thin surfacings

BS sieve size	Percent by mass of total aggregate passing
20 mm	100
14 mm	90–100
10 mm	35–55
6·3 mm	25–40
2 mm	18–30
0·500 mm	12–20
0·063 mm	9–11

A demonstration of this change in approach is that the previous edition of the SHW contained wide grading envelopes whereas there are no grading requirements in the current edition. Table 13.13 gives a typical grading for SMA type proprietary mixtures, to permit comparisons with earlier tables.

In common with all mixtures specified in the Series 900 clauses of the SHW, the manufacture of mixtures for thin surfacings is controlled by a quality management system complying with the 'Sector QA Scheme for the Production of Asphalt Mixes' based on BS EN 9001: 2000[332].

Details of each thin surfacing system approved by BBA/HAPAS are published as a uniquely numbered certificate and copies are available on the internet[333]. A 'Detail Sheet' covers each variant of the approved system; generally aggregate size and binder type are the variables. The Detail Sheet declares typical test results and performance levels for a number of mandatory parameters. Performance levels for selected optional parameters may also be declared – if the producer has chosen so to do.

The mandatory parameters are coarse aggregate properties, wheel-tracking rate, bond strength, sensitivity to water and initial texture depth. The data for each mandatory parameter from different certificates can be compared and evaluated, helping the specifier to choose a system that suits the requirements of the particular works.

Coarse aggregate properties are defined in terms of polished stone value (PSV) and aggregate abrasion value (AAV) in the same way as any other surfacing.

332 BRITISH STANDARDS INSTITUTION. *Quality management systems, Requirements, BS EN ISO 9001: 2000.* BSI, London.

333 *www.bbacerts.co.uk.*

There are four levels of wheel-tracking rate specified by the SHW and declared in BBA/HAPAS certificates – Level 3 to Level 0[334]. Generally, thin surfacing systems will meet the most onerous Level 3 required for heavily trafficked roads, particularly if a polymer modified bitumen is used. Level 3 sets the maximum value of the mean of the wheel-tracking rates as 5·0 mm/h with an upper limit of 7·5 mm/h for mixtures laid at 30 mm thick or more. There are equivalent values for thinner layers. These values are set to give a wheel-tracking performance about the same as that of performance-related design rolled asphalt. However, it should be noted that, perhaps more so than with HRA, the wheel-tracking performance of thin layers can be influenced by the characteristics of the underlying layers. The use of a rut-resistant stone mastic asphalt (SMA) binder course[335] is becoming more common.

Bond strength is a measure of the degree of bonding between the compacted mixture and the layer below. The effectiveness of the bond depends on many factors, particularly the type of bond coat used and the weather conditions at the time of laying. It is tested in the laboratory using a core extracted from the surface. This is a complex test that is usually only done as part of the initial BBA/HAPAS system approval process.

Sensitivity to water is assessed as part of the initial laboratory design of individual mixtures by evaluating the loss (in practice, often a gain) in the stiffness of specimens after vacuum saturation and three cycles in hot (60°C) and cold (5°C) water. The results reflect the affinity between the bitumen and the aggregates used. Polymer modified bitumens can help to improve the results.

A feature of good thin surfacing systems is that the initial texture depth is highly consistent because the composition of the mixture is closely controlled at the asphalt plant. The requirement for untrafficked texture depth is to be measured in accordance with BS EN 13036-1: 2002[336]. For high-speed roads, texture depth is generally prescribed as an average 1·5 mm minimum for each lane kilometre with an individual minimum value of 1·2 mm. For low-speed situations, a lower value will permit the choice of a wider range of systems, including those with a smaller

---

334 HIGHWAYS AGENCY *et al. Manual of contract documents for highway works, table 9/28, Series 900, vol 2, Notes for guidance on the specification for highway works, 2002.* The Stationery Office, London.

335 HIGHWAYS AGENCY *et al. Manual of contract documents for highway works, vol 1, cl 937, Specification for highway works, 2002.* The Stationery Office, London.

336 BRITISH STANDARDS INSTITUTION. *Road and airfield surface characteristics, Test methods, Measurement of pavement surface macrotexture depth using a volumetric patch technique, BS EN 13036-1: 2002.* BSI, London.

size aggregate that may give other important benefits such as greater noise reduction and better resistance to moisture ingress.

Table 9/29[337] gives levels and values for road/tyre noise levels based on a comparison with the noise generated on HRA surface courses. Because road/tyre noise is difficult and expensive to test in a statistically valid way, this is an optional parameter that is not declared on every BBA/HAPAS Detail Sheet. Most thin surfacing systems with 'negative' surface texture compare well against HRA surface course and that is all that is required in most circumstances. Smaller size aggregates will give a lower level of road/tyre noise in urban situations but probably a lower texture depth that may reduce skidding resistance at high speed. Mixtures that are more porous will also reduce the level of road/tyre noise by dissipating the air pressure that builds up in front of the rolling tyre but may not protect the underlying layers from the ingress of moisture.

A feature of the BBA/HAPAS approval system is that it requires the supplier's quality plan to control transport, laying and compaction – a series of steps called installation by the approval body. Preparation of the underlying surface is also to be in accordance with the 'Installation Method Statement' that forms part of the BBA/HAPAS certificate.

Thin surfacings require the contractor to provide a guarantee of two years for performance, surfacing materials and workmanship. This warranty excludes failure due to 'settlement, subsidence or failure of the carriageway on which the material has been laid' but includes 'fretting, stripping and loss of chippings'. Also included is a requirement for trafficked texture depth measured in accordance with BS EN 13036-1: 2002. For high-speed roads, this is usually prescribed as 1·0 mm minimum for each lane kilometre with an average minimum value of 1·2 mm.

13.5 Guidance on the selection of mixtures

A key question for those preparing contracts is the choice of asphalts to be specified for the works. Both BS 4987-1: 2003 and BS 594-1: 2003 contain useful information in this regard[338,339].

337 HIGHWAYS AGENCY *et al. Manual of contract documents for highway works, vol 2, table 9/29, Notes for guidance on the specification for highway works, 2002.* The Stationery Office, London.

338 BRITISH STANDARDS INSTITUTION. *Coated macadam (asphalt concrete) for roads and other paved areas, Specification for constituent materials and for mixtures, annex B, BS 4987-1: 2003.* BSI, London.

339 BRITISH STANDARDS INSTITUTION. *Hot rolled asphalt for roads and other paved areas, Specification for constituent materials and asphalt mixtures, annex A, BS 594-1: 2003.* BSI, London.

Chapter 14

Asphalt production plants

The manufacture of an asphalt involves mixing aggregate in pre-determined proportions with bitumen to produce a homogeneous mixture that complies with the specified composition and temperature or end-product criteria. Before mixing, the aggregate must be blended to ensure that the required grading is achieved. It is then heated to remove any moisture and to raise it to a temperature suitable for coating with bitumen.

The earliest known method of producing coated aggregate (circa 1850) was effected by means of a shallow steel tray mounted over a closed-grate coal fire. Broken and roughly graded aggregate was heated and dried on the tray before hot tar was poured over it and mixed manually[340]. Mechanical means of drying and mixing were developed by the end of the nineteenth century. These machines were based on concrete mixers and cylindrical dryers using coal fires to dry and heat the aggregate. Sub-sequently, the paddle mixer or pugmill was developed and employed a single shaft with blades; the first types of these mixers remained in use until the 1920s.

Three types of aggregate dryer were developed in the UK: the long rotating cylinder; a tower dryer in which the aggregate is slowly passed downwards whilst hot gases are forced upwards through it; and the short drum type of batch dryer which is the forerunner of the batch-heater mixer. In the late 1930s and early 1940s, oil started to replace coal as the fuel for heating. Initially, hot burner gases were ducted into the drums and further development brought the burner inside the drum; a similar system is still in use.

---

340 JONES L. *Recent developments in coating plant technology, Quarry Management, vol 13, no 10, pp 25–30, Oct 1986.*

Fig. 14.1 High-level asphalt plant
Courtesy of Tarmac Ltd

14.1 Types of mixing plant

In general, mixing plants comprise two main sections – the first is the aggregate drier and proportioning unit whilst the second is the mixer itself (Fig. 14.1). Often, the heating and proportioning equipment is common to different types of mixers. There are basically three types of plant used in the UK: asphalt batch, batch heater and drum mix. Each is now described in detail.

14.1.1 Asphalt batch plant

The asphalt batch plant (see Figs. 14.2 and 14.3) is the most common type of asphalt production plant in the UK. It is very flexible, changing from one mixture type to another without difficulty. It is thus ideal for handling the multiplicity of specifications that exist in the UK[341]. These plants have capacities ranging from 100 to over 400 tonnes per hour.

As the description implies, this type of plant produces a single batch of material at a time. This is then repeated to produce the required quantity. A pre-set amount of proportioned aggregate and bitumen is mixed giving batches that may vary from 500 kg to 5 tonnes. Plant capacity is determined by both the type of mixture and the batch size. Mixing

341 CURTIS C R. *Blacktop specifications – should we rationalise, The road ahead for blacktop specifications: Asphalt and Coated Macadam Association Seminar, Nov 1988.*

Fig. 14.2 Asphalt batch plant
Courtesy of Foster Yeoman Ltd

Fig. 14.3 Sectional flow diagram of an asphalt batch plant
Courtesy of Gencor International Ltd

times vary with the type of mixture. A 2·5 tonne batch, for example, requiring a mixing cycle time of 60 s, will give a maximum plant capacity of 150 tonnes per hour.

The cold aggregate feed is the first major component of any mixing plant. The aggregate is collected from a stockpile and loaded into a

269

Fig. 14.4 Cold feed bins
Courtesy of Tarmac Ltd

cold feed holding bin; eight or more cold feed bins are normally used. Some mixing plants have 12 or more bins, each charged with a different aggregate (see Fig. 14.4). Feeder units are located under the bins to control the flow of aggregate onto a conveyor belt. Alternatively, the aggregate can be supplied by feeders mounted directly under crushed stone bins. The feeders can be controlled individually or in combination to provide a proportional arrangement whereby the ratio of one feeder to another is maintained but the total flow rate of all the feeders can be varied. A conveyor belt transports the combined aggregates to the drying and heating unit.

A basic part of any mixing plant is the aggregate dryer/heater. Its function is twofold – to ensure that water is removed from the aggregate before it is mixed with the bitumen and to heat the aggregate to the required mixing temperature. The removal of moisture is necessary to allow adhesion of the bitumen to the aggregate (discussed in Chapter 9). Since the moisture content of the aggregate can be as high as 5% of the dry weight, i.e. around 50 kg of water per tonne of aggregate[342], this part of the plant plays a vital role in assuring the ultimate quality of the finished mixture. It normally consists of a large rotating metal drum that is mounted at an angle of between 3 and 4° along its axis (see

342 GRANT R M. *Flexible pavements and bituminous materials, Residential course at the University of Newcastle upon Tyne, section J, Manufacture of coated materials, pp J1–J12, Sep 1989.*

Fig. 14.5 Aggregate drier with burner

Fig. 14.5). A gas or oil burner at the lower end causes hot gases to rise through the drum and emerge at the upper end. The cold aggregate enters the rotating drum from the upper end and is carried through the hot gas and flame by means of steel angles or flights fitted to the inside, aided by the downward inclination of the drum, finally discharging onto an elevator. The length of time during which the aggregate remains in the drum is known as the 'retention time'. The drying and heating process is very fuel intensive and, consequently, costly. It takes some 10 litres of fuel per tonne of aggregate to remove 5% moisture and increase the temperature of the aggregate by 160°C[343]. At the lower end of the drum, the temperature of the aggregate is measured by a radiation pyrometer. The value is compared to the specified temperature and the burner is adjusted accordingly.

To eliminate dust from the dryer exhaust, dust collectors are fitted. In general, there are three types of dust collector – cyclone, fabric filters or wet collectors (although the last of these is almost redundant). Two or more of these devices – a primary collector and one or more final or secondary collectors – may be fitted. Provision is usually made for the coarse dust to be returned to the hot aggregate. However, if the dust is unsuitable or is produced in excessive quantities, it can be removed from the collector and disposed of elsewhere.

343 GRANT R M. *Flexible pavements and bituminous materials, Residential course at the University of Newcastle upon Tyne, section J, Manufacture of coated materials, pp J1–J12, Sep 1989.*

When the aggregate is discharged from the dryer/heater, it is transported vertically to a screening unit housed at the top of the tower as shown in Fig. 14.2. Here, the aggregate is separated into fractions (usually between 4 and 6 in number) of specified size and stored temporarily in hot storage bins. Material is withdrawn from each bin in predetermined amounts and charged into a weigh hopper. Each aggregate is individually weighed to the quantity determined by the formulation of the mixture as a proportion of the batch size.

The bitumen binder is kept in a storage tank and fed into a recirculating system of pipes that returns it to the tank. When bitumen is needed for the mixer, a valve in the system is opened permitting the required quantity of bitumen to flow from the tank. A weigh bucket is the predominant method of measuring binder quantity.

The dust collected from the dryer exhaust can be reintroduced into the hot aggregate if the gradation of this dust complies with the specified aggregate grading. If not, the dust may be collected in a fines silo that feeds fines into the batch in controlled amounts. Where collected dust is unsuitable for addition to the mixture, mineral filler will be used instead. This is normally limestone filler, supplied in bulk into a silo and added to the mixture in the same way as the collected dust.

The weighed aggregate, fines and bitumen are discharged into the pugmill. This is usually a large semi-circular bowl with two shafts that have angled mixing blades. These blend the hot aggregate and when the bitumen is added the movement of the aggregate ensures that all the aggregate surfaces are coated with a thin film of bitumen. Depending on the efficiency of the pugmill and the material being mixed, the duration of mixing varies from 30 to 60 s. The mixing time should be as short as possible to minimise oxidation of the bitumen. However, it is important that the mixing time is long enough to ensure that uniform distribution of the aggregate sizes and uniform coating of the aggregate is achieved. The mixed material is then discharged directly into the delivery vehicle or transported by skip into a hot storage bin.

In order to meet environmental regulations, the mixing section and mixed material system are generally sheeted with angular profile cladding or positioned inside a large building. The feed hoppers are equipped with a canopy and the conveyor belts are covered to protect the feed aggregates from wind. Emissions from the exhaust stack cannot be seen with the naked eye, particulate emission levels of less than $20 \, \text{mg/m}^3$ can be easily achieved.

Freestanding versions of asphalt batch plants, which need no concrete foundations providing the ground is level and stable, are available. Such plants are designed to be moved quickly from contract to contract.

Mobile versions of the asphalt batch plant are also available. Again these need no foundations and they can be readily relocated. Mobile plants are small in height and are useful in cases where a height restriction is encountered.

A variation of the asphalt batch plant is the multiple hot stone bin version. A diverter chute is positioned under the vibrating screen to enable two different types of aggregate to be stored concurrently. Typically, there are 12 or more hot stone bin compartments. The multiple hot stone bin plant enables a variety of mixtures to be produced without having to clean out the plant when different aggregates are required.

14.1.2 Batch heater

This type of plant is unique to the UK. It is called a batch heater because the aggregates are dried and heated in batches prior to being mixed with bitumen and filler. Batch heaters allow rapid changes in mixture specifications and are not limited by hot aggregate storage capacity. Plant outputs vary from some 50 tonnes per hour to more than 200 tonnes per hour. However, aggregates having moisture contents exceeding 2% cannot be dried economically in these plants and a separate dryer would normally be used. They are better suited to low- and medium-temperature materials, such as coated macadams, than to mixtures with high sand contents, such as rolled asphalt.

The cold aggregate feed is almost identical to that of the asphalt batch plant; the proportional cold feed is charged into a hopper for check weighing before drying. The proportions of aggregate cannot be changed in the process, which means that the feed control of the cold aggregate has to be very accurate. From the weigh hopper, the aggregates are charged into the batch heater, in effect a shortened version of the aggregate dryer, and the batch is dried and heated. The heated aggregate is discharged into a conventional pugmill and is mixed with the bitumen and filler.

14.1.3 Drum mix

Drum mix plants are so called because the aggregates are dried and heated in a large drum on a continuous basis just prior to mixing with the bitumen. This type of plant is not very flexible and requires considerable effort to change the mixture type. However, the increased use of computerised control is beginning to overcome this problem. These plants are relatively simple and the smaller models are ideal for contracts where a portable plant is required (Fig. 14.6). Their simplicity also reduces maintenance costs and improves reliability. However, the overriding advantage of drum mix plants is their production capacity. Plants can produce anything between 100 tonnes per hour to some 700 tonnes

273

Fig. 14.6 Mobile drum mix plant
Courtesy of Lafarge Aggregates Ltd

per hour, making this type of equipment ideal for large contracts where large quantities of the same material are required over a long period.

This type of plant produces asphalts without the use of screens, hot aggregate storage bins, weigh hoppers or pugmill mixers. In principle, the cold aggregate feed system is similar to that which is used in an asphalt batch plant, described above. The aggregate proportioning must be precise as any error in the feed ratio will only be detected when samples of the mixed material are analysed in the laboratory.

The cold feed bins discharge the aggregate onto a collecting conveyor that, in turn, feeds them into a charging conveyor. This is provided with a belt weigher enabling the aggregate flow rate to be maintained (in so doing, a correction is made to take account of the amount of moisture in the aggregate). This allows the calculation of the amount of fine aggregate/filler and bitumen that have to be added to the mixture.

The drum used in a drum mix plant differs from that used in an asphalt batch plant dryer in three respects:

- the burner is at the top end of the drum and the aggregate thus flows away from the flame;
- the configuration of the flights inside the drum is more complex;
- the drum is longer.

274

The drum interior has two zones. The first zone dries and heats the proportioned aggregate and the second zone mixes the hot aggregate with the bitumen. The mixing zone is protected by either a flame shield or a device that increases the density of the aggregate curtain in order to protect the coated aggregate from the radiant heat of the heating unit.

The bitumen is injected into the drum at the start of the second zone. The flow rate of the bitumen is proportional to the aggregate flow rate and is controlled by either variable-speed fixed-displacement pumps or variable orifice control valves.

Exhaust gases from drum mix plants contain less dust than those from other types of mixing plant[344]. This is because most of the airborne dust generated in the heating and drying zone is absorbed by the bitumen-coated aggregate in the second zone. UK environmental legislation, however, still requires dust collection. The collected dust may be blown pneumatically into the drum or simply screwed into the mixing zone. Either method may also be used to introduce controlled amounts of mineral filler into the drum. The final mixture flows continuously from the bottom of the drum; bulk storage is normally in the form of well-insulated high-capacity hot storage silos[345] (discussed in Section 8.2.3).

Developments have led to a new breed of drum mix plants, termed counterflow drum mix plants. The ancillary equipment – such as the feeders, conveyors, bag filter, bitumen system, filter system and hot storage silos – is identical to that used in a drum mix plant. The counterflow drum mix incorporates a dryer similar to the asphalt batch plant dryer with the burner mounted at the discharge end and a separate mixing chamber for the bitumen and filler. The mixing chamber can be mounted directly after the dryer using an extended burner. Alternatively, the mixing chamber can envelope the dryer as an outer skin, this is termed a double drum. The counterflow drum mix is more fuel efficient than the parallel flow drum mix since the incoming aggregates are preheated by the hot exhaust gases.

14.2 The addition of recycled asphalt pavement

Recycled asphalt pavement (RAP) can be added to each of the different types of mixing plants. By superheating the virgin aggregates, percentages between 10 and 30% can be accommodated, dependent upon the

344 GRANT R M. *Flexible pavements and bituminous materials, Residential course at the University of Newcastle upon Tyne, section J, Manufacture of coated materials, pp J1–J12, Sep 1989.*

345 BAXTER J. *Hot storage of bituminous mixes, Problems and advantages, Quarry Management and Products, pp 213–218, Aug 1977.*

moisture content in the RAP. A rejuvenating agent may need to be added to the mixture if the RAP is severely oxidised.

If the RAP is in large sections then a RAP breaker will have to be employed prior to processing. On an asphalt batch plant, or a batch heater plant, the most common method is to convey the RAP over a belt weigher and feed directly into the pugmill. The conveyor operates on a start/stop basis. An alternative operational method in an asphalt batch plant is to feed the RAP into the aggregate weigh hopper. The control system acknowledges it as an additional ingredient. The disadvantage with this method is the potential contamination of the aggregate weigh hopper.

Between 50 and 60% RAP can be added to an asphalt batch plant by incorporating a recycling or 'black' drum. This consists of an inclined steel drum (the angle of inclination is between 3° and 4°) with the burner mounted at the higher end. The RAP is dried and heated to remove moisture, weighed in a separate weigh hopper and then transferred to the pugmill.

A counterflow drum mix plant can accommodate up to 60% RAP by adding it to the isolated mixing chamber. The virgin aggregates are superheated. The RAP is introduced over a separate belt weigher, the control signal being taken from the charging conveyor weigh system.

14.3 Additive systems

All types of mixing plant can be fitted with a variety of additive systems. The mixture can be coloured, most commonly by the addition of red oxide. Blue, green and gold coloured mixtures can also be produced. The employment of a clear binder such as Mexphalte C results in a natural looking mixture. The addition of latex into the pugmill by means of a peristaltic pump is also an option. Deferred-set asphalts can be produced by the addition of flux metered via a spray bar into the pugmill.

Fibres are used in some SMAs and thin surfacings. They can be added to the mixture by volume or weight directly into the pugmill. An alternative method is to weigh the fibres in a separate weigh hopper. The fibres are introduced in small fibre 'strings' or in pellet form.

14.4 Production control testing of asphalts

Production control testing is used to ensure that the asphalt, as manufactured, complies with the specified aggregate grading and bitumen content. Representative samples of the material are taken at prescribed intervals since it is impractical to test every batch of material that is produced. Clearly, it is essential that the sample is representative otherwise the data generated will be meaningless and misleading. It is

Table 14.1 Test methods for the analysis of asphalts[346]

Method	Solvent	Bitumen content	Filler content
Extraction bottle (bitumen by difference)	Trichloroethylene or methylene chloride	By difference or directly	Directly
Extraction bottle (bitumen directly determined)	Methylene chloride	Directly	By difference
Hot extractor	Trichloroethylene	By difference	Directly

equally important that the test methods are carried out by trained and experienced staff using well maintained and properly calibrated equipment.

A procedure for sampling the mixture during discharge at the production plant is included in BS 598-100: 1987[347]. Once the material has been sampled, it requires some preparatory treatment before testing. This treatment is detailed in BS 598-101: 1987[348]. Three test methods for determination of the composition of the mixture with regard to aggregate grading and bitumen content are detailed in BS 598-102: 1996[346]; all three methods involve the separation of the bitumen from the aggregate using chlorinated hydrocarbon solvents as shown in Table 14.1.

Having removed the bitumen from the material, the residual aggregate is graded into its individual fractions. The majority of asphalts can be analysed by any of the methods. Collection and direct determination of filler content and direct determination of bitumen content provides a check on the results but it also increases the duration of the test.

Determining the bitumen content directly involves evaporation of solvent from a known quantity of bitumen/solvent solution (known as an aliquot) and weighing the recovered bitumen. The percentage of bitumen can then be calculated from the mass of the sample and the volume of solvent originally used. The procedure is relatively complicated. The most common sources of error are:

- loss of solvent by evaporation;
- incomplete separation of the insolvable matter from the bitumen/ solvent when centrifugal methods are used;

---

346 BRITISH STANDARDS INSTITUTION. *Sampling and examination of bituminous mixtures for roads and other paved areas, Analytical test methods, BS 598-102: 1996.* BSI, London.

347 BRITISH STANDARDS INSTITUTION. *Sampling and examination of bituminous mixtures for roads and other paved areas, Methods for sampling for analysis, BS 598-100: 1987.* BSI, London.

348 BRITISH STANDARDS INSTITUTION. *Sampling and examination of bituminous mixtures for roads and other paved areas, Methods for preparatory treatments of samples for analysis, BS 598-101: 1987.* BSI, London.

- incomplete removal of the solvent during recovery;
- dust entering the bitumen/solvent solution.

All of the errors cited above are likely to increase the calculated bitumen content.

A pressure filter is used when the filler fraction is determined directly. The disadvantage with this procedure is that it is slow and great care must be taken to avoid leaks from the pressure filter seals and tears in the filter paper. A powder additive is used to aid flow of the solvent/ bitumen mixture.

Correction factors are applied to bitumen and filler contents of base and binder course coated macadam mixtures and to all hot rolled asphalt materials. These factors are included to allow for the experimental errors resulting from variations in aggregate grading.

Ignition furnace methods are used by some laboratories to determine bitumen content and grading. This method involves subjecting the sample to high temperatures that 'burn' the bitumen off the sample. The advantage of this method is that it requires no solvent and uses less equipment. Disadvantages include the need to assess aggregates to determine a correction factor for each aggregate type/source. This value has to be programmed into the software of the machine used to ensure accuracy of results. At present, this method is the subject of a draft European Standard[349].

---

349 EUROPEAN COMMITTEE FOR STANDARDIZATION. *Bituminous mixtures – test methods for hot mix asphalt–binder content by ignition, prEN 12697-39*. CEN, Brussels.

Chapter 15

Transport, laying and compaction of asphalts

15.1 Transportation

There are three distinct phases in transporting the mixed asphalt to the site where it is to be laid. These are loading into a truck, transporting to the site and tipping into the hopper of the paver. In order to minimise disruption, transportation must be well planned and executed efficiently.

During loading, it is important to minimise the risk of segregation. Loading has to be undertaken quickly in order to avoid undue heat loss. Lorries should be loaded as evenly as possible over the entire trailer. If the asphalt in the trailer is loaded with steep sides, this may cause segregation of the asphalt. Covered trailers (Fig. 15.1) are required to keep the mixture hot during transportation and a rounded base in the trailer helps reduce the risk of aggregate and temperature segregation. The phasing and frequency of transport to the site must be well coordinated. If a paver has to stop to await the arrival of the next load, the quality of the surfaced road will suffer. It can lead to unevenness and reduced levels of compaction, both of which may shorten the serviceable life of the road. In contrast, a convoy of lorries waiting at the site should be avoided. The asphalt mixtures may cool whilst waiting which may lead to unsatisfactory compaction results or, in severe cases, having to discard the load. The unloading of the asphalt mass requires skill to avoid both separation and stoppages.

15.2 Use of tack coats

The function of a tack coat is to promote adhesion between two layers of asphalt. An example would be where a new surface course is to be laid on a previously paved road or a binder course that has been opened to traffic. Tack coating is an important stage in road surfacing and is often required by highway specifications.

It is essential in terms of the overall stiffness of the pavement that the layers bind well together. In addition, the improved adhesion afforded by

Fig. 15.1 A modern asphalt delivery lorry
Courtesy of Tarmac Ltd

tack coating means there will be less tendency towards displacement of the mixture or crack formation when rolling takes place.

15.3 Pavers

These machines, occasionally described as 'asphalt finishers', lay the hot asphalt to form the pavement. A number of published texts summarise experiences related to asphalt paving and compaction[350,351].

A paver has to provide homogeneous pre-compaction to give the mixture sufficient stability such that the roller can commence the compaction process. It also has to provide a level surface with a homogeneous texture. The performance of the paver is the single most important factor in achieving these two requirements. All modern asphalt pavers consist of two main units, a tractor and a floating screed.

15.3.1 Tractor unit

The tractor unit provides forward propulsion to the paver and power to other elements of the equipment. It is driven by either pneumatic wheels (Fig. 15.2) or crawler tracks (Fig. 15.3).

350 DYNAPAC. *Compaction and paving, theory and practice, Svedala Dynapac pub no IHCC-CAPEN1, 2000.* Svedala Industri AB, Sweden.
351 FORSSBLAD L and S GESSLER. *Vibratory asphalt compaction, 1975.* Dynapac, Sweden.

Fig. 15.2 Wheeled paver

Fig. 15.3 Tracked paver

Wheeled pavers are easy to transport. Their relatively high travelling speeds allow them to move about the work site rapidly and to move easily between different sites on public roads.

The good traction of tracked pavers makes them suitable for use on unbound surfaces. They are also particularly suited to laying extra wide sections and working on steep sites. Tracked pavers produce

281

Fig. 15.4 Laying a macadam base
Courtesy of Aggregate Industries UK

layers which possess better surface regularity than those placed by wheeled pavers.

Having released the rear gate of the trailer, delivery trucks reverse into the front of the paver. The trailer of the truck is lifted by a ram at the front of the trailer resulting in the asphalt mixture being discharged into the receiving hopper of the paver (Fig. 15.4). The rear tyres of the truck, and therefore the truck itself, are then pushed forward by the paver. Twin or single slat conveyors carry the asphalt from the hopper to the rear of the machine and then onto the auger (screw conveyor) which distributes the asphalt laterally over the entire working width of the screed. The height of the auger is adjustable to allow for different layer thicknesses.

The rate at which the asphalt flows through the paver is controlled by the speed of the slat conveyor and the augers. The conveyor speed is automatically linked to the forward speed of the paver and the height of the material that is spread out ahead of the screed. It is essential that the height is maintained at a constant level.

Figure 15.5 shows how asphalt flows through the paver. The material passes from the hopper (1), via the conveyors (2), on to the auger (3) and under the screed (4). Smooth flow is essential to achieving good paving results.

15.3.2 Screed unit

The screed levels and pre-compacts the asphalt to a specified thickness, grade and cross slope and crown profile. The self-levelling floating

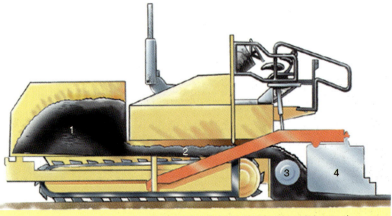

Fig. 15.5 Material flow through a paver – from the hopper (1), via the conveyors (2), on to the auger (3) and under the screed (4) – is essential to good paving results

screed is attached to the tractor by side arms at tow points located on each side of the tractor near its central point. Here, vertical movements caused by any surface unevenness are at a minimum. This allows the screed to produce an even surface despite the fact that the underlying base may be somewhat irregular. As each successive asphalt layer is placed, irregularities become less and less apparent.

The tow points are set to give the required thickness of the mat. Their position may be finely adjusted by electronic systems on a continuous basis as the asphalt is laid.

15.3.3 Angle of attack

The concept of the 'angle of attack' is illustrated in Fig. 15.6. This is the angle between the bottom plate of the screed and the surface being paved. The angle varies from screed to screed according to the weight of the

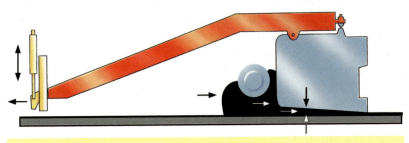

Fig. 15.6 The 'angle of attack' is the angle between the bottom plate of the screed and the surface being paved

screed, the contact area of the bottom plate and the shape of the front part of the screed.

Figure 15.6 illustrates the importance of the angle of attack. Any change in the level of the tow points results in a corresponding adjustment of this angle. The desired surface regularity is obtained if all the forces acting on the screed are in equilibrium, only then will the screed settle into its angle of attack.

The angle of attack can be increased or decreased by raising or lowering the level of the tow point. Any movement of the tow point upsets the equilibrium and results in a rise or fall of the screed. Once the screed has attained the new level, the angle of attack is restored and the forces revert to a state of equilibrium.

15.3.4 Heating of the screed bottom plates
Screeds are heated by diesel, propane gas or electrically powered elements. This is necessary to prevent the bottom plate picking up the hot asphalt.

15.3.5 Fixed or variable paving width
The most common type of screed is the telescopic screed that has a hydraulically variable laying width (Fig. 15.7). In some applications, a fixed screed is the most economical option.

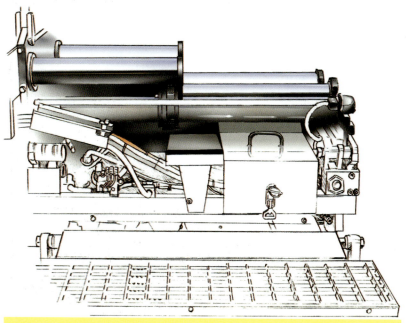

Fig. 15.7 Hydraulic screed extensions

15.4 Additional screed systems

The main factor that dictates the effectiveness of a screed in pre-compacting an asphalt is its weight. The heavier the screed, the greater will be the degree of pre-compaction that is achieved. Additional systems such as tampers and vibration generators are usually attached to improve the flow of material underneath the screed.

The choice of a tamping and/or vibrating screed depends on the application as well as the type/s of asphalt, maximum aggregate size in the asphalt, layer thickness as well as local preferences and specifications.

15.4.1 Tamper

The tamping mechanism uses a vertical, high-amplitude movement at comparatively low frequencies. The main purpose of the tamper is to facilitate the flow of material below the screed plate. A static or vibrating screed plate follows the tamping unit. The width of the tamper and the tamping frequency limit the maximum paving speed since no part of the mat must be left untamped. Figure 15.8 illustrates the action of the tamper. The tamper bar (1) feeds material under the screed plate and the vibrator (2) reduces the friction between the screed plates and the material.

15.4.2 Vibrators

The screed plate is equipped with a vibration generating system. Vibrating the asphalt reduces the friction between the screed plate and

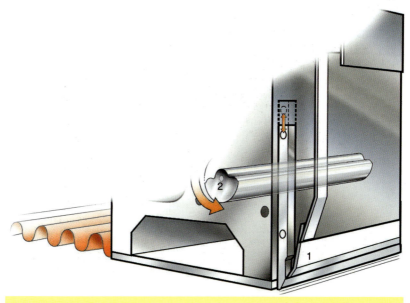

Fig. 15.8 Action of the tamper

the mixture. This allows the screed to float more easily over the asphalt. The vibrations also tend to cause a small amount of bitumen to rise to the surface of the asphalt, providing additional lubrication.

15.4.3 Tamping and vibrating screeds

A screed that is capable of being both tamped and vibrated is extremely versatile since it affords the choice of using either mode independently or both in combination. The weight of the two systems also increases the total weight of the screed which, in turn, results in higher levels of pre-compaction.

15.4.4 High-compaction screeds

For some applications, such as cement-stabilised layers and thicker asphaltic bases, special high-compaction screeds have been developed. These screeds have additional mass and are equipped with double tampers and vibration systems.

15.5 Paving operations

Careful planning of asphalt production and transportation to the site is crucial in maintaining a continuous paving operation. Any stoppages in the operation will result in a pavement of inferior quality. The paver speed should be constant. The rate of supply should be such that the paver can maintain the highest speed that is consistent with laying the asphalt to the required specification. In ensuring that an adequate rate of supply is maintained, key elements will be the capacity of the asphalt manufacturing plant or plants and the number of trucks that are available to deliver material.

In order to meet the requirements of the specification, a number of points must be taken into consideration. The required paving width has to be set and the screed must be heated to prevent the asphalt sticking to the underside of the screed. The tow points must be set to the height that corresponds to the desired thickness of the finished mat. Where necessary, the screed must be adjusted to allow for a crown profile. The height of the auger is also crucial to the outcome. If it is set too low, it will restrict the material flow under the screed which will result in an open texture and cause the mat to tear. If it is set too high, the asphalt may not extend to the outer edges of the screed. Ideally, the distance between the surface of the mat and the lower edge of the auger flights should be equivalent to roughly five times the maximum aggregate size in the mixture.

There are a number of factors that must be controlled during a paving operation. These include:

- head of material in front of the screed

- speed of the paver
- layer thickness
- regularity of the surface being covered
- paving width
- joints in the surface being covered
- laying temperature
- mixture segregation.

15.5.1 Head of material

The height of the asphalt in front of the screed is called the 'head of material'. It should be maintained at a constant level over the entire working width. It has a decisive influence on the vertical position of the screed and, thus, the regularity of the finished mat. The levelling action of a screed relies on there being a state of equilibrium between all the forces acting on it. Any change in these forces causes the screed to move up or down to compensate. If the head of material rises, the resistance to forward travel increases and, in an attempt to overcome this resistance, the screed starts to rise (Fig. 15.9(a)). A ridge will then appear in the mat and/or the layer thickness will increase. Excessive material also accelerates wear on the augers. If, in contrast, the head of material is too low, the screed settles because there is insufficient material to support it (Fig. 15.9(b)). Many pavers can be fitted with an automatic system that monitors and controls the material flow through the paver and, thus, the level of the screed. These systems will significantly reduce the likelihood of the occurrence of the adverse effects described above.

As illustrated in Fig. 15.9(c), if the supply of material to the auger is correct, the forces acting on the screed are in equilibrium and the screed is able to maintain the required level.

15.5.2 Paving speed

The paver should travel at a constant speed since variations will result in an uneven surface. An automatic system to pre-set and maintain speeds under varying load conditions is recommended.

Stoppages are also a problem. Not only are they likely to leave a blemish on the surface, they also result in temperature segregation. Every time the paver stops, the screed tends to sink into the mat. The areas in front of the screed and just behind the paver, both of which are inaccessible to the rollers, cool down while the material below the screed remains hot. When the paver starts moving again, the screed will lift slightly to overcome the cooler material ahead of the screed leaving a ridge in the mat.

If the paver is forced to stop, the screed can be locked in position with a special 'screed stop system', which is powered by hydraulic lift

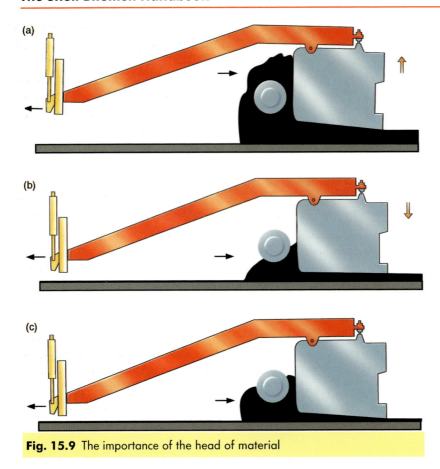

Fig. 15.9 The importance of the head of material

cylinders. This prevents the screed from sinking into the mat and reduces the problems associated with paver stoppages.

Normal paving speeds range from 4 m/min up to 20 m/min depending on the type of asphalt being laid and the performance of the equipment itself. A minimum speed is necessary in order to keep the screed floating. If the paving speed falls below this value, the screed will settle. The layer will then be too thin. When using high compaction screeds, speed should be maintained around 2 to 4 m/min in order to achieve the very high densities for which these screeds were designed.

15.5.3 Layer thickness and surface regularity

In the UK, national requirements for finished tolerances, etc. are contained in the SHW[352]. In terms of the finished road level, requirements

352 HIGHWAYS AGENCY *et al. Manual of contract documents for highway works, Specification for highway works, vol 1, cl 702, 2002.* The Stationery Office, London.

are stated in terms of the maximum number of surface irregularities. In order to achieve the specified surface regularity, the layer thickness may vary to take account of any irregularities in the underlying surface.

Where necessary, electronic levelling devices, such as grade controllers and/or slope controllers should be used to automatically adjust the mat thickness so that a level surface is maintained. If the paver is being operated manually, the surfacing gang must refrain from making frequent corrections to the height of the screed.

15.5.4 Paving width

Most modern pavers are fitted with a telescopic screed that allows the paving width to be varied. Normally, the width is set prior to the start of laying. However some works, e.g. tie-ins at roundabouts, may involve laying a variable width which requires the paver driver to vary the paving width during laying. This has to be undertaken with care and should aim to minimise the amount of asphalt that has to be laid by hand.

15.5.5 Joints

An important contributory factor in achieving good surface regularity and appearance is the quality of the longitudinal and transverse joints. Accordingly, due care is necessary when laying and compacting material that is close to the joints.

When laying asphalt adjacent to an existing lane, the height of the screed above the surface must be carefully adjusted to allow for the reduction in height that occurs as the material is compacted. An allowance of 15 to 20% is not uncommon but the actual value depends on the nature of the material, temperature of the mixture, compactive effort, etc. An automatic grade controller that operates off the adjacent lane is very useful for joint matching.

The side overlap of the joint should be about 25 to 50 mm. There should be as little raking of the joints as possible, so laying must be precise.

To create a smooth transverse joint, the paver screed should be placed on top of the mat placed previously, just in front of the joint. As the forces on the screed need to be in equilibrium when the paver resumes its work, only enough asphalt to cover the auger shaft should be conveyed to the auger box before the paver moves forward. (The 'auger box' is the area immediately in front of the screed.)

In order to ensure a good bond at joints and to seal the area, a coating of bitumen must be applied to the exposed surface. Open joints are a fairly common failure on mats that are only a few years old. Good practice in painting the joints and in compacting them will significantly reduce the possibility of this occurring.

15.5.6 Type of mixture and laying temperature

Asphalts having low workability (sometimes described as 'stiff mixtures'), require heavy screeds. Asphalts that are more workable require relatively light screeds. Stiff mixtures tend to lift the screed above the required level whilst mixtures having high workability very often do not have the resistance to adequately support the weight of the screed.

The load of the screed on high-workability mixtures can be reduced with the help of a 'screed unload system' which transfers the weight of the screed to the tractor. This not only allows heavy screeds to be used on high-workability mixtures, it also improves traction and helps to obtain an even surface and uniform degree of compaction.

Another factor that has a significant effect on the quality of the finished mat is the temperature at which the asphalt is laid (time of day is therefore also important and account should be taken of cooler night temperatures, Fig. 15.10). Variations in temperature can cause irregularities in the surface and result in varying levels of compaction being achieved. As an asphalt cools, it becomes more resistant to compaction. The tractor unit must be able to provide the tractive force necessary to overcome this resistance. Accordingly, successful laying of cooler mixtures requires the use of pavers with good traction and relatively heavy screeds. Colder material may tear under the screed as it is laid.

Fig. 15.10 Night laying on an airfield
Courtesy of Lafarge Contracting Ltd

Fig. 15.11 Segregation in asphalt layers – a potential cause of failure

15.5.7 Segregation of mixtures

Segregation (Fig. 15.11) occurs when the coarser aggregate separates from the asphalt, making the mixture non-homogeneous. Segregation is a common cause of failure in asphalt layers[353].

Segregation may occur as early as when the truck is being loaded with asphalt at the production plant. This is particularly likely if the mixed asphalt is dropped too slowly into the truck. It is often difficult to avoid some concentration of aggregate along the sides of the trailer, particularly in mixtures with large nominal stone sizes, e.g. 0/40 mm mixtures. It is for this reason that the UK coated macadam Standard[354] ceased to include 40 mm dense base and binder course design mixtures in 2003.

Once the asphalt has segregated, it may well remain so as it passes through the paver and is laid and compacted. The appearance of 'hungry' areas (sometimes referred to as lean or dry) in the laid surface is an indication of segregation. This condition may well result in the stiffness of the pavement being reduced and/or the asphalt being open

353 BROCK D and J G MAY. *Hot mix asphalt segregation: Causes and cures, Quality improvement series 110, National Asphalt Pavement Association, 1993.*

354 BRITISH STANDARDS INSTITUTION. *Coated macadam for roads and other paved areas, Specification for constituent materials and for mixtures, p 5, BS 4987-1: 2003.* BSI, London.

to the rigours of the weather. Either of these factors may result in a surface defect in the form of a pothole or cracking appearing relatively early in the life of the pavement.

If the material appears non-homogeneous at the edges of the lanes, this may be caused by segregation along the sides of the trailer or poor mixture distribution by the augers such as may occur if the head of material in the auger box is incorrect. For example, if the material level is too high, it will slope towards the outer edges where the aggregate can separate. The height setting of the auger is another important factor in this respect.

A segregated strip in the middle of the lane occurs if the auger drive unit is located at the centre of the augers. This phenomenon does not occur if the auger drives are located at the outer ends of the drive shafts.

Transverse areas of segregated material normally arise from the separation of materials at the front and rear ends of the trailer.

15.6 Compaction

Proper compaction is necessary if the asphalt is to perform to its full potential. A well-compacted, homogeneous asphalt layer will be resistant to rutting and crack initiation and propagation. It is also, in most cases, substantially impermeable, flexible and resistant to wear. If the compactive effort is not applied equally throughout the mat and the resultant air voids content is variable, there may be problems with the pavement earlier in its life than would otherwise have been the case. Thus, the application of the appropriate level of compaction uniformly throughout the mat is an important determinant of the life of the pavement.

There are a number of types of roller that are currently available for asphalt compaction. The choice of machine depends on the type and size of the job and is often related to local preferences. Factors affecting asphalt compaction have been summarized by the Asphalt Institute[355]. In general terms, factors that affect asphalt compaction can be classified as those relating to the asphalt itself, the laying plant and the environment.

There is also a range of lightweight compaction equipment that is suitable for a variety of asphalt applications including vibratory plate compactors, double-drum walk-behind rollers and lightweight vibratory tandem rollers. The compactive effort applied by a deadweight roller is primarily dependent on its static weight but is also influenced by the diameter of its steel wheels or drums.

355 ASPHALT INSTITUTE. *Factors affecting compaction, Educational Series ES-09, 1980*. AI, Kentucky.

Fig. 15.12 Static three-wheeled roller

Pneumatic-tyred rollers rely on static weight and tyre pressures for their compactive effort. They are often used in combination with static smooth drum or vibratory rollers for finish rolling to remove drum marks and for sealing the surface. Finish rolling is generally used to give a smooth finish rather than to provide any increase in the level of compaction in the mat.

Vibratory rollers combine the static load of the drum with dynamic loads. The effect of the vibration is to reduce internal friction in the asphalt and improve the effect of the compactive effort even when used with comparatively low static linear loads.

Roller capacities are usually expressed in terms of tonnes of asphalt laid per hour. A vibratory asphalt roller will always have a higher capacity than a static roller of the same weight. On asphalts with relatively poor workability, this difference is even more pronounced. A range of rollers is illustrated in Figs. 15.12 to 15.19.

15.6.1 Rolling procedures
Asphalt compaction can be divided into three stages:

- initial rolling
- main compaction
- finish rolling.

293

Fig. 15.13 Static tandem roller

A key factor in dictating the compactability of a hot asphalt is its temperature. The temperature at which it leaves the rear end of the screed is normally somewhere between 130 and 160°C. Within this range, the mixture has a relatively low internal viscosity. As the temperature drops, the viscosity of the bitumen rises, resulting in an increase in the asphalt's resistance to compaction.

Cooling time is defined as the time interval during which an asphaltic layer is available for compaction. It extends from the time when the asphalt has been placed by the paver to the time when the material has reached a temperature where further compactive effort does not increase the density of the material. The calculation of cooling time has been studied by Brown[356] and Hunter[357,358]. Cooling time is a very important parameter in relation to the compaction of asphalt mixtures. Compaction for all layers is critical but it becomes even more so for

356 BROWN J R. *Cooling effects of temperature and wind on rolled asphalt surfacings, Supplementary Report 624, 1980.* Transport and Road Research Laboratory, Crowthorne.

357 HUNTER R N. *The cooling of bituminous materials during laying, Asphalt Technology, J Inst Asph Tech, Sep 1986.* Institute of Asphalt Technology, Dorking.

358 HUNTER R N (R N HUNTER, ed). *Compaction, Asphalts in Road Construction, pp 341–348, 2000.* Thomas Telford, London.

Fig. 15.14 Pneumatic-tyred roller

Fig. 15.15 Vibratory tandem roller

Fig. 15.16 Combination (combi) roller

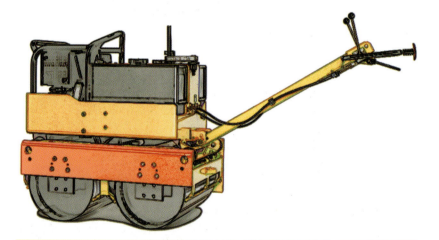

Fig. 15.17 Double-drum walk-behind roller

thin layers where the time available is extremely limited and may be as small as a few minutes. Calculation of this cooling time can most readily be assessed by using CALT[359] (Calculation of Asphalt Laying Temperatures) – a computer program that is part of Shell's suite of analytical computer programs.

The rate at which an asphalt layer cools is a function of layer thickness, ambient temperature, ground temperature and weather conditions; wind will have a pronounced effect on the rate of cooling. Consider the situations illustrated in Fig. 15.20. In a thin surfacing (illustrated in

359 HUNTER R N. *Calculation of temperatures and their implications for unchipped and chipped bituminous materials during laying, PhD Thesis, Heriot-Watt University, 1988.*

Fig. 15.18 Vibratory plate compactor

Fig. 15.19 Single-drum asphalt compactor

Fig. 15.20(a)) in unfavourable conditions, the time available for compaction may be as little as five minutes or less. Under the same conditions, a thick layer may well retain its temperature for over an hour. Accordingly, the thinner the layer, the less time there is available for compaction.

In general, compaction rolling should start as soon as possible after the material has been placed. With a vibratory roller, compaction can usually

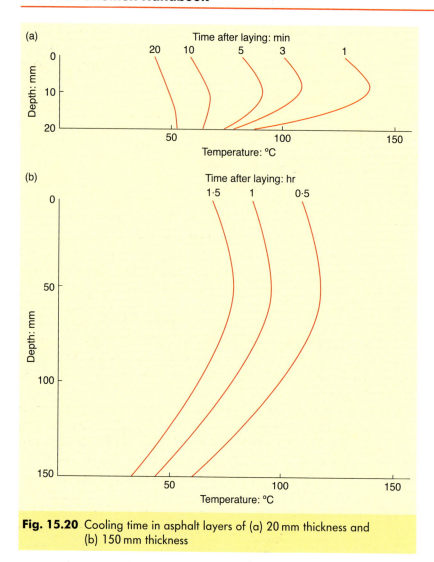

Fig. 15.20 Cooling time in asphalt layers of (a) 20 mm thickness and (b) 150 mm thickness

start with vibrating passes. On mixtures with high workability, it may be more appropriate to start with two static passes that should be made at low rolling speed, say 1 to 2 km/h. The roller should follow as close as possible behind the paver so that compaction can take place whilst the asphalt is above the minimum compaction temperature. This is necessary to ensure that the degree of compaction achieved is adequate. However, if the roller runs repeatedly over the same area at very short intervals when the temperature of the asphalt is high, the surface may crack resulting in a reduction in density.

Finish rolling is effective down to around 60°C. The main purpose of finish rolling is to remove roller marks and other surface blemishes. It improves the appearance of the surface. Finish rolling may also increase the density of the mat especially if the asphalt is comparatively hot.

It is common practice in many countries to use pneumatic-tyred rollers to seal the surface although traffic tends to achieve this in time. However, this effect obviously does not occur on airport runways which explains why pneumatic-tyred rollers are often specified for finish rolling on airport contracts.

15.6.2 *Number and types of roller*

The number and types of rollers required on a particular job are determined by the rate at which the asphalt is placed, normally expressed in the number of square metres per hour. In order to arrive at this figure, a number of elements have to be taken into consideration.

Every paving job can be expressed in terms of the asphalt tonnage that has to be placed on an hourly basis. On larger contracts, this tonnage is usually governed by the capacity of the asphalt production plant.

The combination of the mixture tonnage, the paving width and the layer thickness determines the speed of the paver. The speed multiplied by the paving width gives the laying rate in square metres per hour. This can be used to calculate the compaction plant that will be required for a particular job. Allowance should be made for temporary peaks in the supply of hot material.

Rolling speeds range from 2 to 6 km/h. Low speeds are used on thicker layers and when the material requires substantial compactive effort. The number of roller passes depends on several factors. These are, primarily, the compaction properties of the asphalt and the degree of compaction that is required. Static linear load and vibration characteristics of the rollers also have a decisive influence.

Thin layers with a high stone content are best compacted with a combination of high frequency and low amplitude to reduce the risk of crushing the aggregate. Asphalts having low workability and thick layers are best compacted at high amplitude and low frequency.

It is advisable to compact a test strip in order to determine a suitable rolling procedure that will result in achieving the specified degree of compaction. Nuclear density gauges give an instantaneous reading of the density of the compacted mat. They can be a useful asset in determining the degree of compaction that has been achieved and are particularly helpful at the start of a contract when establishing an effective rolling regime.

15.6.3 *Rolling patterns*

Rolling patterns are illustrated in Figs. 15.21 and 15.22. The paved width in Fig. 15.22 is divided into rolling lanes and zones. The number of rolling lanes depends on the drum width and paving width. The drum

Fig. 15.21 A vibratory tandem roller will achieve uniform compaction over the entire paving area by following a simple pattern

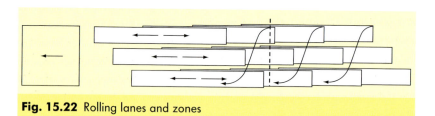

Fig. 15.22 Rolling lanes and zones

width should be related to the paving width so that, for example, three parallel rolling lanes are sufficient to cover the paving width.

When switching rolling lanes, the roller should only travel on material that has been compacted in order to avoid marking the mat. In addition, the roller should never be allowed to stand still on a hot mat.

A rolling pattern normally consists of three parallel rolling lanes divided into zones, these zones being 30 to 50 m in length. In each zone, the roller moves forward and backward until it has executed the required number of passes. However, the roller may start working on the next zone before completing the previous one. The roller strokes (the distance the roller moves from the point at which it changes direction behind the paver to the change in direction at the end of the lane) then cover two zones.

The rolling pattern for conventional deadweight rollers is more complicated than that illustrated in Fig. 15.22. This is because the two rear drums have a significantly different linear load than that exerted by the front drum. In practice, rolling with a conventional deadweight roller often results in higher densities in the middle of a lane than those achieved along the edges of the lane.

When compacting joints, there are two different approaches. Either the joint is compacted with the roller working on the cold lane with a 100 to 200 mm overlap on the hot lane or the drum extends 100 to 200 mm onto the finished lane while the rest of the drum passes over the newly laid mat. As discussed above, efficient joint compaction is particularly important in determining the quality of the finished pavement.

15.6.4 Low-workability mixtures

Increases in traffic loads on major highways in the UK has led to the development of the concept of 'long-life pavements' and materials that are appropriate for such roads. As a consequence, asphalt mixtures containing low penetration bitumen and crushed aggregate with high stone content are now commonplace. Their high mechanical resistance to compaction requires efficient methods of rolling. In this respect, vibratory rollers are the best choice to achieve specified densities.

15.6.5 High-workability mixtures

Asphalts that possess high workability are prone to lateral displacement during compaction which may result in small transverse cracks on the surface some 3 to 5 mm deep. These can normally be closed by suitable finish rolling or by subsequent traffic action. If longitudinal cracks appear, they are often deep and very difficult to close completely.

The rolling of asphalts with high workability requires special measures. Often, they must be allowed to cool down before rolling starts. This means that the roller has to operate a relatively long distance behind the paver. In many cases, it may be best to work with long rolling lanes (100 m or more). In order to stabilise the mixture, it is often advisable to start the compaction with two passes using a static mode or by means of a pneumatic-tyred roller. A large drum diameter and a slow approach also help to prevent shoving or the appearance of cracks. It is often a good idea to select a low amplitude and high frequency on these mixtures. A pneumatic-tyred roller is suitable for finishing the surface.

Figure 15.23 illustrates the difficulties associated with compacting high-workability mixtures. Such materials have very low viscosity in their hot state and may be squeezed out under the drum during rolling, resulting in hairline cracks and the risk of lateral displacement of the asphalt. Adequate compaction can be imparted if the material is allowed to cool.

Fig. 15.23 Rolling mixtures with high workability

15.6.6 Thin layers

Thin layers are placed at fast paving speeds resulting in high surface laying capacities. The compaction capacity should be adjusted to reflect these higher outputs. Simply increasing the speed of rollers is not a satisfactory answer, the danger being that the asphalt is inadequately compacted. In order to achieve sufficient compaction, the number of rollers has to be increased. To avoid crushing the aggregate, a low amplitude and a higher frequency should be employed. In addition, thin layers cool rapidly which is why the rollers must be able to attain specified densities in a timely and efficient manner.

15.6.7 Thick layers

It is possible to achieve high densities on asphalt layers up to 200 mm thick. However, rolling on very thick layers may create undulations in the surface. On thick layers, rolling should start at some distance from the edge of the lane. Successive roller passes should move closer to the edge to prevent it being displaced. A large drum diameter together with a high-amplitude setting is particularly apt in such circumstances. High-amplitude vibration ensures that good compaction is achieved throughout the layer thickness.

15.7 Specification and field control

In general, a method specification or an end-product specification is used for asphalt compaction. A combination of the two is also frequently cited. A method specification lists the type and size of rollers to be used, number of rollers to be used and, sometimes, the way in which they are to be used. In an end-product specification, the client decrees a minimum level of compaction that has to be achieved. This is then checked by laboratory and field tests.

In the UK, the specification is a mixture of method and end-product[360]. End-product specifications are widely used throughout Europe for large projects.

---

360 HIGHWAYS AGENCY *et al. Manual of contract documents for highway works, Specification for highway works, vol 1, 2002.* The Stationery Office, London.

Fig. 15.24 Tandem paving on the M40 Motorway, UK
Photograph courtesy of Tarmac Ltd

On many modern contracts, the client will check that the requirements of end-product density specifications have been met by testing based on Marshall tests[361,362]. The density requirements normally fall in the range of 95 to 98% Marshall (50 or 75 blows on the asphalt sample). Sometimes, requirements also include a given range of air void content and the UK has adopted just such an approach in its national specification[363] (discussed in Section 13.3.1).

Civil engineering contracts, including those that involve an element of asphalt construction, often contain penalty clauses that stipulate sums to be deducted from payment if the pavement as constructed fails to meet specified densities.

The normal method for field density control is to remove a core sample with a diamond-tipped coring bit. Density and air void content are determined on the sample cores in a laboratory.

361 MARSHALL CONSULTING & TESTING LABORATORY. *The Marshall Method for the design and control of bituminous paving mixtures, 1949.* Marshall Consulting & Testing Laboratory, Mississippi.

362 ASPHALT INSTITUTE. *Mix design methods for asphalt concrete and other hot-mix types, Manual series no 2 (MS-2), 1988.* AI, Kentucky.

363 HIGHWAYS AGENCY *et al. Manual of contract documents for highway works, Specification for highway works, vol 1, cl 929, 2002.* The Stationery Office, London.

Nuclear density gauges can be used for rapid density testing on site. As indicated previously, they are very practical when establishing suitable rolling procedures at the start of a job. Modified gauges have now been developed to give more accurate density readings on thin asphalt layers. The final approval of the density level is, however, generally based on the assessment of cores.

Other quality controls that are applied to asphalt layers include checking the surface regularity, texture depth (see Sections 13.4.3 and 19.1.2) and, most importantly, skid resistance (see Section 19.1.2).

Chapter 16

Testing of asphalts

An analytical approach to pavement design is based on a knowledge of the various materials used in the layers of a pavement structure. Laboratory testing has been used extensively in research and this has resulted in a fundamental understanding of the mechanical behaviour of asphalts.

There are two aspects of material properties relevant to pavement design: firstly, the load–deformation or stress–strain characteristics required for the analysis of the structure; secondly, the performance characteristics that determine the mode of failure. The main performance criteria are fatigue cracking and permanent deformation.

Ideally, the mechanical properties of the material in situ are required for design purposes. However, in situ testing of full-scale trial sections is impractical in most cases. Consequently, engineers have to rely on laboratory testing. These tests should reproduce the anticipated in situ conditions as closely as possible, i.e. temperature, loading time, stress conditions, degree of compaction, etc. However, the in situ conditions are continually changing and selection of appropriate testing conditions is therefore extremely difficult.

The stress condition can be reproduced in the laboratory only with great difficulty. Figure 16.1 depicts a representation of a pavement element showing the stresses that occur in it when a wheel load is approaching. This is, of course, a very simplified model. In practice, the stresses are applied three-dimensionally: horizontal and shear stresses occur on planes that are perpendicular to those shown in Fig. 16.1. As the wheel passes over the element, these stresses change with time and this is shown in Fig. 16.2.

It has not yet been possible to reproduce this complex stress regime accurately in the laboratory. The repeated load triaxial test comes close but does not reproduce the shear reversal effect. Hence, simplified tests have been introduced that can reproduce certain aspects of the in situ behaviour. These tests can be divided into the following three groups.

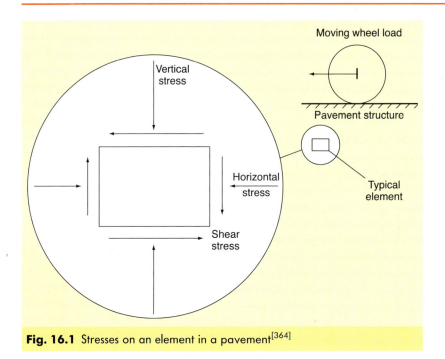

Fig. 16.1 Stresses on an element in a pavement[364]

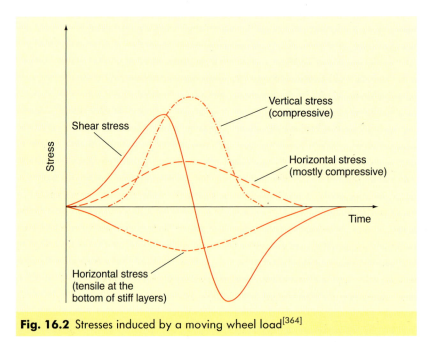

Fig. 16.2 Stresses induced by a moving wheel load[364]

364 PELL P S. *Bituminous pavements: materials, design and evaluation, Residential course at the University of Nottingham, section E, Laboratory test methods, pp E1–E17, Apr 1988.*

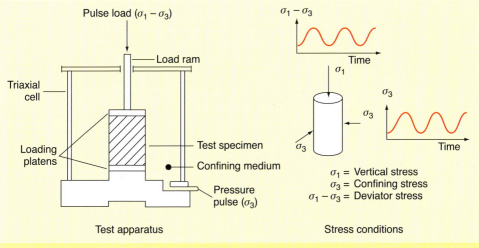

Fig. 16.3 The repeated load triaxial test[365]

(1) Fundamental:
  - repeated load triaxial test
  - unconfined static uniaxial creep compression test
  - repeated load indirect tensile test
  - dynamic stiffness and fatigue tests.

(2) Simulative:
  - wheel-tracking test
  - gyratory compactor
  - durability testing.

(3) Empirical:
  - Marshall test.

16.1 Fundamental tests
16.1.1 Repeated load triaxial test

This test most closely reproduces the stress conditions in a pavement. It is extensively used in research to study permanent deformation as well as the elastic properties of asphalt[366]. It has the advantage that both vertical and horizontal stresses can be applied at the levels predicted in the pavement, as shown in Fig. 16.3. Deformation monitoring devices to measure vertical and horizontal strains induced by the imposed stresses are mounted on the specimens. In view of the complex nature of the

365 PELL P S. *Bituminous pavements: materials, design and evaluation, Residential course at the University of Nottingham, section E, Laboratory test methods, pp E1–E17, Apr 1988.*

366 BROWN S F and K E COOPER. *The mechanical properties of bituminous materials for road bases and basecourses, Proc Assoc Asph Pav Tech, vol 53, pp 415–437, 1984.* Association of Asphalt Paving Technologists, Seattle.

apparatus and testing procedures, this type of test is only suitable for research investigations, although it is the subject of a draft European test method.

16.1.2 Unconfined static uniaxial creep compression testing

The use of a static unconfined uniaxial compression test, termed the 'creep test', for assessing the permanent deformation resistance of asphalt was developed in the 1970s by Shell[367]. This test gained wide acceptance due to the ease of specimen preparation, the simplicity of the test procedure and the low cost of test equipment. The only requirements for the test specimen are that it should be prismatic with flat and parallel ends normal to the axis of the specimen while the test procedure comprises the application of a constant stress to the specimen for one hour at a constant temperature and measurement of the resultant deformation. The test equipment was, therefore, very simple and early versions of the equipment generally applied the load as a dead weight via a mechanical lever arm[368].

Shell developed a rut prediction procedure based on the creep test but it was found that the method underestimated rut depths measured in trial pavements[369]. This was attributed to the effects of dynamic loading producing higher deformations in the wheel-tracking tests[370]. An experimentally derived empirical correction factor that varied according to mixture type was, therefore, subsequently introduced into the creep analysis to account for these effects.

In 1983 Finn *et al.*[371], in an assessment of the creep test and analysis procedures during investigations of pavement rutting, recommended that consideration be given to the development of a repeated load test. The development of plastic strains in the skeleton of an aggregate contributes significantly to permanent deformation. The effect of this form of loading in inducing plastic strains had already been recognised by

---

367 HILLS J F. *The creep of asphalt mixes, J Inst Pet, vol 59, no 570, pp 247–262. Nov 1973*. Institute of Petroleum, London.

368 DE HILSTER E and P J VAN DE LOO. *The creep test: influence of test parameters, Proc. Colloquium 77 Plastic deformability of bituminous mixes, pp 173–215, Institut für Strassen, Eisenbahn und Felsbau an der Eidgenössischen Technischen Hochschule, 1977.* Zürich.

369 HILLS J F, D BRIEN and P J VAN DE LOO. *The correlation of creep and rutting, Paper IP 74-001, 1974.* Institute of Petroleum, London.

370 VAN DE LOO P J. *Creep testing, a simple tool to judge asphalt mix stability, Proc Assoc Asph Pav Tech, vol 43, pp 253–284, 1974.* Association of Asphalt Paving Technologists, Seattle.

371 FINN F N, C L MONISMITH and N J MARKEVICH. *Pavement performance and asphalt concrete mix design, Proc Assoc Asph Pav Tech, vol 52, pp 121–150, 1983.* Association of Asphalt Paving Technologists, Seattle.

Goetz *et al.*[372] who had developed an early version of a rapid cycle repeated load test by 1957. Further concerns over the use of static loading have arisen more recently. Evidence suggests that static loading does not reflect the improved performance of binder modifiers that enhance the elastic recovery properties of an asphalt whereas, in contrast, this can be demonstrated under repeated loading[373].

Triaxial testing on asphalt has shown that the application of a confining pressure has a significant positive effect on the permanent deformation performance of axially loaded specimens[372,374,375]. Results of repeated load (sinusoidal) triaxial tests on various mixtures[376] suggest that the use of confinement is more beneficial for continuously graded materials that rely on the aggregate interlock load transfer mechanism. More recent work in France[377] on güssasphalt, porous asphalt and asphaltic concrete also indicates that confinement is very important for mixture types whose resistance to deformation comes from the aggregate skeleton but less so, if at all, for those whose deformation resistance is based on the stiffness of the binder/filler mortar.

The use of a triaxial cell necessarily makes test equipment more expensive and test procedures more complex. However, it is considered that the facility to confine specimens during testing is important if an axial test is to gain acceptance and be used. This is particularly so as axial testing is being advanced for adoption in European Standards[377]. This has led to the development of a simple system for applying a partial vacuum through the use of latex membrane. This is termed the vacuum repeated load axial test (VRLAT) and is described in detail below.

372 GOETZ, W H, J F MCLAUGHLIN and L E WOOD. *Load–deformation character-istics of bituminous mixtures under various conditions of loading, Proc Assoc Asph Pav Tech, vol 26, pp 237–296, 1957.* Association of Asphalt Paving Technologists, Seattle.

373 VALKERING, C P, D J L LANCON, E DE HILSTER and D A STOKER. *Rutting resistance of asphalt mixes containing non-conventional and polymer modified binders, Proc Assoc Asph Pav Tech, vol 59, pp 590–609, 1990.* Association of Asphalt Paving Technologists, Seattle.

374 MONISMITH C L and A A TAYEBALI. *Permanent deformation (rutting) considera-tions in asphalt concrete pavement sections, Proc Assoc Asph Pav Tech, vol 57, pp 414–463, 1988.* Association of Asphalt Paving Technologists, Seattle.

375 BROWN S F and M S SNAITH. *The permanent deformation characteristics of a dense bitumen macadam subjected to repeated loading, Proc Assoc Asph Pav Tech, vol 43, pp 224–252, 1974.* Association of Asphalt Paving Technologists, Seattle.

376 BROWN S F and K E COOPER. *The mechanical properties of bituminous materials for road bases and basecourses, Proc Assoc Asph Pav Tech, vol 53, pp 415–437, 1984.* Association of Asphalt Paving Technologists, Seattle.

377 BONNOT J. *Discussion paper on uniaxial and triaxial repeated load testing of bituminous mixtures, Paper presented to CEN Committee TC227 (WG1 TG2 – Test Methods), Vienna, 1994.*

Fig. 16.4 Uniaxial creep test and RLAT

Uniaxial creep

The uniaxial creep (UC) test (Figs. 16.3 and 16.4) is the simplest method of assessing the resistance to permanent deformation. Specimens up to 200 mm long are normally tested with axial stresses of up to 500 kPa on a 100 mm diameter specimen. An optional pre-test conditioning period (0 to 60 minutes) is followed by a test period of up to 10 000 s. There is also the option for a post-test relaxation period (10 to 100 minutes). Test data is stored automatically and can be printed out if required. During the test, the relationship between axial strain and elapsed time is normally plotted on the computer screen. Typical results are shown in Fig. 16.5. This method is described in a British Standard[378].

---

378 BRITISH STANDARDS INSTITUTION. *Sampling and examination of bituminous mixtures for roads and other paved areas, Method for determination of resistance to deformation of bituminous mixtures subject to unconfined uniaxial loading, BS 598-111: 1995.* BSI, London.

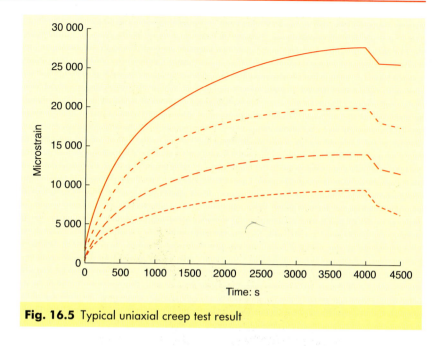

Fig. 16.5 Typical uniaxial creep test result

Standard test conditions and requirements for the UC test are:

conditioning stress	10 kPa
conditioning period	600 s
test stress	100 kPa
test duration	3600 s
test temperature	30 or 40°C

Repeated load axial test

The repeated load axial test (RLAT) is increasingly used in preference to the UC test as the pulsed load is more simulative of traffic loading. The specimen is subjected to repeated load pulses of 1 s duration separated by 1 s duration rest periods (up to 10 000 pulses). An optional pre-test conditioning period (selectable 0 to 60 minutes) is available. As with the creep test, test data are automatically stored and can be printed out, if required. During the test, the relationship between axial strain and number of load pulses is plotted on the computer screen. How the test is set up is as shown in Fig. 16.4 and typical results are shown in Fig. 16.6. This test method is currently being developed[379].

---

379 BRITISH STANDARDS INSTITUTION. *Method for determining resistance to permanent deformation of bituminous mixtures subject to unconfined dynamic loading, DD 226: 1996.* BSI, London.

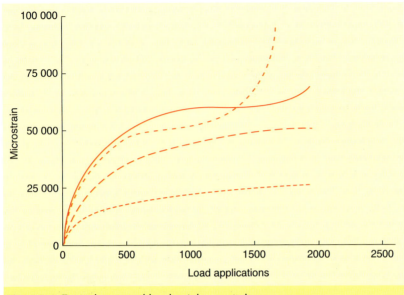

Fig. 16.6 Typical repeated load axial test result

Standard test conditions and requirements for the RLAT are:

conditioning stress	10 kPa
conditioning period	600 s
test stress	100 kPa
test duration	1800 cycles
test cycle	square wave pulse 1 s on, 1 s off
test temperature	30 or 40°C

In the vacuum repeated load axial test (VRLAT), the specimen is evacuated through one of the loading platens (via a porous stone or slots machined into the platen). The loading platen is ported through its body to a vacuum pump. The latex sealing membrane, of the type used in triaxial testing of soils, is fitted over the specimen and both the upper and lower platens. It is then secured using rubber 'O' rings that locate in a groove machined around the circumference of each platen in the plane of the surface of the platen. This configuration is shown in Fig. 16.7.

The test conditions are the same as those used for the RLAT and the confining pressure is, at present, the subject of investigation at TRL Ltd.

16.1.3 Repeated load indirect tensile tests

Various methods have been employed to measure the stiffness of an asphalt. Conventional techniques have principally included uniaxial and

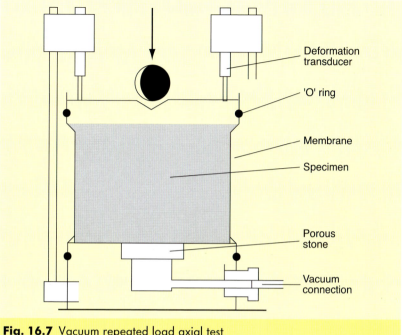

Fig. 16.7 Vacuum repeated load axial test

triaxial compression and/or tension tests and flexural beam tests. However, in the late 1960s, Hudson and Kennedy[380] developed an indirect tension test based on the Brazilian splitting test[381] that is used to measure the tensile strength of concrete. In the early 1980s, the test was accepted as a standard method by the American Society for Testing and Materials[382]. Cooper and Brown[383] introduced the method to the UK with the development of the Nottingham Asphalt Tester (NAT) which has subsequently gained widespread use in much of Europe and has formed the basis of a number of British Standards.

This section presents a brief overview of the theory supporting the indirect tensile test for asphalt.

380 HUDSON W R and T W KENNEDY. *An indirect tensile test for stabilized materials, Research Report 98-1, Center for Highway Research, January 1968.* University of Texas, Austin.

381 CARNIERO F L L B and A BARCELLUS. *RILEM Bulletin, no 13, 1953.*

382 AMERICAN SOCIETY for TESTING and MATERIALS. *Standard test method for indirect tension test for resilient modulus of bituminous mixtures, D4123-82, 1 Jan 1982.* ASTM, Philadelphia.

383 COOPER K E and S F BROWN. *Development of a simple apparatus for the measurement of the mechanical properties of asphalt mixes, Proceedings, Eurobitume Symp, pp 494–498, Madrid, Spain, 1989.*

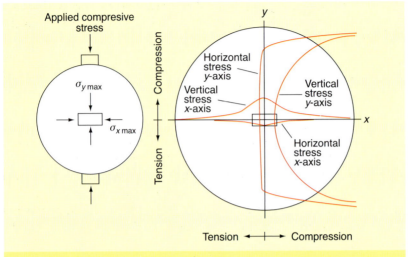

Fig. 16.8 Stress distribution in the indirect tensile mode of test

A repeated line loading is applied across the vertical diameter of a cylindrical specimen as shown in Fig. 16.8. This vertical loading produces both a vertical compressive stress and a horizontal tensile stress on diameters of the specimen. The magnitudes of the stresses vary along the diameters as shown in Fig. 16.8 but are at a maximum in the centre of the specimen.

The situation, as depicted in Fig. 16.8, enables the calculation of strain but depends on the following assumptions.

- The specimen is subjected to plane stress conditions ($\sigma_z = 0$).
- The material behaves in a linear elastic manner.
- The material is homogeneous.
- The material behaves in an isotropic manner.
- Poisson's ratio (ν) for the material is known.
- The force (P) is applied as a line loading.

When all of these conditions are met, or assumed, then the stress conditions in the specimen are given by the closed form solution of the theory of elasticity, that has been derived by various authorities in the field of stress analysis[384–387]. These theories all show that when

384 TIMOSHENKO S. *Theory of Elasticity, pp 104–108, 1934*. McGraw Hill, New York.

385 FROCHT M M. *Photoelasticity, vol 2, pp 121–129, 1948*. John Wiley, New York

386 MUSKHELISHVILI N I. *Some basic problems of the mathematical theory of elasticity, pp 324–328, 1953*. P Noordhoff, Groningen.

387 SOKOLNIKOFF I S. *Mathematical theory of elasticity, pp 280–284, 1956*. McGraw Hill, New York.

the width of the loading strip is less than or equal to 10% of the diameter of the specimen and the distance of the element of material from the centre is very small then:

$$\sigma_{x\,max} = \frac{2P}{\pi dt} \tag{1}$$

$$\sigma_{y\,max} = -\frac{6P}{\pi dt} \tag{2}$$

where P = applied vertical force (in N), d = diameter of the specimen (in m), t = thickness of the specimen (in m), ν = Poisson's ratio, S_m = stiffness modulus of the specimen (in Pa), $\sigma_{x\,max}$ = maximum horizontal tensile stress at the centre of the specimen (in N/m^2), $\sigma_{y\,max}$ = maximum vertical compressive stress at the centre of the specimen (in N/m^2) and $\varepsilon_{x\,max}$ = maximum initial horizontal tensile strain at the centre of the specimen.

As the analysis of the indirect tensile test, for calculation of the strain, is concerned with an element of material at the centre of the specimen, Equations (1) and (2) hold true.

By simple linear elastic stress analysis (Hooke's law):

$$\varepsilon_{x\,max} = \frac{\sigma_{x\,max}}{S_m} - \frac{\nu\sigma_{y\,max}}{S_m} \tag{3}$$

where S_m is the stiffness modulus (in Pa).

Substituting Equations (1) and (2) into Equation (3) gives:

$$\varepsilon_{x\,max} = \frac{2P}{\pi dt S_m} + \frac{\nu 6P}{\pi dt S_m} \tag{4}$$

Substituting Equation (1) into Equation (4) gives:

$$\varepsilon_{x\,max} = \frac{\sigma_{x\,max}}{S_m}(1 + 3\nu) \tag{5}$$

Equation (5) is then used for the calculation of the maximum tensile strain ($\varepsilon_{x\,max}$).

Hadley et al.[388] and Anagnos and Kennedy[389] developed equations that permit the calculation of the tensile strength, tensile strain, modulus of elasticity and Poisson's ratio for a homogeneous, isotropic, linear elastic cylinder subjected to a diametrically applied static 'line' load. The equations were extended to allow calculation of the 'instantaneous

388 HADLEY W O, W R HUDSON and T W KENNEDY. *A method of estimating tensile properties of materials tested in indirect tension, Research Report 98-7, Center for Highway Research, Jul 1970.* The University of Texas, Austin.

389 ANAGNOS J N and T W KENNEDY. *Practical method of conducting the indirect tensile test, Research report, Center for Highway Research, Aug 1972.* The University of Texas, Austin.

resilient Poisson's ratio' and the 'instantaneous resilient modulus of elasticity' for a cylinder subjected to repeated loading as follows:

$$\nu = \frac{\dfrac{\delta_v}{\delta_h}c_1 + c_2}{\dfrac{\delta_v}{\delta_h}c_3 + c_4} \tag{6}$$

where ν = instantaneous resilient Poisson's ratio, δ_v = instantaneous resilient (elastic) vertical deformation (in m), δ_h = instantaneous resilient (elastic) horizontal deformation (in m) and c_1, c_2, c_3 and c_4 = constants depending on the specimen diameter and loading strip width and

$$E_R = \frac{P(c_5 - \nu c_6)}{\delta_h t} \tag{7}$$

where E_R = instantaneous resilient modulus of elasticity (in Pa), P = applied load (in N), ν = instantaneous resilient Poisson's ratio, δ_h = instantaneous resilient (elastic) horizontal deformation (in m), t = thickness (height) of specimen (in m) and c_5 and c_6 = constants dependent on the specimen diameter and loading strip width.

Calculation of the above elastic properties for an asphalt requires measurement of the load and elastic (recoverable) deformations. In theory, determination of the load and deformations appears to be quite simple and straightforward. However, in practice, this may not be the case. The difficulty lies in determination of the inflection point on the unloading section of the deformation curves, which may not be readily discernible, or could be virtually non-existent, depending on the frequency and duration of the load, test temperature, material type, etc. Therefore, it could be reasoned that the indirect tensile test is not a reliable method for the determination of the modulus of elasticity of an asphalt. As a result, the theory has been adapted to allow measurement of the load and total instantaneous horizontal deformation. The vertical deformation is not measured so a value of Poisson's ratio has to be assumed. As the deformation consists of both the elastic (recoverable) and viscous (time-dependent) components, a resilient modulus of elasticity cannot be calculated. Instead, a stiffness modulus is calculated as follows:

$$S_m = \frac{P(c_5 - \nu c_6)}{\delta_h t} \tag{8}$$

where S_m = stiffness modulus, resilient modulus or total modulus (in Pa), P = applied repeated load (in N), ν = resilient Poisson's ratio, δ_h = resilient total horizontal deformation (in m), t = thickness (height) of specimen (in m), c_5 and c_6 = constants dependent on the

Fig. 16.9 Indirect tensile stiffness modulus test

specimen diameter and loading strip width. For example, for a 101·6 mm diameter specimen and a 12·7 mm loading strip width, $c_5 = 0·2692$ and $c_6 = −0·9974$.

Indirect tensile stiffness modulus

The indirect tensile stiffness modulus (ITSM) test, defined in DD 213[390], is the most commonly used test method in the NAT (Fig. 16.9) and is used for the determination of the stiffness modulus of a specimen. The test is simple and can be completed quickly. The operator selects the target horizontal deformation and a target load pulse rise time (time from start of load application to peak load). The force applied to the specimen is then automatically calculated by the computer and a number of conditioning pulses are applied to the specimen. These conditioning pulses are used to make any minor adjustments to the magnitude of the force needed to generate the specified horizontal

---

390 BRITISH STANDARDS INSTITUTION. *Method for determination of the indirect tensile stiffness modulus of bituminous mixtures, DD 213: 1993.* BSI, London.

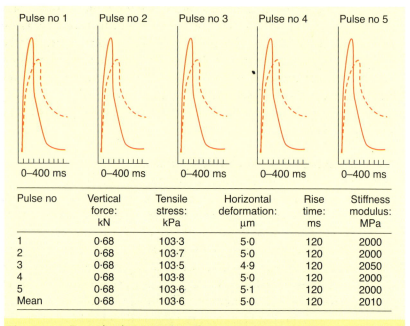

Pulse no	Vertical force: kN	Tensile stress: kPa	Horizontal deformation: μm	Rise time: ms	Stiffness modulus: MPa
1	0·68	103·3	5·0	120	2000
2	0·68	103·7	5·0	120	2000
3	0·68	103·5	4·9	120	2050
4	0·68	103·8	5·0	120	2000
5	0·68	103·6	5·1	120	2000
Mean	0·68	103·6	5·0	120	2010

Fig. 16.10 Typical indirect tensile stiffness modulus test result

deformation and to seat the loading strips correctly on the specimen. Once the conditioning pulses have been completed, the system applies five load pulses. This generates an indirect movement on the horizontal diameter and, since the diameter of the specimen is known beforehand, the strain can be calculated. As the cross-sectional area of the specimen is also known and the force applied is measured, the applied stress can be calculated. Thus, since the stress and the strain are now known, the stiffness modulus of the material can be calculated. A typical test result is shown in Fig. 16.10.

Standard test conditions and requirements for the ITSM test are:

horizontal strain	0·005% of the specimen diameter
rise time	124 ms – equivalent to a frequency of 1·33 Hz
specimen diameter	100 mm, 150 mm and 200 mm
specimen thickness	between 30 mm and 70 mm
test temperature	20°C

Resilient modulus

The resilient modulus (RESMOD) test, defined in an ASTM publication[391], details how instantaneous and total Poisson's ratio and modulus

391 AMERICAN SOCIETY for TESTING and MATERIALS. *Standard test method for indirect tension test for resilient modulus of bituminous mixtures, D4123-82, 1 Jan 1982.* ASTM, Philadelphia.

may be calculated from the recovered vertical and horizontal deformations. In the NAT, two vertical and two horizontal deformation transducers are used for the RESMOD test. To the operator, the software appears to be very similar to that used in the ITSM test but a comprehensive data acquisition routine is employed with five test pulses continuously scanned over a period of 12 s. The software then analyses each test pulse, calculating Poisson's ratio and stiffness modulus for loading, instantaneous recovery and total recovery.

16.1.4 Dynamic stiffness and fatigue tests

For the measurement of the dynamic stiffness and fatigue characteristics of asphalt, researchers have developed a number of tests that are illustrated diagrammatically in Fig. 16.11. Bending tests using beams or cantilevers involving repeated applications of load have been used to determine both dynamic stiffness and fatigue resistance[392–395]. In bending tests, the maximum stress occurs at a point on the surface of the specimen and its calculation, using the standard beam bending formula, depends on the assumption of linear elasticity.

In the USA, a repeated load uniaxial test involving only compressive stresses has been adopted by ASTM as a standard method[396] for evaluating the dynamic stiffness or 'resilient modulus' of asphalt.

Figure 16.12 shows a three-point bending test where a beam of asphalt 200 mm long by 40 mm wide by 30 mm deep is held at its ends and the mid-point is vibrated at relatively low strains or deflections. By measuring the force required to achieve the level of strain, the dynamic stiffness of the material can be calculated using the following standard beam bending formula:

$$\delta = \frac{WL^3}{48EI} \rightarrow E = \frac{WL^3}{48\delta I}$$

392 BROWN S F. *Practical mechanical tests for the design and control of asphaltic mixes, RILEM 3rd Int Symp on the testing of hydrocarbon binders and materials, Belgrade, 1983.*

393 MONISMITH C L, J A EPPS and F N FINN. *Improved asphalt mix design, Proc Assoc Asph Pav Tech, vol 54, pp 347–391, 1985.* Association of Asphalt Paving Technologists, Seattle.

394 BONNAURE F, G GEST, A GRAVOIS and P A UGE. *A new method of predicting the stiffness of asphalt paving mixtures, Proc Assoc Asph Pav Tech, vol 46, pp 64–104, 1977.* Association of Asphalt Paving Technologists, Seattle.

395 COOPER K E and P S PELL. *The effect of mix variables on the fatigue strength of bituminous materials, Laboratory Report 633, 1974.* Transport and Road Research Laboratory, Crowthorne.

396 AMERICAN SOCIETY FOR TESTING and MATERIALS. *Standard test method for indirect tension test for resilient modulus of bituminous mixtures, D4123-82, 19 Jan 1982.* ASTM, Philadelphia.

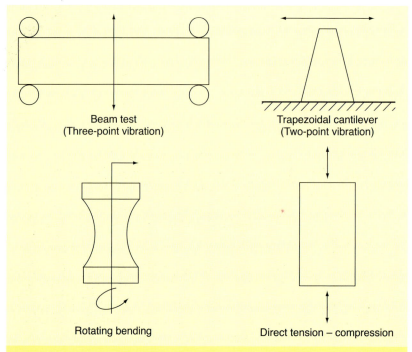

Fig. 16.11 Tests for measuring dynamic stiffness and fatigue

where E = dynamic stiffness, L = length of beam, δ = deflection, I = second moment of area = $bd^3/12$ (b = breadth, d = depth) and W = applied load.

Most laboratory fatigue tests are carried out under uniaxial conditions, either bending or direct loading[397,398]. Fatigue tests are carried out by applying a load to a specimen in the form of an alternating stress or strain and determining the number of load applications required to induce 'failure' of the specimen. 'Failure' is an arbitrary end point, not where the sample literally fails. In a constant strain test, the sample is usually deemed to have 'failed' when the load required to maintain that level of strain has fallen to 50% of its initial value. Because of the scatter of test results associated with fatigue testing, it is normal to test several specimens at each stress or strain level and the results are plotted as stress or strain against cycles to failure on a log–log graph as shown in Fig. 16.13[398].

397 PELL P S. *Characterization of fatigue behaviour, Special Report 140, pp 49-64, 1973.* Transportation Research Board, Washington.

398 READ J M. *Fatigue cracking of bituminous paving mixtures, PhD Thesis, University of Nottingham, 1996.*

Fig. 16.12 The three-point bending test

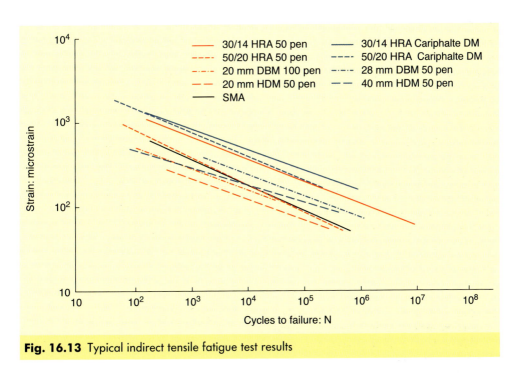

Fig. 16.13 Typical indirect tensile fatigue test results

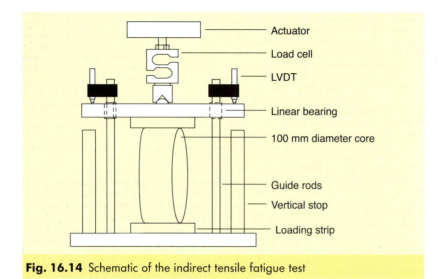

Fig. 16.14 Schematic of the indirect tensile fatigue test

Indirect tensile fatigue test

The indirect tensile fatigue test (ITFT) uses a repeated controlled stress pulse to damage the specimen and the accumulation of vertical deformation against number of load pulses is plotted. As with other indirect tensile tests the duration of the load pulse and also the magnitude of the stress is specified by the operator, however, this test is presently being considered by a British Standard committee. The first draft test method was published in 1996[399] and is currently being revised. A typical set of results is shown in Fig. 16.13. A schematic of the test is shown in Fig. 16.14.

Standard test conditions and requirements for the ITFT are:

stress level	between 50 and 600 kPa (determined by the user)
target rise time	120 ms – equivalent to a frequency of 1·33 Hz
failure criterion	9 mm vertical deformation
test temperature	20°C

16.2 Simulative tests

As the stress conditions in a pavement loaded by a rolling wheel are extremely complex and cannot be replicated in a laboratory test on a sample of asphalt with any precision, simulative tests have been used to compare the performance of different materials.

399 BRITISH STANDARDS INSTITUTION. *Method for the determination of the fatigue characteristics of bituminous mixtures using indirect tensile fatigue, Draft for Development ABF, 1996.* BSI, London.

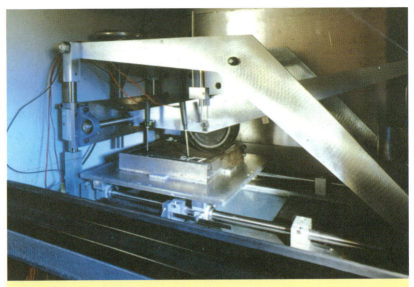

Fig. 16.15 UK wheel-tracking apparatus

16.2.1 *Wheel-tracking tests*

The equipment ranges from large full-scale pavement test facilities such as those at the TRL[400] and the Laboratoire Central des Ponts et Chaussées (LCPC) Test Track at Nantes or the mid-scale pavement test facility at the University of Nottingham[401] to the smaller laboratory-scale wheel-tracking test (see Fig. 16.15). Full-scale and mid-scale facilities can be used to test entire pavement structures whereas laboratories can evaluate the resistance of asphalt to permanent deformation[402]. In the UK, the laboratory-scale test is usually carried out at either 45°C or 60°C, the wheel applying a load of 520 N to the surface of the sample. The performance of the material is assessed by measuring the resultant rut depth after a given number of passes or rate of tracking (in millimetres per hour). Work carried out by the TRL has shown that a good correlation exists between permanent deformation on open stretches of road and the performance in the laboratory-scale wheel-tracking test[403].

400 OECD. *Full-scale pavement tests: report by the OECD Scientific Expert Group, Paris, 1985.* OECD Scientific Expert Group, Paris.

401 BROWN S F and B V BRODRICK. *Nottingham pavement test facility, Transportation Research Record 810, 1981, pp 67- 72.* Transportation Research Board, Philadelphia.

402 ROAD RESEARCH LABORATORY. *The wheel-tracking test, Leaflet LF 50, Issue 2, 1971.* RRL, Crowthorne.

403 SZATKOWSKI W S and F A JACOBS. *Dense wearing courses in Britain with high resistance to deformation, Proc Int Symp on plastic deformability of bituminous mixes, pp 63–76, Zurich, 1977.*

The method for the determination of the wheel-tracking rate in cores of asphalt surface courses was developed at the TRL. The equipment was originally devised for testing 305 mm square by 50 mm deep slabs manufactured using a laboratory roller compactor. This method has now been adapted for testing 200 mm diameter cores cut from the carriageway and is the subject of a full British Standard[404].

The wheel-tracking test is normally carried out on surface courses although there is no reason why other layers cannot be tested. Cores of diameter 200 mm are cut and the surface course is separated from the underlying material, by sawing if necessary. When dealing with a chipped surface course such as hot rolled asphalt, the material is tracked on the underside.

Standard test conditions and requirements for the UK wheel-tracking test are as follows:

Tyre	The tyre should have an outside diameter between 200 and 205 mm. The tyre should be of rectangular section 50 ± 1 mm, treadless and 10 to 13 mm thick. It should be made of solid rubber with a hardness number of 80 IRHD units.
Wheel load	The wheel load should be 520 ± 5 N.
Test temperature	The test temperature should be 45 or 60°C.
Test duration	The test is continued until a rut depth of 15 mm is achieved or for 45 minutes, whichever is the sooner.
Results	The vertical displacement of the loaded wheel is plotted against time. The mean rate of rutting during the last third of the test is determined and expressed in mm/hour to the nearest 0·1 mm/hour.

However, the conditions for such tests vary from country to country. The UK method[404] is not applicable to laboratory-prepared and compacted specimens for compliance testing although some other countries permit such testing.

Clearly, the method used to prepare and compact specimens will have a significant effect on the test results. Typical test conditions are, for example, set out in a German procedure called the Hamburg wheel-tracking test[405].

---

404 BRITISH STANDARDS INSTITUTION. *Sampling and examination of bituminous mixtures for roads and other paved areas, Method of test for the determination of wheel-tracking rate and depth, BS 598-110: 1998.* BSI, London.

405 FORSCHUNGSGESELLSCHAFT FÜR STRAßEN- UND VERKEHRSWE-SEN. *Technische Prüfvorschriften für Asphalt im Straßenbau (TP A-StB), 1997.* FGSV, Cologne, Germany.

Either laboratory-produced test specimens (320 mm × 260 mm) or large cores (diameter >300 mm) may be used for the test. A steel wheel (diameter 203·5 mm, breadth 47 mm) passes the asphalt slab under water at 50°C. The load is 700 N, the wheel pass distance is 230 mm and the frequency is 53 cycles/min. After 20 000 passes, the test is halted. The deformation curve (rut depth) during the test is recorded. The test is carried out simultaneously with two asphalt slabs.

Provided that the mixture design of the asphalt is stable, the deformation of the specimen generally correlates with the viscosity of the binder. In the case of polymer modified bitumens, the zero shear viscosity of the binder gives a good correlation with rutting behaviour. However, stripping of the binder can also lead to failure in wheel-tracking testing carried out in the presence of water.

16.2.2 *Gyratory compaction*

This tool (Fig. 16.16) is a routine test for mixture design in France and the USA and is becoming standard in the rest of Europe. The test

Fig. 16.16 Gyratory compactor

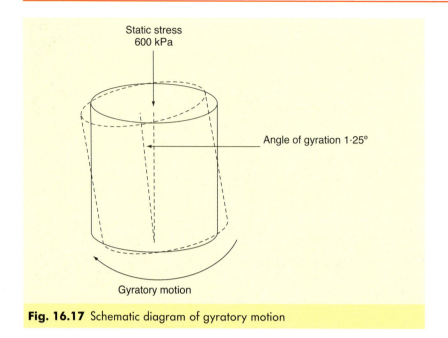

Static stress
600 kPa

Angle of gyration 1·25°

Gyratory motion

Fig. 16.17 Schematic diagram of gyratory motion

itself consists of placing a sample of hot asphalt in a mould and applying a static pressure of a controlled magnitude. The mould is then gyrated through a small angle in order to allow the aggregate particles to re-orientate themselves under the loading. A schematic diagram of the gyratory motion is shown in Fig. 16.17. This method is considered to simulate, to a reasonable degree, compaction that actually takes place in service.

The gyratory compactor also allows a measure of the compactability of a mixture to be assessed by monitoring the vertical movement of the loading ram and the number of gyrations of the mould. Hence, if one mixture takes fewer gyrations in comparison to another mixture for the same vertical movement of the loading ram then the former would be said to be more easily compacted.

Standard test conditions and requirements for the gyratory compactor are:

stress level   600 kPa
angle of gyration   1·25°
speed of gyration   30 revs/min

Additionally, slab compaction is becoming more widespread and a photograph for reference purposes is given as Fig. 16.18.

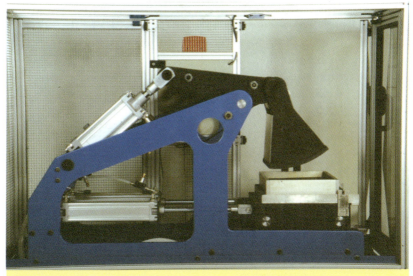

Fig. 16.18 Slab compactor

16.2.3 Durability testing

In order to develop tests to characterise durability, an appropriate definition has to be established. Scholz[406] suggested the following.

> Durability as it applies to bituminous paving mixtures is defined as
> the ability of the materials comprising the mixture to resist the
> effects of water, ageing and temperature variations, in the context
> of a given amount of traffic loading, without significant deteriora-
> tion for an extended period.

This definition, therefore, implies the need to be able to artificially age
materials in order to assess the effects on the mechanical properties. In
addition, there is also the need to be able to simulate the effect of water
damage on the specimens. Many attempts have been made to establish
methods that do one or other of these but none has been widely adopted
and Bell[407] stated that 'Compared to research on asphalt cement, there has
been little research on the ageing of asphalt mixtures, and, to date, there is

406 SCHOLZ T V. *Durability of bituminous paving mixtures, PhD thesis, University of
Nottingham, 1995.*
407 BELL C A. *Summary report on aging of asphalt-aggregate systems, SHRP-A/IR-89-004,
Strategic Highway Research Program, 1989.* National Research Council, Washington
D C.

no standard test'. With this in mind, the LINK Bitutest project[408] undertook to establish standard methods for simulating the ageing that occurs in an asphalt subsequent to compaction and for assessing the water sensitivity of asphalts. This research culminated in a series of Standards[408] being written that addressed the issue of durability. In essence, durability is dealt with by two procedures.

(1) Standard practice for long-term oven ageing of compacted asphalts. This standard practice is used to simulate the long-term ageing of compacted asphalt. Long-term ageing considers the hardening of the bitumen in the mixture subsequent to construction. The practice should result in ageing that is representative of 5 to 10 years in service for dense, continuously graded mixtures.

(2) Test method for measurement of the water sensitivity of compacted asphalts. This test method determines the water sensitivity of compacted asphalt under warm and cold climatic conditions. This method is applicable to laboratory-moulded specimens and core specimens obtained from existing roads.

The first procedure defines a method that allows a specimen to be artificially aged. This permits the properties of the specimen to be assessed before and after ageing in order to determine any changes. The second procedure is a stand-alone method in which the stiffness modulus of a specimen is measured after which it is vacuum saturated in water. The specimen is then subjected to thermal cycles of 60°C (6 hours) and 5°C (16 hours) with the stiffness modulus being measured between each complete cycle of hot and cold. This permits the calculation of a ratio that gives an indication of the water sensitivity of the asphalt. The lower the ratio, the more sensitive is the asphalt to the effects of water.

These protocols are now gaining widespread use in the UK, particularly as they form the backbone of the durability testing in the BBA/HAPAS (British Board of Agrément/Highway Authorities Product Approval Scheme) for thin surfacings[409].

408 BROWN S F, J M GIBB, J M READ, T V SCHOLZ and K E COOPER. *Design and testing of bituminous mixtures, vol 3, Experimental procedures, Report to DOT/ EPSRC LINK programme on transport infrastructure and operations, Jan 1995.* University of Nottingham.

409 BRITISH BOARD OF AGRÈMENT. *Guidelines document for the assessment and certification of thin surfacing systems for highways, SG3/00/173, SG3, 2000.* BBA/ HAPAS, Watford.

16.3 Empirical tests

The current UK empirical pavement design procedure specifies three tests for the evaluation of the mechanical properties of pavement materials. These are the Marshall test, the California bearing ratio (CBR) test (discussed in Section 18.4.3), which is used for soils and unbound granular material, and the uniaxial compression test, which is used for the determination of the crushing strength of cement bound materials[410].

The Marshall and CBR tests are carried out to failure under continuous loading. In both cases, the complex stress system set up in the material is quite unrelated to the actual pavement condition in situ under traffic loading. Neither of these two tests provides basic information on the stress–strain characteristics of the material because each has its origin in an empirical method – mixture design as far as the Marshall test is concerned, and pavement thickness in the case of the CBR test.

16.3.1 The Marshall test

The concepts of the Marshall test were developed by Bruce Marshall, formerly Bituminous Engineer with the Mississippi State Highway Department. In 1948, the US Corps of Engineers improved and added certain features to Marshall's test procedure and ultimately developed mixture design criteria[411]. Since 1948, the test has been adopted by organisations and government departments in many countries, sometimes with modifications either to the procedure or to the interpretation of the results. Asphaltic concrete mixtures for airfield runways and taxiways have been designed using the Marshall test for the last 20 years and the option to design the mortar fraction of hot rolled asphalt was first published in BS 594: 1973[412]. This was extended to an option for designing the entire mixture in BS 598-3: 1985[413].

The Marshall test entails the manufacture of cylindrical specimens 102 mm in diameter × 64 mm high using a standard compaction hammer (see Fig. 16.19) and a cylindrical mould. The specimens are tested for their resistance to deformation at 60°C at a constant rate of 50 mm/min. The jaws of the loading rig confine the majority but not

410 BRITISH STANDARDS INSTITUTION. *Testing concrete, Method for determination of compressive strength of concrete cubes, BS 1881-116: 1985.* BSI, London.

411 ASPHALT INSTITUTE. *Mix design methods for asphalt concrete and other hot-mix types, Manual series no 2 (MS-2), 1988.* AI, Kentucky.

412 BRITISH STANDARDS INSTITUTION. *Hot rolled asphalt for roads and other paved areas, BS 594: 1973.* BSI, London.

413 BRITISH STANDARDS INSTITUTION. *Sampling and examination of bituminous mixtures for roads and other paved areas, Methods for design and physical testing, BS 598-3: 1985.* BSI, London.

Fig. 16.19 Impact compactor

all of the circumference of the specimen, the top and bottom of the cylinder being unconfined. Thus, the stress distribution in the specimen during testing is extremely complex. Two properties are determined: the maximum load carried by the specimen before failure ('Marshall stability') and the amount of deformation of the specimen before failure occurred ('Marshall flow'). The ratio of stability to flow is known as the 'Marshall quotient'.

Although the Marshall test is very widely used, it is important to recognise its limitations. Research at the University of Nottingham[414] compared the mechanical properties of various mixtures, using repeated load triaxial tests, triaxial creep tests, uniaxial unconfined creep tests and Marshall tests. The result of this work, shown in Fig. 16.20, suggests that

414 BROWN S F, K E COOPER and G R POOLEY. *Mechanical properties of bituminous materials for pavement design, Proc Eurobitume Symp, Cannes, France, pp 143–147, 1981. Reprinted in a special publication of Asphalt Technology, Jan 1982.* Institute of Asphalt Technology, Dorking.

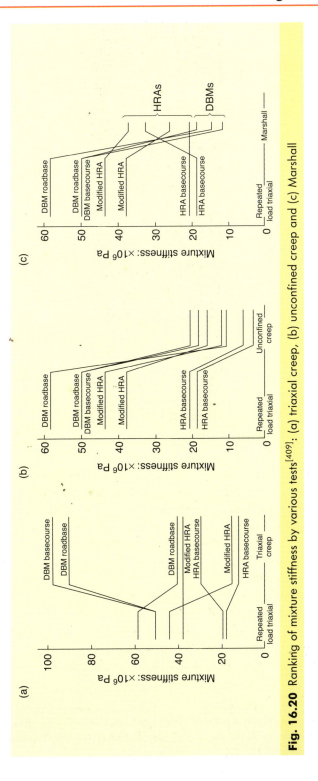

Fig. 16.20 Ranking of mixture stiffness by various tests[409]: (a) triaxial creep, (b) unconfined creep and (c) Marshall

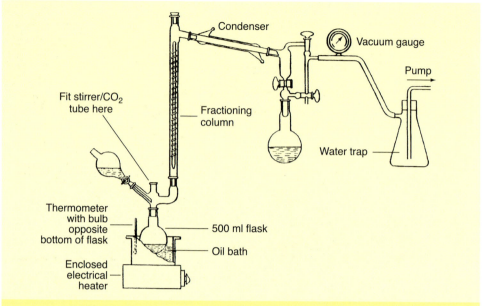

Fig. 16.21 Distillation apparatus used for recovery of bitumen[415]

the Marshall test is a poor measure of resistance to permanent deformation and does not rank mixtures in order of their deformation resistance. Repeated load triaxial tests give more realistic results.

16.4 Determination of recovered bitumen properties

In addition to routine production control tests, there are a number of procedures available to assist highway engineers in the investigation of unsatisfactory performance. One useful test is to recover the bitumen from the mixture and determine its properties, usually the penetration and softening point.

In this test, a sample of the asphalt is soaked in dichloromethane (methylene chloride) to remove the bitumen from the aggregate into solution. The bitumen/solvent solution is separated from the fine mineral matter by filtration and centrifuging. The solvent is then evaporated under controlled conditions, using the apparatus shown in Fig. 16.21, ensuring that all of the solvent is removed without taking away any of the lighter components of the bitumen. When the recovery procedure has been completed, the penetration and softening point of the recovered bitumen can be determined using the standard tests. The method is

---

415 BRITISH STANDARDS INSTITUTION. *Sampling and examination of bituminous mixtures for roads and other paved areas, Method of recovery of soluble bitumen for examination, BS 598-103: 1990.* BSI, London.

Fig. 16.22 Rotary evaporator

described in detail in BS EN 12697-4: 2000[416]. Alternatively, the solvent can be removed using a rotary evaporator (Fig. 16.22), described in BS EN 12697-3: 2000[417], which has the advantage that it removes the solvent very rapidly from the solution.

The precision for the BS EN 12697-4: 2000[416] test method is as follows.

- Repeatability
    penetration   5 dmm
    softening point   2·5°C
- Reproducibility
    penetration $0·19x^{1/2}$ dmm
    softening point   3·5°C

where x is the mean of the two results.

These precision values only apply to bitumens with a penetration up to 120 dmm although indications are that they would apply to bitumens with a higher penetration value.

416 BRITISH STANDARDS INSTITUTION. *Bituminous mixtures, Test methods for hot mix asphalt, Bitumen recovery: Fractionating column, BS EN 12697-4: 2000.* BSI, London.

417 BRITISH STANDARDS INSTITUTION. *Bituminous mixtures, Test methods for hot mix asphalt, Bitumen recovery: Rotary evaporator, BS EN 12697-3: 2000.* BSI, London.

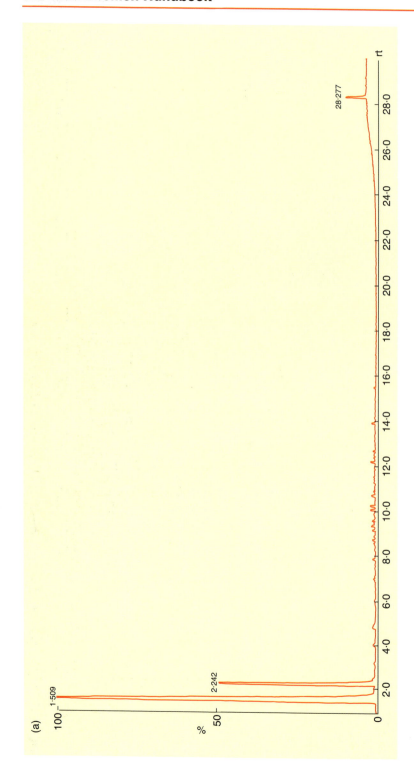

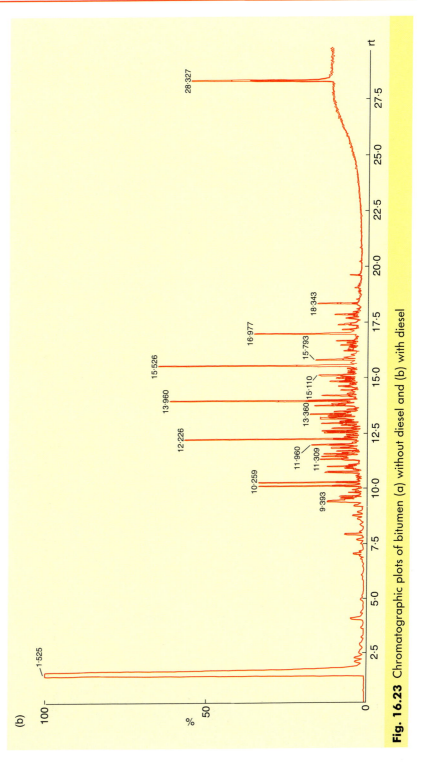

Fig. 16.23 Chromatographic plots of bitumen (a) without diesel and (b) with diesel

The test is very operator sensitive and deviation from the standard method will almost certainly result in misleading data[418]. For these reasons, the test is not suitable for contractual purposes but is a very useful tool for investigating defects in service. With sensible interpretation of test results, the bitumen recovery method can be used to determine if the bitumen in the mixture is abnormally hard or soft.

In addition to penetration and softening point determination, chromatographic examination of the recovered bitumen can be useful to detect the presence of any contaminants. Figures 16.23(a) and (b) show chromatographic plots, without and with the presence of diesel respectively, generated using a combination of gas chromatography and mass spectroscopy. The presence of the contaminant is very clear. Identification, however, is less precise because diesel can weather losing lighter fractions and the composition can change between summer and winter. Although the technique is only qualitative and not quantitative, it is a useful tool for examining recovered bitumens that are abnormally soft.

418 WADELIN F A. *Workshop on binder recovery from asphalt and coated macadam, Asphalt Technology, no 32, pp 28–42, Aug 1982.* Institute of Asphalt Technology, Dorking.

Chapter 17

Properties of asphalts

The analytical design of asphalt or flexible pavements involves consideration of two aspects of material properties. These are (i) the load–deformation or stress–strain characteristics used to analyse critical stresses and strains in the structure and (ii) the performance characteristics of the materials that show the mode, or modes, of failure. The two principal structural distress modes are cracking and permanent deformation. The former applies to bound materials only in the pavement whilst the latter is valid for all materials in the pavement, both bound and unbound.

17.1 Stiffness of asphalts

As discussed previously, asphaltic mixtures behave visco-elastically, i.e. they respond to loading in both an elastic and viscous manner. The proportion of each depends on the time of loading and on the temperature at which the load was applied. The complexity of this behaviour is increased by the heterogeneity of the mixture components, the bitumen being responsible for the visco-elastic properties whilst the mineral skeleton influences elastic and plastic properties. Mixture components and compositions can be extremely diverse, which makes prediction of the properties of a particular mixture difficult.

Asphalt stiffness can be divided into elastic stiffness that occurs under conditions of low temperatures or short loading times; and viscous stiffness that occurs at high temperatures or long loading times. The former is used to calculate critical strains in the structure in analytical design. The latter is used to assess the resistance of the material to deformation. It has also been shown that, at intermediate temperatures where the stiffness has both an elastic and viscous component, the stiffness is stress dependent[419]. High stresses result in lower stiffness and low stresses

---

419 READ J M. *Fatigue cracking of bituminous paving mixtures, PhD Thesis, University of Nottingham, 1996.*

result in higher stiffness making the assessment of performance even more difficult. However, this stress dependency is less important compared to the effects of both the time of loading and temperature.

Stiffness at a particular temperature and time of loading can be measured by a variety of methods:

- bending or vibration tests on a beam specimen (see Section 16.1.4);
- direct uniaxial or triaxial tests on cylindrical specimens (see Section 16.1.1);
- indirect tensile tests on cylindrical specimens (see Section 16.1.3).

Different types of loading can be used in the tests but for elastic stiffness relevant to moving traffic, sinusoidal or pseudo-sinusoidal repeated loading at high frequency is most appropriate[420].

17.1.1 The prediction of asphalt stiffness

Asphalt stiffness can be measured quickly and easily using tests such as the indirect tensile stiffness modulus[421] (ITSM) test. However, when testing is not feasible, such as in the design office, then the stiffness of a particular mixture at any temperature and time of loading can be estimated by empirical methods to an accuracy that is acceptable for most purposes.

In 1977, Shell produced a nomograph (shown in Fig. 17.1) for predicting the stiffness of an asphalt[422]. The data required for this nomograph are:

- the stiffness modulus of the bitumen (in Pa)
- the percentage volume of bitumen
- the percentage volume of the mineral aggregate.

The University of Nottingham has also developed a method for calculating mixture stiffness[423] and the data required are:

- the stiffness modulus of the bitumen (in Pa $\times 10^6$)
- the voids in the mineral aggregate, VMA (%).

420 RAITHBY K D and A B STERLING. *Some effects of loading history on the fatigue performance of rolled asphalt, Laboratory Report 496, 1972.* Transport and Road Research Laboratory, Crowthorne.
421 BRITISH STANDARDS INSTITUTION. *Method for the determination of the indirect tensile stiffness modulus of bituminous mixtures, DD 213: 1993.* BSI, London.
422 BONNAURE F, G GEST, A GRAVOIS and P A UGE. *A new method of predicting the stiffness of asphalt paving mixtures, Proc Assoc Asph Pav Tech, vol 46, pp 64–104, 1977.* Association of Asphalt Paving Technologists, Seattle.
423 BROWN S F and J M BRUNTON. *An introduction to the analytical design of bituminous pavements, 3rd ed, University of Nottingham, 1986.*

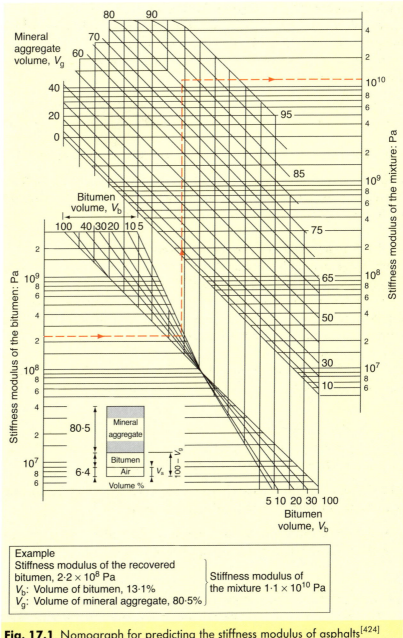

Fig. 17.1 Nomograph for predicting the stiffness modulus of asphalts[424]

Example
Stiffness modulus of the recovered
bitumen, $2 \cdot 2 \times 10^8$ Pa
V_b: Volume of bitumen, 13·1%
V_g: Volume of mineral aggregate, 80·5%
Stiffness modulus of
the mixture $1 \cdot 1 \times 10^{10}$ Pa

424 BONNAURE F, G GEST, A GRAVOIS and P A UGE. *A new method of predicting the stiffness of asphalt paving mixtures, Proc Assoc Asph Pav Tech, vol 46, pp 64–104, 1977.* Association of Asphalt Paving Technologists, Seattle.

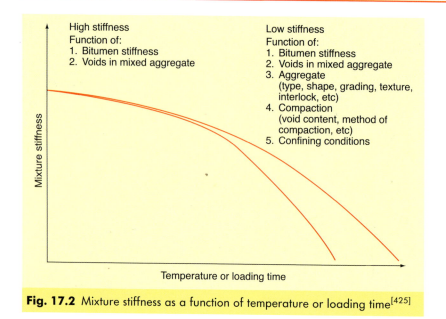

Fig. 17.2 Mixture stiffness as a function of temperature or loading time[425]

These two procedures can only be applied when the stiffness modulus of the bitumen exceeds 5 MPa, i.e. under high-stiffness conditions appropriate to moving traffic where the response is predominantly elastic and, for the Nottingham method, values of VMA between 12% and 30%. These two methods assume that the grading, type and characteristics of the aggregate affect only the elastic stiffness of the mixture since they influence the packing characteristics of the aggregate and, thus, the state of compaction of the material.

17.2 Permanent deformation of asphalts

In order to determine the permanent deformation characteristics of an asphalt, the low-stiffness response of the material, i.e. its response at high temperature or long loading times, must be analysed. When the stiffness of the bitumen is <0·5 MPa, mixture behaviour is much more complex than it is in the elastic zone. Under these conditions, the stiffness of the mixture not only depends on that of the bitumen and the volume of aggregate and bitumen, but also on a variety of other factors. These include the aggregate grading, its shape, texture and degree of interlock, and the degree of compaction. This is illustrated in Fig. 17.2.

The simplest test used to study the deformation behaviour of asphalts is the creep test (described in Section 16.1.2). Figure 17.3(a) shows the

425 BROWN S F. *Bituminous pavements: materials, design and evaluation, Residential course at the University of Nottingham, Section 1, Bituminous materials: elastic stiffness and permanent deformation, pp 1–13, Apr 1988.*

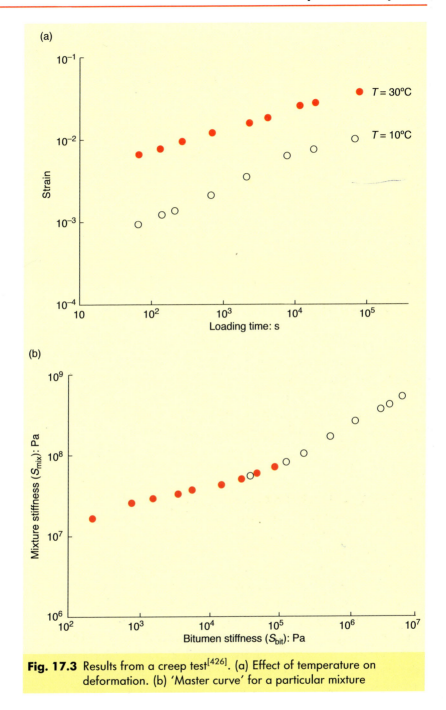

Fig. 17.3 Results from a creep test[426]. (a) Effect of temperature on deformation. (b) 'Master curve' for a particular mixture

426 HILLS J F, D BRIEN and P J VAN DE LOO. *The correlation of rutting and creep tests on asphalt mixes, J Inst Pet, Paper IP74-001, 1974.* Institute of Petroleum, London.

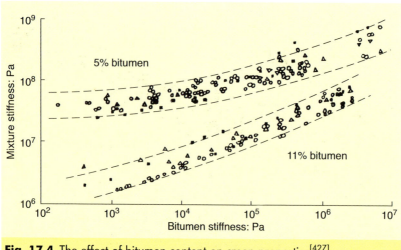

Fig. 17.4 The effect of bitumen content on creep properties[427]

deformation behaviour of a mixture tested at different temperatures. If these same results are plotted on a graph of mixture stiffness (S_{mix}) against bitumen stiffness (S_{bit}), the test results form a single continuous 'master curve' for the mixture as shown in Fig. 17.3(b). Thus, the effect of testing at different temperatures is combined in a single curve. Similarly, the effect of using different grades of bitumen or the application of different stress levels can also be combined.

Thus, the S_{mix} versus S_{bit} curve is a means of assessing the resistance to permanent deformation that is independent of arbitrarily chosen test variables for a particular asphalt. For example, Fig. 17.4 shows the deformation characteristics of two asphalt mortar mixtures having identical aggregate gradings but containing 5% and 11% bitumen respectively. It can be seen that the stiffness of the leaner mixture levels out with decreasing bitumen stiffness. In contrast, the stiffness of the richer mixture continues to decrease with decreasing bitumen stiffness. Clearly, therefore, maintaining a high value of S_{mix} when S_{bit} is decreasing is a desirable characteristic for long-term resistance to permanent deformation. Similarly, Fig. 17.5 demonstrates the effect of aggregate shape on an asphalt mortar having the same aggregate grading and bitumen content[428]. As would be expected, crushed aggregate increases the degree of aggregate interlock and this results in an increase in the value of S_{mix} and higher resistance to permanent deformation.

427 SHELL INTERNATIONAL PETROLEUM CO LTD. *The Shell pavement design manual, 1978.* Shell International Petroleum Company Ltd, London.

428 HILLS J F. *The creep of asphalt mixes, J Inst Pet, vol 59, no 570, Nov 1973.* Institute of Petroleum, London.

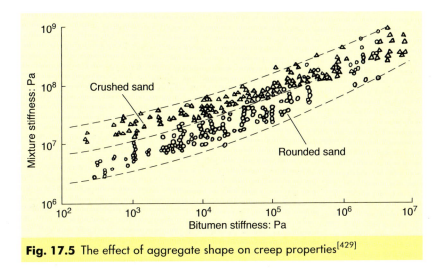

Fig. 17.5 The effect of aggregate shape on creep properties[429]

Tests

The resistance of mixtures to permanent deformation can be determined by tests such as the unconfined creep test and the repeated load uniaxial or triaxial test; their results being plotted as S_{mix} versus S_{bit}. These tests reproduce stress conditions on the road more accurately and, as a result, are gaining widespread popularity. National Standards have been developed[430] and end product performance-specifications have been trialled in the UK[431].

17.2.1 The prediction of permanent deformation

When attempting to calculate deformation on a highway, account has to be taken of a range of wheel load spectra, contact areas and pressures, lateral distribution of the wheel loads and ambient temperature gradients within the asphalt layers. Clearly, the situation is very complex. The final step in the *Shell Pavement Design Manual*[432,433] when layer thicknesses

429 HILLS J F. *The creep of asphalt mixes, J Inst Pet, vol 59, no 570, Nov 1973.* Institute of Petroleum, London.

430 BRITISH STANDARDS INSTITUTION. *Method for determining resistance to permanent deformation of bituminous mixtures subject to unconfined dynamic loading, DD 226: 1996.* BSI, London.

431 NUNN M E and SMITH T. *Evaluation of a performance specification in road construction, Project Report 55, 1994.* Transport Research Laboratory, Crowthorne.

432 SHELL INTERNATIONAL PETROLEUM CO LTD. *The Shell pavement design manual, 1978.* Shell International Petroleum Company Ltd, London.

433 CLAESSEN A I M, J M EDWARDS, P SOMMER and P C UGE. *Asphalt pavement design – the Shell method, Proc 4th Int. Conf. on the structural design of asphalt pavements, vol 1, pp 39–74, 1977.*

have been determined, is to predict the rut depth from:

$$\text{Rut depth} = C_\text{m} \times h \times \frac{\sigma_\text{ave}}{S_\text{mix}}$$

where C_m = empirical correction factor, h = thickness of the layer, σ_ave = average stress and S_mix = stiffness of the mixture.

However, as temperature gradients occur within the asphalt layers and different mixture types are often used in the surface course, binder course and base, the total asphalt thickness must be subdivided and each sub-layer considered individually. The rut depth is the sum of the deformations in each of these sublayers. An alternative method for predicting the total rut depth in a pavement is to take the results from simple repeated load uniaxial testing[434] and to use the permanent deformation program[435] (PDP) developed at the Transport Research Laboratory.

17.3 Fatigue characteristics of asphalts

Fatigue can be defined as: 'The phenomenon of fracture under repeated or fluctuating stress having a maximum value generally less than the tensile strength of the material'[436]. However, this has been generally accepted as referring to tensile strains induced by traffic loading and, since other means of generating tensile strains in a pavement exist, a better definition may be: 'Fatigue in bituminous pavements is the phenomenon of cracking. It consists of two main phases, crack initiation and crack propagation, and is caused by tensile strains generated in the pavement by not only traffic loading but also temperature variations and construction practices'[437].

Under traffic loading, the layers in a flexible pavement are subjected to flexing that is virtually continuous. The size of the strains is dependent on the overall stiffness and nature of the pavement construction, but analysis confirmed by in situ measurements has suggested that tensile strains of the order of 30×10^{-6} to 200×10^{-6} for a standard wheel load occur. Under these conditions, the possibility of fatigue cracking exists[436].

As described in Section 16.1.4, dynamic bending tests are normally used to measure the fatigue characteristics of asphalts. The constant-stress

434 BRITISH STANDARDS INSTITUTION. *Method for determining resistance to permanent deformation of bituminous mixtures subject to unconfined dynamic loading, DD 226: 1996.* BSI, London.

435 NUNN M E. *Prediction of permanent deformation in bituminous pavement layers, Research Report 26, 1986.* Transport and Road Research Laboratory, Crowthorne.

436 PELL P S. *Bituminous pavements: materials, design and evaluation, Residential course at the University of Nottingham, Section J, Bituminous materials: fatigue cracking, pp J1–J13, Apr 1988.*

437 READ J M. *Fatigue cracking of bituminous paving mixtures, PhD Thesis, University of Nottingham, 1996.*

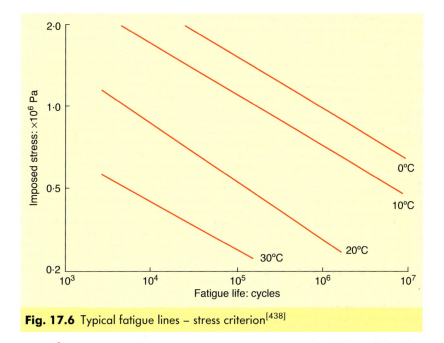

Fig. 17.6 Typical fatigue lines – stress criterion[438]

fatigue life characteristics for the same mixture at different temperatures are shown in Fig. 17.6. The lines are essentially parallel and show longer fatigue lives at lower temperatures. If the tests are carried out at a different frequency, the result is similar. Thus, fatigue lives are longer at higher frequencies. If the fatigue results are replotted in terms of constant strain, ε_t, results from different stiffnesses coincide and this is shown in Fig. 17.7. Thus, the criterion of failure is strain and temperature and time of loading affect stiffness. This effect is known as 'the strain criterion'[439].

The general relationship defining the fatigue life, in terms of initial tensile strain, is:

$$N_f = c \times \left(\frac{1}{\varepsilon_f}\right)^m$$

where N_f = number of load applications to initiate a fatigue crack, ε_f = maximum value of applied tensile strain, c and m = factors depending on

438 PELL P S. *Bituminous pavements: materials, design and evaluation, Residential course at the University of Nottingham, Section J, Bituminous materials: fatigue cracking, pp J1–J13, Apr 1988.*

439 PELL P S and I F TAYLOR. *Asphaltic road materials in fatigue, Proc Assoc Asph Pav Tech, vol 38, pp 371–422, 1969.* Association of Asphalt Paving Technologists, Seattle.

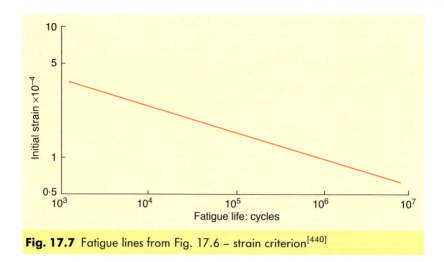

Fig. 17.7 Fatigue lines from Fig. 17.6 – strain criterion[440]

the composition and properties of the mixture; m is the slope of the strain/fatigue life line.

However, the strain criterion does not account for the differences in fatigue characteristics predicted using controlled stress and controlled strain experiments. Accordingly, researchers have tried to find alternative failure criteria that do take account of the differences in fatigue characteristics.

It has been suggested by Van Dijk et al.[441–443] that the differences in the fatigue life of an asphalt determined under conditions of stress and strain control can be explained by the dissipated energy concept. This is the amount of energy that is lost from the system, due to fatigue damage, per cycle summed for the entire life of the specimen. Van Dijk stated that, for a given mixture, the relationship between dissipated energy and the number of load repetitions to failure is valid, independent of testing method and temperature. This work has been progressed by

440 PELL P S. *Bituminous pavements: materials, design and evaluation. Residential course at the University of Nottingham, Section J, Bituminous materials: fatigue cracking, pp J1–J13, Apr 1988.*

441 VAN DIJK W. *Practical fatigue characterization of bituminous mixes, Proc Assoc Asph Pav Tech, vol 44, pp 38–74. 1975.* Association of Asphalt Paving Technologists, Seattle.

442 VAN DIJK W and W VISSER. *The energy approach to fatigue for pavement design, Proc Assoc Asph Pav Tech, vol 46, pp 1–41, 1977.* Association of Asphalt Paving Technologists, Seattle.

443 VAN DIJK W, H MOREAUD, A QUEDIVILLE and P UGE. *The fatigue of bitumen and bituminous mixes, 3rd Int Conf on structural design of asphalt pavements, vol 1, pp 354–366, London, 1972.*

Himeno et al.[444], who developed the dissipated energy concept for three-dimensional stress conditions and applied it to the failure of an asphaltic layer in a pavement, and by Rowe[445], who has shown that dissipated energy can be used to predict the life to crack initiation accurately. The dissipated energy concept shows considerable promise as a failure criterion that completely encompasses all the variables of fatigue. However, widespread support for the strain criterion still exists as research into dissipated energy is still in its infancy.

The fatigue characteristics of a mixture can be influenced significantly by its composition. Mixture stiffness is also influenced by the composition of an asphalt and distinguishing between the two is important. For example, adding filler or reducing the void content will increase mixture stiffness, resulting in increased fatigue life at a given level of stress since the resulting strain is smaller. In other words, a point has been reached lower down on the strain–fatigue life line. At a particular strain level however, it has been found that the entire line moves to the right[446], i.e. the fatigue life is improved if, for example, the volume of bitumen is increased. Basic fatigue performance is influenced by other mixture variables only insofar as these affect the volume of bitumen in the mixture. If rounded aggregate is used, for instance, a denser mixture with a lower void content will result and this will increase the relative volume of bitumen and, thus, improve the fatigue performance.

17.3.1 The prediction of fatigue performance

Traditionally, the establishment of the fatigue line for a particular mixture required involved and specialised testing equipment. A simpler procedure to predict fatigue performance with sufficient accuracy for pavement design purposes was clearly needed. A method for predicting the fatigue life of asphalts was developed by Shell[447] using the nomograph shown in Fig. 17.8. The required data are:

444 HIMENO K, T WATANABE and T MARUYAMA. *Estimation of fatigue life of asphalt pavements, 6th Int Conf on structural design of asphalt pavements, Ann Arbor, pp 272–288, 1987.*

445 ROWE G M. *Performance of asphalt mixtures in the trapezoidal fatigue test, Proc Assoc Asph Pav Tech, vol 62, pp 344–384, 1993.* Association of Asphalt Paving Technologists, Seattle.

446 PELL P S and K E COOPER. *The effect of testing and mix variables on the fatigue performance of bituminous materials, Proc Assoc Asph Pav Tech, vol 44, pp 1–37, 1975.* Association of Asphalt Paving Technologists, Seattle.

447 BONNAURE F, A GRAVOIS and J UDRON. *A new method for predicting the fatigue life of bituminous mixes, Proc Assoc Asph Pav Tech, vol 49, pp 499–529, 1980.* Association of Asphalt Paving Technologists, Seattle.

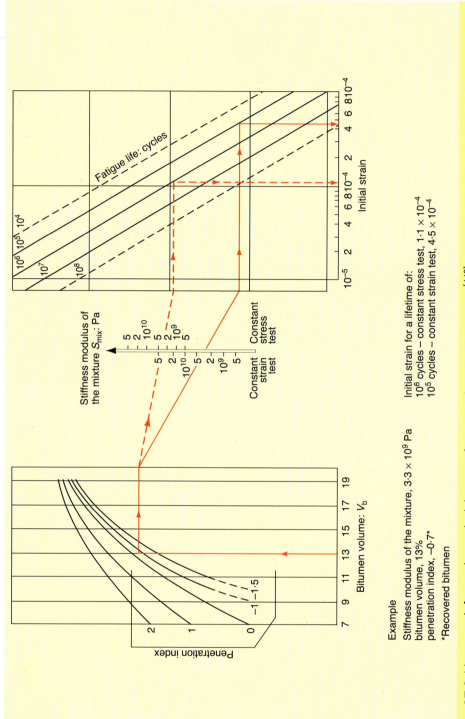

Fig. 17.8 Nomograph for predicting the laboratory fatigue performance of asphalts[448]

Example

Stiffness modulus of the mixture, $3 \cdot 3 \times 10^9$ Pa
bitumen volume, 13%
penetration index, $-0 \cdot 7^*$
*Recovered bitumen

Initial strain for a lifetime of:
10^6 cycles – constant stress test, $1 \cdot 1 \times 10^{-4}$
10^5 cycles – constant strain test, $4 \cdot 5 \times 10^{-4}$

- the percentage volume of bitumen
- the penetration index of the bitumen
- the stiffness modulus of the mixture (in Pa)
- the initial strain level.

Other prediction techniques based on some or all of the above input data[449,450] have also been developed but these empirical techniques are unable to deal with material developments such as polymer modified mixtures as they are based on historical data. This has led to the need for a quick and simple fatigue test. The indirect tensile fatigue test[451], discussed previously, has been developed to the status of a national Standard[452] at the University of Nottingham and offers an economical method of determining the life to crack initiation.

The application of laboratory-determined fatigue lives to predict actual pavement performance in practice is a complex problem that is likely to yield conservative results. This happens because simple continuous cycles of loading neglect the beneficial effects of rest periods that occur in practice between axle loads. Longer lives are also likely in practice because of the lateral distribution of wheel loads in the wheel-track and the fact that a degree of crack propagation will occur before the performance of the pavement is adversely affected. As a result of these problems, laboratory fatigue lives have to be calibrated to correlate with actual pavement performance and the calibration factor is likely to depend on environmental and loading conditions. However, these factors have been incorporated into a new prediction technique[451] for use in analytical pavement design, based around work carried out at the University of Nottingham.

448 BONNAURE F, A GRAVOIS and J UDRON. *A new method for predicting the fatigue life of bituminous mixes, Proc Assoc Asph Pav Tech, vol 49, pp 499–529, 1980.* Association of Asphalt Paving Technologists, Seattle.

449 ASPHALT INSTITUTE. *Thickness design, Asphalt pavements for highways & streets, Manual Series No 1 (MS-1), Sep 1981.* AI, Kentucky.

450 COOPER K E and P S PELL. *The effect of mix variables on the fatigue strength of bituminous materials, Laboratory Report 633, 1974.* Transport and Road Research Laboratory, Crowthorne.

451 READ J M. *Fatigue cracking of bituminous paving mixtures, PhD Thesis, University of Nottingham, 1996.*

452 BRITISH STANDARDS INSTITUTION. *Method for the determination of the fatigue characteristics of bituminous mixtures using indirect tensile fatigue, DD ABF, 1996.* BSI, London.

Chapter 18

Design of flexible pavements

The objective of pavement design is to produce an engineering structure that will distribute traffic loads efficiently whilst minimising the whole-life cost of the pavement. The term 'whole-life' when applied to a pavement, refers to all costs incurred in connection with a pavement and throughout its life. Thus, it includes the works costs (construction, maintenance and residual value) and user costs (traffic delays, accidents at roadworks, skidding accidents, fuel consumption/tyre wear and residual allowance). Pavement design is essentially a structural evaluation process, needed to ensure that traffic loads are distributed such that the stresses and strains developed at all levels in the pavement and the subgrade are within the capabilities of the materials at those levels. It involves the selection of materials for the different layers and the calculation of the required thickness. The load-carrying capacity of a pavement is a function of both the thickness of the material and its stiffness. Consequently, the mechanical properties of the materials that constitute each of the layers in a pavement are important for designing the structure. Since moisture may affect the subgrade and the sub-base (and also the base if it is unbound) and temperature affects the bitumen-bound layers, it is essential that the design process takes account of the climatic conditions.

The loading exerted by a car does not contribute to any structural deterioration in a pavement. Only commercial vehicles, which are defined as vehicles having an unladen weight $>15\,\mathrm{kN}$, cause any structural deterioration. Figure 18.1 diagrammatically illustrates the two classical modes of failure caused by trafficking of a pavement. It assumes two failure modes:

- fatigue cracking at the underside of the base
- cumulative deformation.

It is the modes illustrated in Fig. 18.1 that pavement engineers have traditionally sought to hold to acceptable limits within the design life.

351

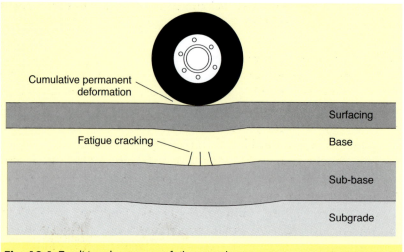

Cumulative permanent deformation

Surfacing

Fatigue cracking

Base

Sub-base

Subgrade

Fig. 18.1 Traditional pavement failure modes

The concept of a design life was regarded as being particularly important for pavements since it was thought that they do not fail suddenly but gradually deteriorate over a period of time. In fact, revolutionary research carried out in the UK and published as TRL Report 250[453] towards the end of the last century has proven that this approach was not borne out by fact in thicker pavements.

The TRL research project examined a substantial number of pavements that had been in service for several decades. It was from one of the many important findings of this project that the concept of 'long-life' pavements emerged. TRL 250 describes its main conclusions as follows[454].

> A well-constructed, flexible pavement, built above a defined threshold strength, will have a very long structural life provided that distress, in the form of cracks and ruts appearing at the surface, is detected and remedied before it begins to affect the structural integrity of the road. There is no evidence that structural deterioration due to fatigue or cracking of the asphalt base, or deformation originating deep within the pavement structure exists in roads that conform with the criteria for long-life.

In practical terms, this translates into a pavement that is designed to cope with a loading of 80 msa (million standard axles) or more having at least 300 mm thickness of asphalt.

453 NUNN M E, A BROWN, D WESTON and J C NICHOLLS. *Design of long-life flexible pavements for heavy traffic, Report 250, 1997.* TRL, Crowthorne.
454 NUNN M E, A BROWN, D WESTON and J C NICHOLLS. *Design of long-life flexible pavements for heavy traffic, Report 250, p 4, 1997.* TRL, Crowthorne.

This chapter considers the development and current status of empirical and analytical methods for designing pavements.

18.1 The importance of stiffness

Stiffness may be defined as a measure of the load-spreading ability of a material and applies whether it is granular, asphaltic or cementitious. It is a fundamental and important parameter that must be fully understood by the Pavement Engineer. It can be readily understood by considering the situations depicted in Figs. 18.2(a) and (b).

In Fig. 18.2(a), assume that Material 1 has a lower stiffness value than Material 2. If both are laid at the same thickness then the pressure resulting from a particular load will be higher at the underside of the layer of Material 1. This, in turn, subjects the layer underneath to a higher value of stress. If this value exceeds that which can be tolerated by the underlying layer then it will fail. In order to get the same pressure on the underside it will be necessary to lay a greater thickness of Material 2 which has a lower stiffness. Compare that situation with that shown in Fig. 18.2(b), where a thicker layer of the material having a lower stiffness is required to produce an acceptable value of resultant pressure at the underside.

This explains why different materials require different thicknesses for a particular design. Another important facet of material behaviour is explained by considering the situation where the underlying layer is granular and is moisture susceptible. If the area becomes saturated, then the stiffness of the underlying layer may be reduced. The outcome may be that the reduced value of stiffness is such that the imposed stresses normally taken by the dry material cause the layer to fail. This problem will not occur if drainage is installed such that groundwater never reaches the pavement layers.

18.2 The structural elements of a flexible pavement

It is convenient to consider a pavement as consisting of three components:

- the foundation, made up of the capping (if present) and the sub-base
- the base
- the surfacing, made up of the binder course and the surface course.

These layers are depicted in Fig. 18.3.

18.2.1 The foundation

The combination of capping and sub-base is described as the 'foundation'. The foundation is the platform upon which the more expensive layers are placed. The function of the foundation is to transmit the stresses generated by the traffic loading to the subgrade without causing any

353

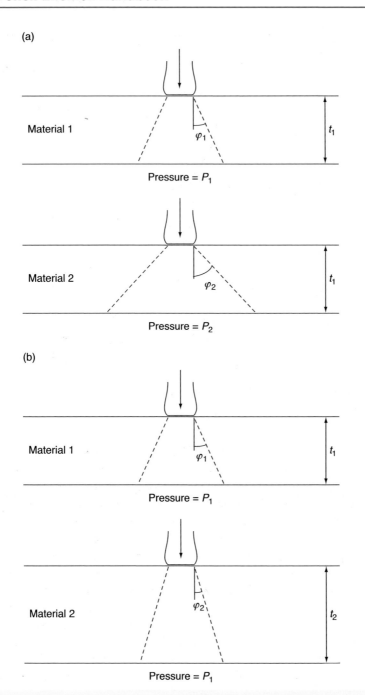

Fig. 18.2 Stiffness of pavement layers. (a) Equal thicknesses, different material stiffnesses, different resultant pressures. (b) Different thicknesses, different material stiffnesses, equal resultant pressures.

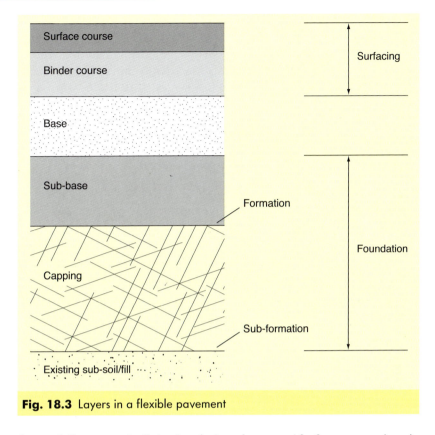

Fig. 18.3 Layers in a flexible pavement

form of distress to it. It is also designed to provide frost protection, in those countries that require it, so that frost cannot penetrate the subgrade causing catastrophic failure. This is necessary not only during its service life but also during the construction phase. Although the frequency of stress application is relatively lower during construction, the values of stress can often be higher than those generated in service.

A capping will not be required if the subgrade has adequate stiffness and shear strength to carry the anticipated construction traffic. Indeed, if the subgrade has sufficient stiffness and shear strength then a sub-base may not be required. Most modern pavements include a capping and a sub-base when constructed. The underside of the sub-base is called the 'formation' and the underside of the capping is called the 'sub-formation'. Further information on foundations including the design of capping and sub-base thicknesses and types can be found in the UK's national design standard[455].

455 HIGHWAYS AGENCY *et al. Design manual for roads and bridges, Pavement design and maintenance, Foundations, vol 7, HD 25/94, 1994.* The Stationery Office, London.

18.2.2 The base

The base is the main structural element in the pavement. Its function is to distribute the imposed loading so that the underlying materials are not overstressed. It must resist permanent deformation and cracking caused by fatigue through repeated loading. It must also be capable of withstanding stresses induced by temperature gradients through the structure. The base is usually a dense asphalt material but wet lean concrete or cement-bound material may also be used. For lightly trafficked applications, an unbound granular base may be perfectly adequate.

18.2.3 The surfacing

The surfacing consists of two layers. The lower layer is termed the 'binder course'. Its primary function is to distribute the stresses from the surface course to the base without overstressing. It also provides a layer with good surface regularity upon which the surface course can be constructed. The tolerance at the top of the surface course is often ±6 mm from the design levels.

The upper layer of the surfacing is termed the surface course. It is this layer and this layer only that is visible to the road user. The surface course has to meet a formidable list of requirements. It must:

- resist deformation by traffic
- be durable, resisting the effects of weather, abrasion by traffic and fatigue
- provide a skid-resistant surface
- generate acceptable levels of noise
- provide a surface of acceptable riding quality
- except in the case of porous asphalt, be impervious thus protecting the lower layers of the pavement
- contribute to the strength of the pavement structure
- generate acceptable levels of spray.

18.3 Factors involved in pavement design
18.3.1 Pavement loading

Loads are applied to the pavement through contact between vehicle tyres and the surface course. Obviously the degree and nature of the traffic loading is one of the major factors affecting the design and performance of pavements. Both the magnitude and the number of loadings contribute to the overall damage in the pavement. The tyre pressure primarily affects the stresses (and strains) developed at the surface and within the upper layers of the pavement. (In road pavement design, a typical commercial vehicle tyre pressure would be 0·5 MPa, whereas aircraft tyres inflict pressures up to 3·0 MPa. These values have a significant influence on the design of the materials used as surfacings of road and airfield pavements.) The structural effects of multiple-wheel configurations are also primarily confined to the upper layers of the pavement.

Although a great deal of publicity has been given to the physical size and gross weight of commercial vehicles, it is the double wheel or axle load that is critical in pavement design and performance. Heavier vehicles are usually carried on a larger number of axles, thereby maintaining or reducing the axle load.

Most countries limit the axle loadings that can be carried on their carriageways. However, these limits are difficult to enforce and overloaded vehicles are commonplace.

Normal traffic on conventional roads is mixed in composition and magnitude of axle loads and it is therefore necessary for design purposes to simplify the real situation by converting actual axle loads to an 'equivalent' loading system. This idea originated with the American Association of State Highway Officials (AASHO) road test in the USA in the early 1960s[456] and was adopted in the UK with the publication of the third edition of Road Note 29 in 1970[457]. The mixed axle load spectrum is converted to an equivalent number of standard axles, the standard adopted in the UK being 80 kN. This is based on a concept of equivalent damage to the pavement. (The concept of a 'standard axle' was introduced in Laboratory Report 833[458] as that exerting or applying a force of 80 kN.) The loading exerted (and thus the damage) for each vehicle can then be computed in terms of a number of equivalent standard axles by using the 'fourth power law' given below. Although the relative damage inflicted by wheel loads of different magnitudes depends on a variety of factors, for design calculations, the 'fourth power law' is normally used, viz:

$$\text{Equivalence factor} = \left(\frac{W}{80}\right)^4$$

where W = axle load in kN.

456 LIDDLE W J. *Application of AASHO road test results to the design of flexible pavement structures, Proc 1st Int Conf on the structural design of asphalt pavements, Ann Arbor, Michigan, pp 42–51, 1962.*

457 DEPARTMENT OF THE ENVIRONMENT. *A guide to the structural design of pavements for new roads, Road Note 29, 3rd ed, 1970.* The Stationery Office, London.

458 KENNEDY C K and N W LISTER. *Prediction of pavement performance and the design of overlays, Laboratory Report 833, 1978.* Transport and Road Research Laboratory, Crowthorne.

It has been shown[459] that on weak pavements the exponential power can be as high as 6 if trafficked by heavy axle loads at or above the current legal limit. However, the value of 4 is accepted in the UK's national design standard[460] and has proven to lead to the construction of highways that have successfully withstood the enormous growth in traffic in the UK.

Thus, under normal conditions, it takes the application of sixteen 40 kN axle loads to cause the same damage as a single application of an 80 kN axle load. Furthermore, one application of a 160 kN axle load induces 16 times the damage of a standard 80 kN axle load.

18.3.2 Environmental effects

The two environmental effects that influence pavement design are moisture, in particular the water table which influences subgrade performance, and temperature, which strongly affects the properties of asphalts.

The properties of subgrade soils should be determined for the existence of critical conditions after the construction of the pavement. The relationship between elastic modulus and moisture content of the subgrade is important. In addition, frost can significantly affect soils and unbound pavement materials. By choosing materials that have a low susceptibility to the effects of frost and ensuring that any which might be so affected are located below the depth to which frost is likely to penetrate, frost damage can be minimised. A more effective approach is to construct the pavement such that water is kept out of the pavement by the judicious construction of effective drainage. The effect of temperature on the stiffness of asphalt materials has been discussed in depth in previous chapters.

18.4 Empirical and semi-empirical pavement design

There are two approaches to pavement design, empirical and analytical. Most design methods in current practice around the world are empirical, being based on experience accumulated in practice and from specially constructed test sections. This approach to pavement design is somewhat removed from engineering principles and cannot cope with circumstances beyond those included in the trial section upon which the method is based. In contrast, the analytical approach uses theoretical analysis of the mechanical properties of materials and is capable, in principle, of dealing with any design situation.

---

459 CURRER E W H and M G D CONNOR. *Commercial traffic: its estimated damaging effect, 1945–2005, Laboratory Report 910, 1979.* Transport and Road Research Laboratory, Crowthorne.

460 HIGHWAYS AGENCY *et al. Design manual for roads and bridges, Pavement design and maintenance, Traffic assessment, vol 7, HD 24/96, 1996.* The Stationery Office, London.

18.4.1 Development of empirical and semi-empirical pavement design methods in the UK

An early UK attempt to provide a basis for designing pavements by analysing layer stresses was published in 1948[461]. However, formal advice on the structural design of pavements for use on Trunk Roads first appeared in Road Note 29, published in three editions in 1960, 1965 and 1970[462]. Its approach was based on measurements that were taken on a number of experimental highways constructed in the public highway network. Design curves related traffic loading to the strength of the subgrade. As traffic levels grew, the approach propagated by Road Note 29 was found to be flawed. This led to work by the Transport and Road Research Laboratory that enhanced the empirical approach adopted previously with theoretical analysis. This culminated in the publication in 1984 of Laboratory Report 1132[463].

Advice on all aspects of the preparation and design of road schemes for Trunk Roads has been provided to agent authorities since 1965. This took the form of technical memoranda, advice notes and standards that were issued as and when the Department of Transport deemed necessary. The problem was that the system was uncontrolled and not all offices or individual engineers had copies of or access to all documents that had been published. Thus, contracts were prepared to differing procedures. In order to overcome this major shortcoming, a composite system, called the *Design Manual for Roads and Bridges* (DMRB), was issued in 1992. It was achieved very simply by combining all the technical memoranda, etc. together in one suite of documents under a unified system of control.

The DMRB is divided into 15 volumes each of which is accorded a different title. Volume 7 is entitled *Pavement Design and Maintenance* and is the document that is of greatest interest to Pavement Engineers. Those elements of Volume 7 that dictate the design of a pavement are:

- HD 24 Traffic assessment[464]

---

461 FOX L. *Computation of traffic stresses in a simple road structure, Dept of Scientific and Industrial Research, Road Research Technical Paper no 9, 1948.* The Stationery Office, London.

462 DEPARTMENT OF THE ENVIRONMENT. *A guide to the structural design of pavements for new roads, Road Note 29, 3rd ed, 1970.* The Stationery Office, London.

463 POWELL W D, J F POTTER, H C MAYHEW and M E NUNN. *The structural design of bituminous roads, Laboratory Report 1132, 1984.* Transport and Road Research Laboratory, Crowthorne.

464 HIGHWAYS AGENCY *et al. Design manual for roads and bridges, Pavement design and maintenance, Traffic assessment, vol 7, HD 24/96, 1996.* The Stationery Office, London.

- HD 25 Foundations[465]
- HD 26 Pavement design[466].

The design method contained in Volume 7 is the most common approach used in the UK for the design of pavements.

18.4.2 Assessment of the load on a pavement

The size of any structural element is, amongst other factors, a function of the loading that is applied to that structural element. When it comes to the empirical design of pavements, the total traffic loading over the anticipated life of the pavement is assessed and expressed in terms of millions of standard axles (msa). The pavement necessary to support that loading is designed as a unit, that unit being the base plus binder course plus surface course. Thus, the pavement is designed as if it was a single layer. Since the thicknesses of the binder course and the surface course are generally known, the thickness of the base can be determined by deduction.

HD 24 deals with assessing the loading on the structure to be applied to the pavement throughout its useful life, i.e. the cumulative commercial vehicle traffic (the 'design traffic'). It confirms that lighter vehicles, i.e. cars, contribute very little to the loading of a pavement. Thus, the assessment is restricted to a calculation of the number of commercial vehicles that are defined as vehicles having an unladen vehicle weight that exceeds 15 kN (kN is, of course, a unit of force, 15 kN is equivalent to an unladen weight of 1·5 tonnes). There are three classifications of commercial vehicles – PSV (public service vehicle), OGV1 (other goods vehicle) and OGV2. These are defined as:

- PSV   Buses and coaches
- OGV1   2 axle rigid
          3 axle rigid
          3 axle articulated
- OGV2   4 axle rigid
          4 axle articulated
          5 axle or more

Assessment of the design traffic is achieved by one of two methods:

(1) the 'standard method' in the case of new roads although it can be used, in certain circumstances, for maintenance schemes or realignment; or

---

465 HIGHWAYS AGENCY *et al. Design manual for roads and bridges, Pavement design and maintenance, Foundations, vol 7, HD 25/94, 1994.* The Stationery Office, London.
466 HIGHWAYS AGENCY *et al. Design manual for roads and bridges, Pavement design and maintenance, Pavement design, vol 7, HD 26/01, 2001.* The Stationery Office, London.

(2) the 'structural assessment and maintenance method' for maintenance schemes although it may be appropriate for certain new roads where non-standard traffic exists.

The standard method determines the design traffic by means of traffic studies for new roads. For maintenance schemes or realignment, a 12, 16 or 24 hour traffic count is carried out. The results are converted to an average annual daily flow (AADF). Using this value, the design traffic can then be read from one of four charts in HD 24[467] and these are used to obtain design traffic for:

- flexible (20 year life) and flexible composite – single carriageway
- flexible (20 year life) and flexible composite – dual carriageway
- rigid, rigid composite and flexible (40 year life) – single carriageway
- rigid, rigid composite and flexible (40 year life) – dual carriageway.

The thinner lines on the graphs are used for higher percentages of OGV2. Note that a minimum value of design traffic of 1 msa on lightly trafficked roads is to be assumed. These figures include a bias to reflect the increased proportion of traffic in lane 1 (lanes should be described as lane 1, lane 2, etc. and not nearside lane, middle lane, etc).

The structural assessment and maintenance method caters for both the assessment of past traffic and the prediction of future design traffic. It allows a more detailed assessment than the standard method for new roads.

A simple formula is then applied to each category of traffic. The total design traffic is the sum of the cumulative traffic in each category.

In the case of dual carriageways, the proportion of vehicles in the most heavily trafficked lane (usually, but not always, lane 1) is used in design. This value of design traffic is to be applied to all lanes including the hard shoulder. In other words, the pavement layer thicknesses are to be the same in all lanes.

It is sometimes necessary to estimate the traffic in other lanes for maintenance purposes and a method of distributing traffic between lanes is described.

It should be noted that due to the application of a cut-off line to the charts in the standard method, the two methods will not always produce comparative results.

---

467 HIGHWAYS AGENCY *et al. Design manual for roads and bridges, Pavement design and maintenance, Traffic assessment, vol 7, HD 24/96, figs 2.1–2.4, 1996.* The Stationery Office, London.

18.4.3 *Designing the foundation*

HD 25[468] is concerned with the design of the foundation. As explained above, the foundation consists of the capping and the sub-base.

The properties of the subgrade, in particular the stiffness modulus and shear strength, are seldom known at all or with any certainty. Thus, it is necessary to use other simple test methods that, with experience, can be used to design a foundation. Such tests are often called 'index tests'. The California bearing ratio (CBR) test is an example. The CBR value is usually quoted as an integer percentage (although some documents suggest that it should be quoted to two figures of decimals, which is wholly impractical in most cases). One method of obtaining a CBR value is to drive a piston, 1932 mm^2 in area, into the material at a specified rate. The load acting on the piston is recorded and that corresponding to a penetration of 2·5 mm is determined. The ratio of that load to 1360 kg (the value obtained with a standard crushed stone sample) expressed as a percentage is designated as the CBR value of the material under test. The total thickness of the pavement is related to the CBR of the subgrade for particular wheel loads by curves of the type shown in Fig. 18.4.

HD 25[468] suggests that assessment of the true value of CBR on site or in the laboratory is the preferred approach. If that is not possible then it provides two very simple tables that allow a value to be read off if the soil type is known. Once the CBR has been assessed, the thickness of the sub-base and, if necessary, the capping can be determined from a single chart[469]. The CBR of a weak soil would be in single figures. If the subgrade has a CBR of 15% or more then no capping is necessary. If the value is less than 15% then it will be necessary to add a capping or to thicken the sub-base. (The sub-base is normally 150 mm thick.) In order to ensure that the pavement can be constructed with confidence, the value of CBR at the top of the sub-base should be ≥30%. Where cracking of the sub-base occurs, the material (and perhaps the material in the layer below) should be removed and replaced with uncontaminated sub-base or some other measure taken to ensure the integrity of the sub-base.

The casual observer may conclude that the condition of the subgrade and the thickness of the foundation need be no more than adequate since they will have little effect on the life of a pavement. Such a conclusion would be singularly incorrect. Research published late in the 1990s

---

468 HIGHWAYS AGENCY *et al. Design manual for roads and bridges, Pavement design and maintenance, Foundations, vol 7, HD 25/94, 1994.* The Stationery Office, London.

469 HIGHWAYS AGENCY *et al. Design manual for roads and bridges, Pavement design and maintenance, Foundations, vol 7, HD 25/94, fig 3.1, 1994.* The Stationery Office, London.

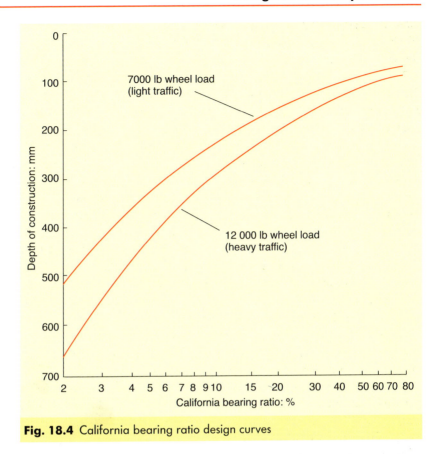

Fig. 18.4 California bearing ratio design curves

Table 18.1 Comparison of rates of rutting[470]

Base type	Rate of rutting (mm/msa)			Life for 10 mm rut: msa
	Mean	Standard deviation	Sample size	
HRA or DBM (subgrade CBR <5%)	0·58	0·18	20	17
HRA or DBM (subgrade CBR >5%)	0·36	0·12	21	28

demonstrates the importance of constructing a firm foundation by comparing the traffic taken on roads where the CBR was <5% and that where the CBR was >5% before a rut of depth of 10 mm was generated (see Table 18.1). Accordingly, the usually relatively modest costs of ensuring that a foundation is adequate is money that is likely to enhance the useful life of the pavement.

470 NUNN M E, A BROWN, D WESTON and J C NICHOLLS. *Design of long-life flexible pavements for heavy traffic, table A1, Report 250, 1997.* TRL, Crowthorne.

18.4.4 *Designing the pavement*

HD 26[471] sets out the UK Highways Agency's method for designing pavements. It is based on the approach that is contained in LR 1132[472].

LR 1132 contains design recommendations, derived from the structural performance of experimental roads, for pavements with different types of binder course. The extrapolation from about 20 msa carried by full-scale road trials to 200 msa has been based on analysis. The basic designs provide reference structures, having a 0·85 probability of survival, which can be modified, using the analytical approach, to cater for new materials or alternative designs. Thus, the LR 1132 approach to pavement design recognises the variability in the road structure and the uncertainties of traffic forecasts. The concept of probability of survival certainly offers a challenge to the engineer who is aiming to achieve a minimum whole-life cost.

In LR 1132, the estimation of the cumulative number of equivalent standard 80 kN axles to be carried during the design life is similar to the modified methods used as part of Road Note 29 but allowances are made for further increases in vehicle damage factors.

The 'design traffic', which is the only output from HD 24, is used to calculate the total thickness of the layers above the sub-base in four different generic types of pavement:

- *flexible* – surfacing and base are bound with a bitumen
- *flexible composite* – surfacing and upper base (if used) are bound with a bitumen whereas the lower base is bound with cement
- *rigid* – surfacing and base are concrete which may be jointed unreinforced (URC) or jointed reinforced (JRC)
- *rigid composite* – Continuously reinforced (CRCP) or continuously reinforced concrete base (CRCR) with asphaltic surface course/surfacing.

Figure 2.1 of HD 26/01 is reproduced here as Fig. 18.5. The parameter 'design traffic' enables the 'design thickness of asphalt layers' to be ascertained. This design thickness is the total thickness of the pavement, i.e. the sum of the thicknesses of the base, binder course and surface course.

This is the first attempt to include an element of an end-product specification into pavement construction. It includes asphalt 'grades'.

471 HIGHWAYS AGENCY *et al. Design manual for roads and bridges, Pavement design and maintenance, Pavement design, vol 7, HD 26/01, 2001.* The Stationery Office, London.

472 POWELL W D, J F POTTER, H C MAYHEW and M E NUNN. *The structural design of bituminous roads, Laboratory Report 1132, 1984.* Transport and Road Research Laboratory, Crowthorne.

473 HIGHWAYS AGENCY *et al. Design manual for roads and bridges, Pavement design and maintenance, Pavement design, vol 7, HD 26/01, fig 2.1, 2001.* The Stationery Office, London.

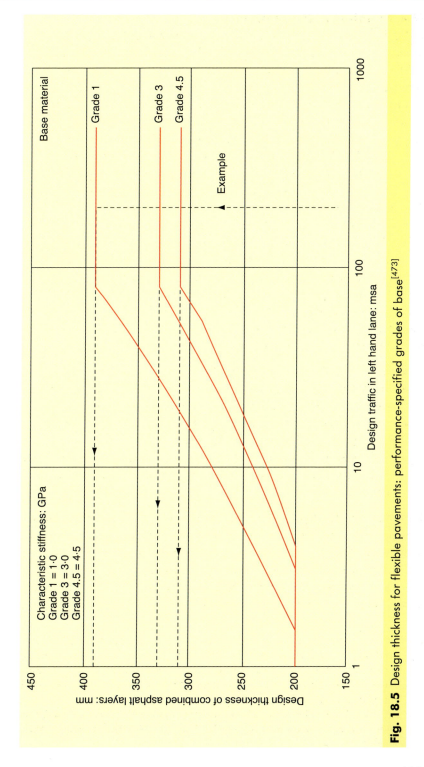

Fig. 18.5 Design thickness for flexible pavements: performance-specified grades of base[473]

These grades indicate the 'characteristic stiffness' of the mixture which is a key determinant of the thickness of the asphalt which is required to support the calculated loading.

Attached to Fig. 2.1 in HD 26 is a series of Notes. These contain important matters affecting the design process and it is vital that users apply the points therein. Advice on the materials that can be used in the pavement layers is given in HD 36[474], HD 37[475] and HD 38[476], all of which can be found in Volume 7 of the DMRB.

18.5 Analytical pavement design using the Shell Pavement Design Method

The philosophy of the analytical approach to pavement design is that the structure should be treated in the same way as other civil engineering structures. The basic procedure is as follows:

(1) assume a form for the structure; this will usually involve rational simplification of the actual structure to facilitate analysis
(2) specify the loading
(3) estimate the size of the components
(4) carry out a structural analysis to determine the stresses, strains and deflections at critical points in the structure
(5) compare these values with the maximum allowable values to assess whether the design is satisfactory
(6) adjust the materials or geometry repeating steps (3), (4) and (5) until a satisfactory design is achieved
(7) consider the economic feasibility of the result.

In the late 1950s, there was great interest in analytically-based design procedures. Accordingly, in 1963 and subsequently, Shell published sets of design charts[477–479] that permitted engineers to calculate stresses

474 HIGHWAYS AGENCY *et al. Design manual for roads and bridges, Pavement design and maintenance, Surfacing materials for new and maintenance construction, vol 7, HD 36/99, 1999.* The Stationery Office, London.

475 HIGHWAYS AGENCY *et al. Design manual for roads and bridges, Pavement design and maintenance, Bituminous surfacing materials and techniques, vol 7, HD 37/99, 1999.* The Stationery Office, London.

476 HIGHWAYS AGENCY *et al. Design manual for roads and bridges, Pavement design and maintenance, Concrete surfacing and materials, vol 7, HD 38/97, 1997.* The Stationery Office, London.

477 SHELL INTERNATIONAL PETROLEUM COMPANY LTD. *Shell design charts for flexible pavements, 1963.* Shell International Petroleum Company Ltd, London.

478 SHELL INTERNATIONAL PETROLEUM COMPANY LTD. *Shell pavement design manual, 1978.* Shell International Petroleum Company Ltd, London.

479 SHELL INTERNATIONAL PETROLEUM COMPANY LTD. *Addendum to the Shell pavement design manual 1985.* Shell International Petroleum Company Ltd, London.

and strains in a structure based upon a multi-layer linear elastic analysis. In 1978, this system was extended to incorporate all relevant major design parameters that were derived from empirical design methods, the AASHO Road Test and laboratory data and published as the *Shell Pavement Design Manual* (SPDM). SPDM allowed for the effects on the pavement of temperature, traffic density and the physical properties of the bitumen and aggregates whilst standardising the asphalt mixtures in regard to stiffness and fatigue properties. It was presented in the form of graphs, charts and tables. In order to keep the quantity of graphs, etc. down to a manageable level, the number of variable parameters used was limited and engineers were expected to interpolate where necessary.

In keeping with the substantial increase in computing power that is now available to every individual within an organisation, the manual procedures were updated to provide user-friendly computer-based packages that are integrated with other computer-based applications. Accordingly, the *Shell Pavement Design Manual* became the Shell Pavement Design Method to recognise the fact that it is no longer a paper-based system.

In 1992, SPDM-PC (a version of SPDM designed for use on a desktop or laptop) was published. The package followed the same design method as the 1978 version but was enhanced to permit the use of a wide variety of data without the need for cumbersome interpolations.

In 1998, Shell launched SPDM 3.0 which operates within Windows©. It is described in a paper published subsequently[480]. Again, this program followed the same design method as the 1978 manual but with an improved graphical user interface.

SPDM is based on a three-layered structure consisting of an asphalt pavement on a base of unbound granular material overlying the existing subgrade, as depicted in Fig. 18.6.

For convenience, the traffic loads imposed on the structure are converted to an equivalent number of standard 80 kN axle loads. The life of the structure is expressed as the number of standard axle loads applied to the structure before a given level of damage occurs. In this context, the design method distinguishes between structural damage, i.e. fatigue failure of the asphalt, and permanent deformation resulting from deformation of the subgrade and permanent (plastic) deformation of the asphaltic layers. These mechanisms are illustrated in Figs. 18.7(a) and (b).

The principal criterion for asphalt fatigue is the value of horizontal strain at the bottom of the asphaltic layer; for subgrade deformation it is the vertical strain at the top of the subgrade. Cracking of the asphaltic

---

480 STRICKLAND D. *Shell pavement design software for Windows, 2000.* Shell International Petroleum Company Ltd, London.

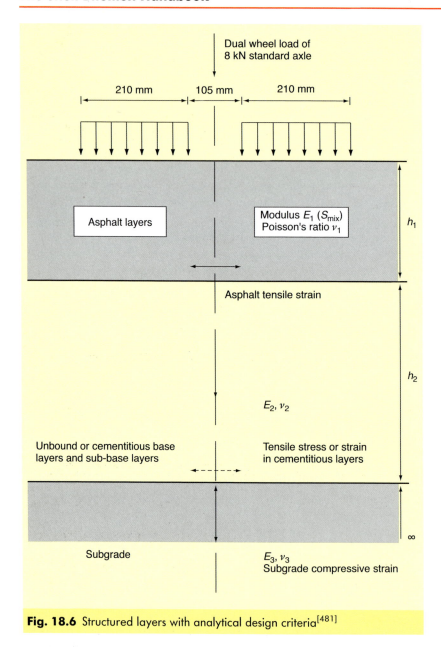

Dual wheel load of
8 kN standard axle

210 mm   105 mm   210 mm

Asphalt layers

Modulus E_1 (S_{mix})
Poisson's ratio ν_1

h_1

Asphalt tensile strain

h_2

E_2, ν_2

Unbound or cementitious base
layers and sub-base layers

Tensile stress or strain
in cementitious layers

Subgrade

∞

E_3, ν_3
Subgrade compressive strain

Fig. 18.6 Structured layers with analytical design criteria[481]

layer arises from repeated tensile strains, the maximum value of which
occurs at the bottom of the base, as shown in Fig. 18.7(b). In thinner
pavements, the crack, once initiated, propagates upwards causing gradual
weakening of the structure. Development of a rut arises from the

481 SHELL INTERNATIONAL PETROLEUM COMPANY LTD. *Shell pavement design manual, 1978.* Shell International Petroleum Company Ltd, London.

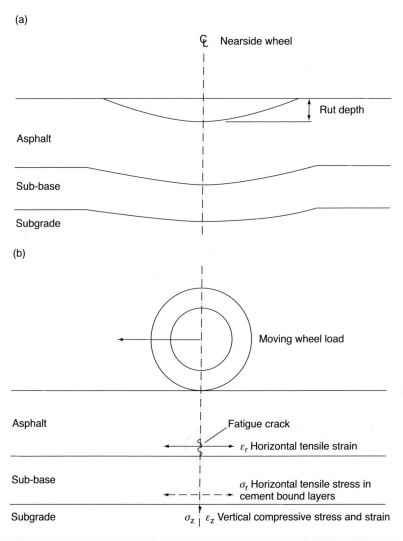

Fig. 18.7 (a) Permanent deformation and (b) fatigue cracking in asphalt pavements

accumulation of permanent strain throughout the structure. The calculation of rut depth using analytical procedures is a complicated process. However, if the vertical strain in the subgrade is kept below a specific level, experience has shown that excessive rutting will not occur, unless the material is poorly designed or inadequately compacted.

Accordingly, the design problem is to proportion the pavement structure so that the critical level of strain will not be exceeded in the design life for the chosen material. To achieve this, the mechanical properties of the materials must be known to analyse the structure. The simplified three-layered structure used as part of SPDM is regarded as a linear elastic multi-layered system in which the materials are characterised by Young's modulus of elasticity (stiffness modulus) and Poisson's ratio. The materials are assumed to be homogeneous and isotropic. It was on the basis of this model that, in the late 1960s, Shell Global Solutions developed BISTRO (BItumen STructures in ROads), a computer program that calculates stresses, strains and displacements at any position in a multi-layer system. This program has been enhanced and renamed BISAR[482,483] (BItumen Stress Analysis in Roads). In 1998, Shell launched BISAR 3.0 which works within Windows©. It is described in a paper published subsequently[484]. This program followed the same principles as previous versions of BISAR but again has an improved graphical user interface.

Using BISAR, the horizontal tensile strain at the bottom of an asphalt layer and the vertical compressive strain at the surface of a subgrade can be determined for various pavement configurations and these two values compared with permissible values.

Considerable effort has been made to reduce the number of variables involved in the analysis of pavements in order to keep the process as simple as possible. Therefore, a number of assumptions have been made. The first assumption is based on results from materials testing which demonstrated that Poisson's ratio does not vary greatly and a value of 0·35 can be assumed for the asphalt layer, sub-base and subgrade. The second assumption is that the thickness of the subgrade is infinite. These assumptions leave only five variables for any particular system. These are:

- the elastic modulus of the asphalt layer
- the thickness of the asphalt layer
- the elastic modulus of the granular sub-base layer
- the thickness of the granular sub-base layer
- the elastic modulus of the subgrade.

Therefore, as these five variables are linked by analytical relationships, if any four are known, or assumed, the fifth can be calculated with relative ease.

482 PEUTZ M G F, H P M VAN KEMPEN and A JONES. *Layered systems under normal surface loads, Highway Research Record no 228, pp 34-45, 1968.*

483 DE LONG D L, M G F PEUTZ and A R KORSWAGEN. *Computer program BISAR, Layered systems under normal and tangential surface loads, Koninklijke/Shell Laboratorium Amsterdam, external report AMSR.0006.73, 1973.*

484 STRICKLAND D. *Shell pavement design software for Windows, 2000.* Shell Bitumen, Wythenshawe, UK.

Chapter 19

Surface dressing and other specialist treatments

19.1 Surface dressing

Surface dressing is a very useful and cost effective process for restoring skid resistance to a road surface that is structurally sound. In a new road, adequate surface texture is designed into the running surface by specifying requirements for both aggregate properties and texture depth. However, during its service life, the surface becomes polished under the action of traffic and the skid resistance will eventually fall below the minimum specified value. A major benefit of surface dressing is that the process restores skid resistance to a surface that has become smooth under traffic.

Surface dressing is the most extensively used form of surface treatment. In the UK, data suggest that local authorities spend approximately 20% of their maintenance budget on this process[485]. In 2001, approximately 75 million square metres of road were surface dressed in the UK. With care, surface dressing can be used on roads of all types from the country lane carrying only an occasional vehicle to Trunk Roads carrying tens of thousands of vehicles per day. However, in some locations, particularly those where vehicles undertake sudden or sharp turning manoeuvres, surface dressing may not be the appropriate surface treatment. This is due to the relatively poor ability of the process to resist tangential forces. Additionally, the durability of a surface dressing does not match that of a road resurfaced with a traditional asphalt. Some large cities, Hamburg for example, prohibit the use of surface dressings on their carriageways.

Surface dressing as a maintenance process has three major aims:

- to provide both texture and skid resistance to the road surface
- to stop the disintegration and loss of aggregate from the road surface

---

485 DEPARTMENT OF TRANSPORT. *Local road maintenance in England and Wales 1994/95, Transport Statistics Report, 1996.* DTp, London.

- to seal the surface of the road against the ingress of water and, thus, protect its structure from damage resulting therefrom.

In addition, it can be used as a treatment to provide:

- a distinctive colour to the road surface
- a more uniform appearance to a patched road.

However, surface dressing cannot restore evenness to a deformed road nor will it contribute to the structural strength of the road structure.

Basic surface dressing consists of spraying a film of binder onto the existing road surface followed by the application of a layer of aggregate chippings as shown in Fig. 19.1. The chippings are then rolled to

Fig. 19.1 Surface dressing
Courtesy of Colas Ltd

promote contact between the chippings and the binder and to initiate the formation of an interlocking mosaic.

19.1.1 The design of surface dressings

The design of surface dressings primarily addresses the retention of surface texture. The skill lies in designing a system that, taking account of the hardness of the existing road, is adequately embedded by rolling and subsequently incident traffic.

The design of surface dressings in Germany

In Germany, a set of guidelines[486] prescribe the type of road maintenance that is to be applied and this depends on the condition of the original pavement. These guidelines invariably form part of the contract between employer and contractor in German contracts. Surface dressings compete with thin surfacings (both hot and cold applied), in reshaping and wholesale replacement of the surface course. The German guidelines suggest which surface dressing should be applied in terms of the type, volume of binder and quantity of chippings. This depends on the structure of the existing road and the nature of the surface dressing. Three different surface dressing systems are listed:

- single application of binder and one layer of chippings;
- single application of binder and two layers of chippings of different sizes (racked-in surface dressing);
- application of a layer of chippings, subsequent single application of binder followed by a further application of a layer of chippings.

Recommendations for binder and chipping rates are given according to the condition of the road surface.

The design of surface dressings in the UK

In the UK, Road Note 39[487] and the Code of Practice for Surface Dressing[488] are commonly used to provide guidance on the appropriate type of dressing, its design and execution. The principles in Road Note 39 are as follows:

---

486 FORSCHUNGSGESELLSCHAFT FÜR STRABEN- UND VERKEHRS-WESEN. *Zusätzliche Technische Vertragsbedingungen und Richtlinien für die Bauliche Erhaltung von Verkehrsflächen – Asphaltbauweisen (ZTV BEA-StB 98) (Bauliche Erhaltung von Verkehrsflächen)*, 1998. FGSV, Cologne, Germany.

487 NICHOLLS J C et al. *Design guide for road surface dressing, Road Note 39, 5th ed, 2002.* TRL Ltd, Crowthorne.

488 ROAD SURFACE DRESSING ASSOCIATION. *RSDA code of practice for surface dressing, 3rd ed, Oct 2002.* RSDA, Colchester.

- select the size and type of dressing that is appropriate for the particular site and the time of application;
- identify the chipping size and basic binder application rate and type of binder;
- refine the binder application rate to match the source of chippings selected;
- further refine the binder application rate on site immediately before work commences.

Road Note 39 is set out in such a way that the application rate of the binder is only obtainable after the above stages have been completed. The tables in Road Note 39 provide coefficients for application rates for binders for single, racked-in, double, sandwich, inverted double surface dressings. These coefficients are then adjusted by reference to tables that take into account the type of aggregate and its shape, the condition of the road surface, the gradient, the degree to which the site is shaded by trees and structures and for sites where traffic volumes are exceedingly light. The coefficients are then translated into actual target application rates of the binder expressed in litres per square metre.

19.1.2 Factors that influence the design and performance of surface dressings

The principal factors that influence the design, performance and the service life of a surface dressing are:

- traffic volumes and speeds
- condition and hardness of the existing road surface
- size and other chipping characteristics
- surface texture and skid resistance
- type of binder
- adhesion
- geometry of site, altitude, latitude and local circumstances
- time of year.

Traffic volumes and speeds
In the UK, a commercial vehicle is defined as having an unladen weight exceeding 1·5 tonnes for design purposes. As the rate of chipping embedment is dependent on the number of commercial vehicles using each lane, it is necessary to establish the number of such vehicles using a road that is to be surface dressed and the speed of traffic using the carriageway.

Condition and hardness of the existing road surface
If the existing road surface is rutted, cracked or in need of patching, these defects must be corrected before surface dressing can be

374

undertaken[489]. Other surface conditions that require special consideration are those where the binder has flushed up resulting in a binder-
rich surface and, conversely, surfaces that have become very dry, lacking
in binder.

The hardness of the road surface affects the extent to which the applied
chippings become embedded into the road surface during the life of the
dressing. The choice of chipping size is directly related to the hardness of
the existing road surface. The use of chippings that are too small will
result in early embedment of the chippings into the surface leading to
a rapid loss of texture depth and, in the worst cases, 'fatting up' of the
binder which may cover the entire surface of the road. The use of
chippings that are too large may result in immediate failure of the treatment due to stripping of the binder from the aggregate under the applied
stresses of the traffic and can also result in excessive surface texture and an
increase in the noise generated between the tyre and the road surface. The
hardness of the road surface also influences the rate of application of
binder required for a given size of chipping. The rate of spread must
be decreased where the road surface is soft in order to compensate for
the greater embedment of chippings into the road surface under the
action of traffic.

In the UK, the categories of hardness of the existing road surfaces for
the purposes of surface dressing design are described in Road Note 39.
Four graphs take account of the geographical location within the UK
and the altitude of the site. The geographical location is not strictly
related to latitude but also takes account of climatic conditions partly
caused by the Gulf Stream. The method of measuring hardness on the
site is by use of a road hardness probe[490]. This device measures the
depth of penetration of a probe with a 4 mm semi-spherical head after
the exertion of a force of 340 N for a period of 10 s. Ten or more
probe tests taken at approximately 0·5 m intervals in the nearside
wheel-track of the lane under consideration are taken and the average
calculated. The road surface temperature is also recorded. Experience
has shown that tests are only significant within the temperature range
of 15 to 35°C.

Having measured the depth of penetration and the surface temperature
and identified whether the site lies in the southern, central or northern
region, the hardness category of the road can be established by reference
to the appropriate graph, as shown in Fig. 19.2.

489 ROAD SURFACE DRESSING ASSOCIATION. *RSDA guidance note on preparing roads for surface dressing, Oct 2002*. RSDA, Colchester.
490 BRITISH CARBONISATION RESEARCH ASSOCIATION. *The CTRA road
hardness probe, Carbonisation research report 7, Jun 1974*.

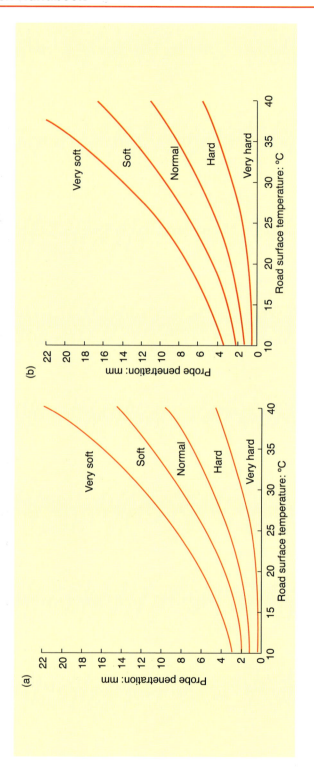

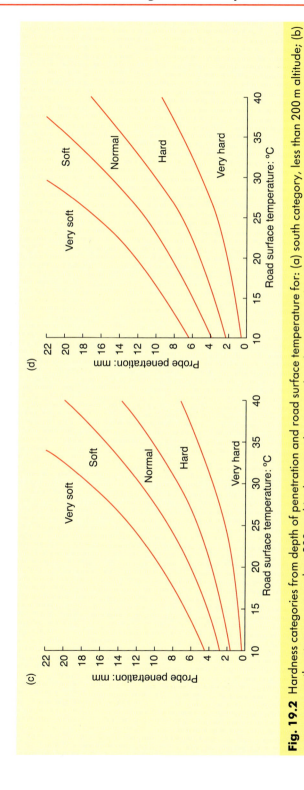

Fig. 19.2 Hardness categories from depth of penetration and road surface temperature for: (a) south category, less than 200 m altitude; (b) south category, more than 200 m altitude; (c) north category less than 200 m altitude; (d) north category more than 200 m altitude[491]

491 NICHOLLS J C et al. *Design guide for road surface dressing, Road Note 39, 5th ed, fig 6.2.3*, 2002. TRL Ltd, Crowthorne.

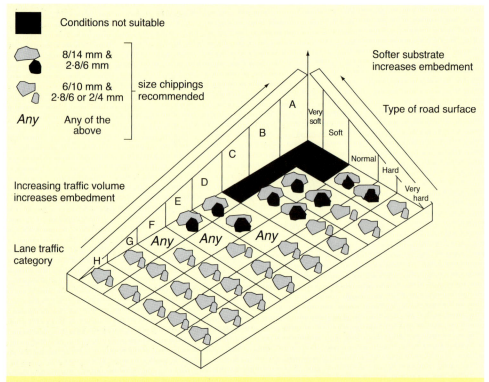

Fig. 19.3 Recommended sizes of chippings for racked-in and double surface dressings[492]

Size and other chipping characteristics

The selection of the size and type of chipping should take into account the following factors.

- The size of chipping has to offset the gradual embedment into road surfaces of different hardness caused by traffic.
- Maintaining both microtexture and macrotexture during the life of the surface dressing by selecting a chipping of appropriate size, polished stone value (PSV) and aggregate abrasion value (AAV).
- Relating the traffic category for each lane and the road surface hardness category to the size of chipping required. Figure 19.3 illustrates this relationship for a racked-in surface dressing.
- Ensuring that the rate of spread of chippings is adequate to cover the binder film after rolling without excess chippings lying free on the road surface.

492 NICHOLLS J C et al. *Design guide for road surface dressing, Road Note 39, 5th ed, fig 8.2.2.1, 2002*. TRL Ltd, Crowthorne.

- Binder type and viscosity must be appropriate for the time of year, traffic category and degree of stress that is likely to be encountered. Similarly, the rate at which it is applied must be appropriate for the traffic category, road hardness and nominal size of chipping.

In the UK, chippings for surface dressing should meet the requirements of BS EN 13043: 2002[493]. Usually, chippings have a nominal size of 2·8/6, 6/10 or 8/14 mm. Extreme caution should be exercised when using chippings larger than 8/14 mm nominal size due to the risk of premature failure.

The presence of dust can delay or even prevent adhesion and is particularly acute at low temperatures and with the smaller sizes of chipping. As well as being of nominal single size and dust free, surface dressing chippings need to satisfy other characteristics. They should:

- not crush under the action of traffic or shatter on impact
- resist polishing under the action of traffic
- not have an extreme value of 'flakiness index' (the ratio of length to thickness).

Chippings are rarely cubical and when they fall onto the binder film and are rolled, they tend to lie on their longest dimension. If the chippings used for a surface dressing have a large proportion of flaky chippings, less binder will be required to hold them in place and the excess binder may 'flush up' onto the surface of the dressing. Such a dressing will have a reduced texture and life. It is for this reason that the UK sets a maximum limit of flakiness for single-size chippings for use in surface dressing. Typical requirements for surface dressing chippings are shown in Table 19.1.

Gritstones, basalts, quartzitic gravels and artificial aggregates such as slag constitute the majority of chippings used for surface dressing. However, not all aggregates with a high PSV are suitable for surface dressing due to insufficient strength. The PSV of the aggregate defines its resistance to polishing. In the UK, its determination is defined in BS EN 1097-8: 2002[494] and the relationship between PSV, traffic and skid resistance is detailed in HD 36[495].

---

493 BRITISH STANDARDS INSTITUTION. *Aggregates for bituminous mixtures and surface treatments for roads, airfields and other trafficked areas, BS EN 13043: 2002.* BSI, London.

494 BRITISH STANDARDS INSTITUTION. *Tests for mechanical and physical properties of aggregates. Determination of the polished stone value, BS EN 1097-8: 2000.* BSI, London.

495 HIGHWAYS AGENCY *et al. Design manual for roads and bridges, Pavement design and maintenance, Surfacing materials for new and maintenance construction, vol 7, HD 36/99, 1999.* The Stationery Office, London.

Table 19.1 Typical properties of surface dressing chippings

Property	Value
Los Angeles coefficient (LA)	LA_{30}
Flakiness index (FI)	FI_{20}
Polished stone value (PSV)	$PSV_{Declared}$[a]
Aggregate abrasion value (AAV)	AAV_{10}

[a] The required PSV value is usually stated in individual contracts. Because the PSV categories in BS EN 13043 do not always use values that correspond with established UK practice, the Standard allows an aggregate supplier to declare an intermediate value.

Coated chippings have a thin film of bitumen applied at an asphalt plant. This bitumen film eliminates surface dust and promotes rapid adhesion to the bitumen. Coated chippings are used to improve adhesion with cutback bitumens particularly in the cooler conditions that occur at the extremes of the season. They should not be used with emulsions as the shielding effect of the binder will delay 'breaking' of the emulsion.

Surface texture and skid resistance

In the UK, two machines are used for monitoring the skid resistance of roads – SCRIM and the GripTester.

SCRIM is the acronym for the 'sideways-force coefficient routine investigation machine' that was introduced into the UK in the 1970s. It provides a measure of the wet road skid-resistance properties of a bound surface by measurement of a sideways-force coefficient at a controlled speed. A freely rotating wheel fitted with a smooth rubber tyre, mounted mid-machine in line with the nearside wheel-track and angled at $20°$ to the direction of travel of the vehicle, is applied to the road surface under a known vertical load. A controlled flow of water wets the road surface immediately in front of the test wheel which rotates but also slides in the direction of travel onto the wetted surface. The force set up by the resistance to sliding is proportional to the skid resistance and measurement of this force allows the sideways force coefficient (SFC) to be calculated.

The GripTester was introduced in the 1980s and is a braked-wheel fixed-slip trailer. It has two drive wheels and a single test wheel that is mounted on a stub axle and is mechanically braked to give a relative slip speed between the drive wheels and the test wheel of just over 15%. A controlled flow of water from a tank in the tow vehicle wets the road surface immediately in front of the test wheel, which skids over the wetted surface. The resulting horizontal drag and the vertical load on the measuring wheel are continuously measured and the coefficient of friction, known as the GripNumber, is calculated.

Both SFC and the GripNumber vary with speed. Testing is normally carried out at 50 km/h; corrections may be made to allow for variations in speed but for these to be reliable it is necessary that the macrotexture of the surface is known. Tight radii, such as roundabouts, are normally tested at 20 km/h.

SFC and GripNumber also vary with traffic flow, temperature and time of year. Network testing is restricted to May to September, with three tests carried out at fairly similar intervals throughout the season. The average is defined as the 'mean summer SCRIM coefficient' (MSSC) and skid resistance is normally expressed in terms of the MSSC.

In 1973, the TRL[496] introduced the concept of 'risk rating', which is a measure of the relative skidding accident potential at an individual site and is determined by its geometry and location. Within each category, there is a range of target SFC values with the target value for each site depending on its risk rating. A road will need to be surface dressed if the measured skid resistance (MSSC value) is inadequate for the risk rating of the site. The 'investigatory levels' of skid resistance for Trunk Roads including Motorways are given in HD 28[497]. In this document, the investigatory levels are those at which the measured skid resistance of the road has to be reviewed. In the review process, it may be decided to restore skid resistance by surface dressing or, in certain circumstances, the risk rating of the site may warrant change.

The skidding resistance of a road surface is determined by two basic characteristics, the microtexture and macrotexture, as shown in Fig. 19.4. Microtexture is the surface texture of the aggregate. A significant level of microtexture is necessary to enable vehicle tyres to penetrate thin films of water and thus achieve dry contact between the tyre and the aggregate on the carriageway. Macrotexture is the overall texture of the road surface. An adequate value of macrotexture is necessary to provide drainage channels for the removal of bulk water from the road surface.

TRL Research Report 296[498] considered the relationship between accident frequency and surface texture on roads. It concluded that, on

496 SALT G F and W S SZATKOWSKI. *A guide to levels of skidding resistance for roads, Laboratory Report 510, 1973.* Transport and Road Research Laboratory, Crowthorne.

497 HIGHWAYS AGENCY *et al. Design manual for roads and bridges, Pavement design and maintenance, Skidding resistance, vol 7, HD 28/94, 1994.* The Stationery Office, London.

498 ROE P G, D C WEBSTER and G WEST. *The relation between the surface texture of roads and accidents, Research Report 296, 1991.* Transport and Road Research Laboratory, Crowthorne.

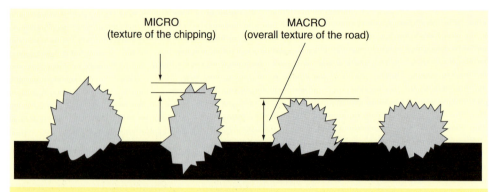

Fig. 19.4 Microtexture and macrotexture[499]

asphalt road surfaces with speed limits greater than 64 km/h and probably on roads with lower speed limits:

- skidding and non-skidding accidents, in both wet and dry conditions, are less if the microtexture is coarse;
- the texture level below which accident risk begins to increase is a sensor-measured texture depth (SMTD, see later) of around 0·7 mm;
- all major types of surfacing provide texture depths across the practical range;
- macrotexture has a similar influence on accidents whether they occur near hazards such as junctions or elsewhere.

Microtexture is the dominant factor in determining the level of skid resistance at speeds up to 50 km/h. Thereafter, macrotexture predominates, particularly in wet conditions. Traditionally, macrotexture has been measured using the sand patch method[500]. A standard volume of a specified type of sand is spread by circular motion on the road surface. On contracts carried out under the SHW, which require a two year guarantee period, the test is generally carried out along the wheel track[501]. The texture depth can then be calculated by dividing the volume of sand by the cross sectional area of the patch. The test cannot be carried out on a wet or sticky surface.

499 NICHOLLS J C *et al. Design guide for road surface dressing, fig 2.1.2, Road Note 39, 5th ed, 2002.* TRL Ltd, Crowthorne.

500 BRITISH STANDARDS INSTITUTION. *Road and airfield surface characteristics, Test methods, Measurement of pavement surface macrotexture depth using a volumetric patch technique, BS EN 13036-1: 2002.* BSI, London.

501 HIGHWAYS AGENCY *et al. Manual of contract documents for highway works, Specification for highway works, vol 1, cl 922, 2002.* The Stationery Office, London.

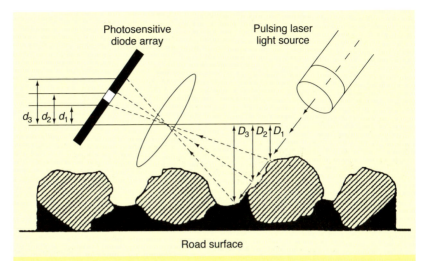

Fig. 19.5 Principle of the laser-based contactless sensor for measuring texture depth[502]

Mature surface dressings in good condition typically have texture depths between 1·5 mm and 2·0 mm; double dressings usually have slightly lower values. Double surface dressings, however, usually stay above intervention levels particularly on hard surfaces such as those surfaced with hot rolled asphalt surface course.

Laser texture measurement (technically, sensor-measured texture depth (SMTD)) is a technique that covers areas much more quickly than sand patch testing. Accordingly, it is used as a screening method. The technique is illustrated in Fig. 19.5. This method is not directly comparable with measurements resulting from sand patch tests. Thus, when quoting texture depth values, the method used to measure the parameter should be stated. For compliance testing on contracts prepared under the SHW, it is usual to quote sand patch texture, correlate the sand patch test results to the values returned by SMTD, check as required during the two year guarantee period by SMTD and carry out additional sand patch tests if necessary.

It is the petrographic characteristics of the aggregate that largely influence its resistance to polishing and this property is measured using the PSV test[503]. This test subjects the aggregate to simulated trafficking

502 ROE P G, L W TUBEY and G WEST. *Surface texture measurements on some British roads, fig 1, Research Report 143, 1988.* Transport and Road Research Laboratory, Crowthorne.

503 BRITISH STANDARDS INSTITUTION. *Tests for mechanical and physical properties of aggregates. Determination of the polished stone value, BS EN 1097-8: 2000.* BSI, London.

Table 19.2 The effect of macrotexture on the change in skidding resistance with speed[504]

Drop in skidding resistance with speed change from 50 to 130 km/h: %	Average texture depth: mm	
	Flexible	Concrete[a]
0	2·0	0·8
10	1·5	0·7
20	1·0	0·5
30	0·5	0·4

[a] When textured predominantly transversely.

in an accelerated polishing machine. A high value of PSV (say 65 or higher) indicates good resistance to polishing. The majority of road aggregates have a good microtexture prior to trafficking and, consequently, most road surfaces have a high skid resistance when new, particularly after the bitumen has worn off the aggregate. The time required for the bitumen to be abraded from the aggregate at the road surface depends upon a number of factors including the type of mixture and, in particular, the thickness of the film of bitumen. Until the latter has been removed by traffic, the microtexture of the aggregate will not be fully exposed to vehicle tyres. Subsequently, the exposed aggregate surfaces become polished and, within a year, the skid resistance normally falls to an equilibrium level.

The macrotexture is determined by the nominal size of aggregate used and the nature of the asphalt mixture. However, resistance to abrasion is also important. Aggregates with poor abrasion resistance, determined by the AAV test[505], will quickly be worn away with a consequent loss of macrotexture. The lower the value of AAV, the greater the resistance to abrasion. High-speed roads normally require an AAV of 14 or less. Surface dressings generally give better value, i.e. last longer, where the chippings have a lower value of AAV.

The rate at which skid resistance falls with increasing vehicle speed is influenced by the macrotexture. The TRL has shown that there is a correlation between macrotexture, expressed in terms of average texture depth, and skid resistance, measured by percentage reduction in braking force coefficient (BFC) between 130 km/h and 50 km/h, as shown in Table 19.2.

504 SALT G F and W S SZATKOWSKI. *A guide to levels of skidding resistance for roads, table 4, Laboratory Report 510, 1973.* Transport and Road Research Laboratory, Crowthorne.

505 BRITISH STANDARDS INSTITUTION. *Tests for mechanical and physical properties of aggregates. Determination of the polished stone value, BS EN 1097-8: 2000.* BSI, London.

The skid resistance necessary at any site will depend on the stresses likely to be induced. Sections of road with minor road junctions and with low traffic volumes are unlikely to require a high value of skid resistance whereas sharp bends on roads carrying heavy volumes of traffic will require significantly higher skid resistance characteristics. It is important to note that roads carrying high volumes of traffic will require aggregates with higher values of PSV than roads where traffic levels are low or moderate.

Type of binder
The essential functions of the binder are to seal surface cracks and to provide the initial bond between the chippings and the road surface. The viscosity of the binder must be such that it can wet the chippings adequately at the time of application, prevent dislodgement of chippings when the road is opened to traffic, not become brittle during periods of prolonged low temperature and function effectively for the design life of the dressing.

The subsequent role of the binder is dependent on the traffic stresses applied to the surface dressing from low-stressed, lightly trafficked country roads to highly stressed sites carrying large volumes of traffic. Clearly a range of binders is necessary if optimal performance is to be achieved under these widely differing circumstances.

The following five types of surface dressing bitumen binders are available; each is described in more detail below.

- conventional bitumen emulsions
- fluxed or cutback bitumens
- polymer modified emulsions
- polymer modified fluxed or cutback bitumens
- epoxy resin thermoset binders.

The relative amounts of the modified and unmodified emulsion and fluxed/cutback bitumen used in Germany and the UK are given in Table 19.3.

Table 19.3 Usage of surface dressing binder by type in Germany and the UK

Type of binder	Use in Germany, %	Use in UK, %
Conventional emulsion	35	22
Modified emulsion	40	67
Conventional fluxed/cutback	0	6
Modified fluxed/cutback	25	5
Total usage in 2001, tonnes	>90 000	113 000

- *Conventional bitumen emulsions.* In the UK, conventional bitumen emulsions are produced to BS 434-1: 1984[506]. The most common material is K1-70 bitumen emulsion, which is a cationic emulsion containing approximately 29% of water, 70% of 160/200 pen or 250/330 pen bitumen to which, approximately, 1% by weight of kerosene has been added. The relatively low viscosity of bitumen emulsions enables them to be sprayed at temperatures between 75 and 85°C through either swirl or slot jets.

  K1-70 emulsions will flow on road surfaces at the time of spraying because they have a relatively low viscosity. Whilst this is helpful in some respects, migration of the binder from the centre to the side of cambered roads, or downhill on gradients, requires the application of chippings to the binder film as quickly as possible after spraying. After spraying, the water migrates from the emulsion leaving a film of bitumen, a process known as 'breaking'. This can take from as little as 10 minutes to over an hour depending on ambient temperature, humidity, aggregate water absorption, wind speed and the extent to which the dressing is subject to slow-moving traffic. Humidity is one of the most important of these factors and the use of emulsions when humidity is above 80% should only be undertaken on very minor roads and where traffic speeds can be kept within the range of 10 to 20 mph until the emulsion has fully broken. In conditions of high humidity, work can continue with emulsions, particularly where the binder is polymer modified, by varying the type of dressing. Good results can be obtained with double and sandwich surface dressings. Pre-coated chippings should not be used with emulsions as the coating delays breaking and serves no tangible benefit but carries additional cost.

  Binder manufactures are able to control the rate at which emulsions break within certain limits, but these have to be compatible with the storage, transport and application of the emulsion. On site, breaking can be accelerated by spraying the binder film with a mist of a breaking agent before the application of the chippings. However, this usually requires specially adapted spray bars and the fitting of storage tanks for the breaking agent.

- *Conventional fluxed or cutback bitumens.* Fluxed or cutback bitumen for surface dressing is usually 70/100 pen or 160/220 pen bitumen that has been diluted with kerosene. In the UK, the use of a suffix X in the description of cutback bitumens, e.g. 100X, indicates that they have been doped with a specially formulated, thermally

---

506 BRITISH STANDARDS INSTITUTION. *Bitumen road emulsions (anionic and cationic), Specification for bitumen road emulsion, BS 434-1: 1984.* BSI, London.

Table 19.4 Advantages and disadvantages of fluxed/cutback binders

Advantages	Disadvantages
Rapid adhesion	High spraying temperature
Low risk of brittle failure	Early-life bleeding
	Environmental & health and safety issues
	Requires dry chippings

stable, passive adhesion agent. This additive assists wetting of the aggregate and the road surface during application and resists stripping of the binder from the aggregate in the presence of water. Either the immersion tray test[507] or the total water immersion test (a test carried out by Shell in house, see Section 9.3.1) can demonstrate the effectiveness of this additive. If surface dressing is being carried out in marginal weather conditions or if the aggregate is wet, it is recommended that an active wetting agent is added to cutback bitumens immediately prior to spraying.

Cutback bitumens are typically sprayed at temperatures in the range of 130 to 170°C. During spraying, between 10 and 15% of the kerosene evaporates and a further 50% dissipates from the surface dressing in the first few years after application. At these temperatures, wetting of the road surface and the applied chippings is achieved rapidly. Cutback bitumens, once applied to the road surface, cool quickly and it is important to ensure that the chippings are applied as soon as possible after spraying to ensure satisfactory wetting of the chippings.

During the surface dressing season (see Section 19.1.7), as temperatures rise, binder suppliers may recommend the use of different viscosity cutback binders in order to promote adhesion and avoid the fatting-up of new dressings. Both cutback and emulsion manufacturers may also make minor changes to their formulations within the ranges covered by the appropriate standards to meet changes in road and ambient temperatures.

The advantages and disadvantages of fluxed/cutback and emulsion binders are listed in Tables 19.4 and 19.5 respectively.

There has been a trend with the introduction of 'best value' assessments based on whole-life costing to opt for modified binders rather than using binders that are cheaper in operational terms alone.

- *Proprietary polymer modified binders.* The relentless increase in the volume of all types of vehicles on roads throughout Europe has

---

507 NICHOLLS J C *et al. Design guide for road surface dressing, appendix D, Road Note 39, 5th ed, 2002.* TRL Ltd, Crowthorne.

Table 19.5 Advantages and disadvantages of emulsion binders

Advantages	Disadvantages
Low spraying temperature	Low initial adhesion
Low risk of early bleeding	Skinning[a]
Environmentally friendly	Risk of brittle failure in first winter[b]
May be used on higher-viscosity base bitumen	
May be used on damp (not wet) surfaces	
Do not require pre-coated chippings	

[a] The use of a sandwich or, preferably, a double surface dressing will reduce the film thickness and, thus, the possibility of skinning.
[b] Employing modified binders will reduce this risk considerably.

led to the development of modified binders with improved adhesivity and substantially higher cohesive strengths than traditional fluxed/ cutback and emulsion binders. These characteristics are particularly important immediately after chippings have been applied to the binder film and mean that they are less likely to be plucked from the road surface by passing traffic than would be the case with conventional unmodified binders. As well as greater adhesive and cohesive strength, modified binders perform satisfactorily over a greater temperature range than is the case with traditional binders. The result is that they do not flush up as rapidly in hot conditions as traditional binders and are not subject to the same degree of brittleness in winter conditions.

A high proportion of the European road network can be classified as easy or average and many of these roads can be surface dressed using bitumen emulsion or cutback bitumen. However, enhanced treatment may be necessary on certain highly stressed sections such as at sharp bends, road junctions and, in shaded areas, under trees and bridges. At these locations, it is probable that the same binder can be used but the specification will change in respect of one or more of the following:

- rate of application of binder
- chipping size and rate of spread
- double chip application
- double binder, double chip application.

These latter techniques are widely practised in France on relatively heavily trafficked roads, and with correct procedures, their use can significantly reduce the risk of flying chippings and consequential windscreen breakage. Conventional emulsions and cutbacks have the lowest cost and with careful workmanship, compliance with specification and appropriate aftercare, good results can be obtained.

As a further aid to consistent performance under variable weather conditions, a number of proprietary adhesion agents are available that act in two basic ways. Some types (active) modify surface tension at the stone/binder interface to improve the preferential wetting of chippings by the binder in the presence of water. Other types (passive) improve the adhesion characteristics of the binder and give an improved resistance to the subsequent detachment of the binder film in the presence of water throughout the early life of the dressing. The effectiveness of the former type of adhesion agent reduces after a few hours at elevated temperature and so it is generally mixed with the binder a few hours before spraying. The latter type of adhesion agent is more temperature-stable and is generally added to the binder by the supplier.

Bearing these performance requirements in mind, it is possible to examine the ways in which the various polymers can be used with bitumen to modify its performance. The bitumen properties where improvement would be sought through the addition of a modifier are set out below.

- Reduced temperature susceptibility, which can loosely be defined as the extent to which a binder softens over a given temperature range. It is clearly desirable for a binder to exhibit a minimum variation in viscosity over a wide range of ambient temperatures.
- Cohesive strength enables the binder to hold chippings in place when they are subject to stress by traffic. This property can also be coupled with the elastic recovery properties of the binder, to enable a surface dressing to maintain its integrity even when it is subjected to high levels of strain.
- Adhesive power or 'tack' is an essential property of binders but it is difficult to define. This is especially true in the early life of a surface dressing before any mechanical interlock or embedment of the chippings has taken place.

In addition, it is desirable for the modified binder to:

- maintain its premium properties for long periods in storage and subsequently during its service life;
- be physically and chemically stable at storage, spraying and in-service temperatures;
- be applied using conventional spraying equipment;
- be cost effective.

Choosing between an emulsion or cutback formulation for modified binders is not easy. Generally speaking, emulsions are more vulnerable during the early life of the dressing before the emulsion has fully broken. The rate at which breaking occurs is influenced by a number of factors including emulsion formulation, weather and mechanical agitation during rolling and trafficking. As discussed above, under certain conditions the emulsion goes through a state during cure where the bitumen droplets have agglomerated but not coalesced and the dressing is extremely vulnerable. The timing of this condition is variable and is generally worsened by the presence of high polymer contents that can cause skinning and retard breaking of the emulsion. It is essential that the emulsion is completely broken before uncontrolled traffic is allowed on the dressing. This can be a serious disadvantage on heavily trafficked roads or where work is executed under lane rental contracts. Chemical after-treatment sprays have been shown to be effective in promoting breaking.

Cutback bitumens have benefits in terms of immediate cohesive grip of the road and chippings, but generally the residual binder properties are more variable due to the differences in the evolution of the diluent and its absorption into the road surface. Nevertheless, polymer modified formulations are the preferred choice for high-speed and heavily trafficked roads because of their better cohesion properties and resistance to bleeding.

The development of polymer modified surface dressing binders to complement conventional and epoxy resin systems offers the Highway Engineer a range of binders suitable for all categories of site. Regardless of the improved properties of a particular binder, it is the performance of the aggregate binder system and application mode that will dictate the success or otherwise of the dressing.

- *Epoxy resin thermoset binders.* These are bitumen-extended epoxy resin binders used with high PSV aggregates such as calcined bauxite. Epoxy resin based binders are used in high-performance systems such as Spraygrip® that are fully capable of resisting the stresses imposed by traffic on the most difficult sites, e.g. roundabouts or approaches to traffic lights and designated accident black spots. It is these systems that are often described as 'Shellgrip' but are more correctly generically described as 'high-friction surfacings'. Binders used in such systems are classified as 'thermosetting' (see Section 5.6) as the epoxy resin components cause the binder to cure by chemical action and harden and it is not subsequently softened by high ambient temperatures or by the spillage of fuel. The dressing thus acts as an effective seal against the ingress of oil and fuel, which is particularly important on roundabouts where spillages regularly occur. Very little embedment of the chippings into the road surface takes place with this type of binder and the integrity of the surface dressings is largely a function of the cohesive strength of the binder.

The extended life of this binder justifies the use of a durable aggregate with an exceptionally high value of PSV. The initial cost of these surface dressing systems is high compared with conventional binders. However, their exceptional wear-resistant properties and the ability to maintain the highest levels of skid resistance throughout its life make it a cost effective solution for very difficult sites by significantly reducing the number of skidding accidents. Statistics suggest that the average value of prevention per serious accident in the UK is £141 490 whilst that for a fatal accident is £1 207 670[508].

Adhesion

Some aggregates adhere more rapidly to binder than others and it is wise to avoid problems by testing the compatibility of the selected binder and chippings. When using cutback binder, this can be checked using the total water immersion test (carried out by Shell in house). In this test, clean 8/14 mm chippings are totally coated with cutback bitumen. After 30 minutes of curing at ambient temperature, the chippings are immersed in demineralised water at 40°C for 3 hours. After soaking, the percentage of binder that is retained on the binder is assessed visually. The extent to which the base of the chipping has been coated with the binder after 10 minutes is then assessed visually. If the coated area is considered to be below 70%, serious consideration should be given to lightly pre-coating the chippings as it is much easier to obtain a bond between the binder film and a light coating of binder on the chippings than is the case with uncoated chippings.

The amount of binder required to lightly coat surface dressing chippings varies according to the type of binder and the size of the chipping but it is typically around 1% by weight. A thick coating of binder, e.g. as used in chipped hot rolled asphalt surface course, will make the chippings sticky and will inhibit free flow through chipping machines. When lightly coated chippings are used, this free-flowing property is critical to a satisfactory surface dressing. Minor pinholes in the coating of the chippings are not detrimental. It is particularly important that the amount of filler is less than the specified level (usually $\leq 1\%$ passing the 0·063 mm sieve).

Once the chippings have been lightly coated, it is important that they are not allowed to come into contact with dust as this will impair the

508 *www.roads.dft.gov.uk/roadsafety/hen198/index.htm* table 4a accessed on 19 Feb 2003.

bond between the chippings and the applied binder film on the road surface.

Geometry of site, altitude and latitude and local circumstances
Sections of road that include sharp bends induce increased traffic stresses on a surface dressing. Two categories of bend are identified in Road Note 39[509] – less than 250 m radius and less than 100 m radius. Gradients that are steeper than 10% will affect the rate at which the binder should be applied due to the fact that on the ascent side of the road, traffic loading will be present for longer than on the descent side. Typically, there is a difference of $0.2 \, l/m^2$ in rate of binder application between the ascent and the descent sides.

Time of year
See Section 19.1.7.

19.1.3. Surface dressing specifications
Surface dressing specifications in Germany
Polymer modified binders for surface dressing in Germany are specified according to a set of material guidelines[510]. These guidelines invariably form part of the contract between employer and contractor in German contracts.

Included therein are specifications for:

- polymer modified bitumen
- polymer modified fluxed bitumen
- cationic emulsions of polymer modified bitumen
- cationic emulsions of polymer modified fluxed bitumen.

In all cases, special attention is given to the adhesion properties of the binder. Recommendations for the exact composition of the surface dressing and for the properties of the chippings are given in another set of national guidelines[511].

509 NICHOLLS J C et al. Design guide for road surface dressing, Road Note 39, 5th ed, 2002. TRL Ltd, Crowthorne.
510 FORSCHUNGSGESELLSCHAFT FÜR STRABEN- UND VERKEHRS-WESEN. Technische Lieferbedingungen für gebrauchsfertige polymermodifizierte Bindemittel für Oberflächenbehandlungen (TL PmOB), 1997. FGSV, Cologne, Germany.
511 FORSCHUNGSGESELLSCHAFT FÜR STRABEN- UND VERKEHRS-WESEN. Zusätzliche Technische Vertragsbedingungen und Richtlinien für die Bauliche Erhaltung von Verkehrsflächen – Asphaltbauweisen (ZTV BEA-StB 98) (Bauliche Erhaltung von Verkehrsflächen), 1998. FGSV, Cologne, Germany.

Surface dressing specifications in the UK

The role of the *Manual of Contract Documents for Highway Works* (MCHW) and the involvement of the Highways Agency, etc. was explained in Section 11.3. Volumes 1 (SHW) and 2 (NG) thereof contain provisions relating to surface dressing. Two clauses in the SHW are of particular interest to those involved in surface dressing. They are Clause 919 (Surface Dressing: Recipe Specification) and Clause 922 (Surface Dressing: Design, Application and End Product Performance). The equivalent Clauses in the NG (NG 919 and NG 922) are also worthy of consultation.

In the past, individual clauses have tended to call up British Standard tests and then give some guidance as to their use. However, a number of new end-performance specifications have recently been developed. The principal differences between Clause 919 and Clause 922 are that the responsibility for the design of the dressing is transferred from the Highways Agency, etc. to the contractor and performance measurement of the surface dressing is undertaken at time intervals specified in the contract.

19.1.4 Types of surface dressing

The original concept of a normal single-layer surface dressing has been developed over the years and there are now a number of techniques available that vary according to the number of layers of chippings and applications of binder. Each of these techniques has its own particular advantages and associated cost and it is quite feasible that along the length of a road, a variety of techniques would be used depending on the stresses at any particular feature. Figure 19.6 provides diagrams of the different types of surface dressing that are available.

Pad coats

A pad coat consists of a single dressing with small chippings and is applied to a road that has uneven surface hardness possibly due to extensive patching by the utilities. The pad coat is used to provide a more uniform surfacing that can subsequently be surface dressed. The chipping size for a pad coat is traditionally 2·8/6 mm with a slight excess of chippings. Pad coats can also be used on very hard road surfaces (such as concrete or heavily chipped asphalts) to reduce the effective hardness of the surface although using a racked-in system with a polymer modified binder is now the preferred option.

After laying and compaction by traffic, excess chippings should be removed before the road is opened to unrestricted traffic. Pad coats may be left for several months before the application of the main dressing that may be either a single dressing or a racked-in system. Either system will embed into the pad coat and have immediate significant mechanical

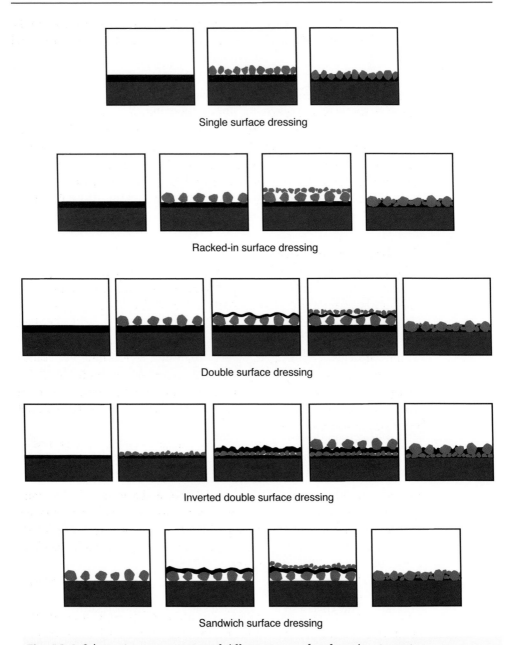

Single surface dressing

Racked-in surface dressing

Double surface dressing

Inverted double surface dressing

Sandwich surface dressing

Fig. 19.6 Schematic representation of different types of surface dressing prior to embedment[512]

512 NICHOLLS J C *et al. Design guide for road surface dressing, fig. 2.2, Road Note 39, 5th ed, 2002.* TRL Ltd, Crowthorne.

strength, reducing the risk of failure. All loose material should be swept from the surface of the road prior to the application of the final dressing.

Single surface dressing

The single surface dressing system is still the most widely used and cheapest type of surface dressing. It consists of one application of binder followed by one application of single-size chippings. Its advantages are that it has the least number of operations, uses the least amount of material and is sufficiently robust for minor roads and some main roads where excessive braking and acceleration are unlikely to occur and where speeds are unlikely to exceed 100 km/h.

Racked-in surface dressing

This type of dressing uses a single application of binder. A first application of 6/10 mm or 8/14 mm chippings (called the 'principal chippings') is spread at approximately 80 to 90% cover which leaves a window of binder between the chippings. This window is filled with small chippings of size 2/4 mm (for the 10 mm principal chipping) or 2·8/6 mm (for the 14 mm principal chipping) to achieve high mechanical interlock (called a 'mosaic'). Generally speaking, 3 mm chippings are not readily available in the UK. The rate of application of binder on this type of surface dressing is usually slightly higher than that which would be required for a single-size dressing.

The initial surface texture and mechanical strength of a racked-in dressing are high. The configuration of the principal chippings is different from that which is achieved with a single dressing. The mosaic, referred to above, cannot be formed because the principal chippings are locked in place. In time, small chippings that are not in contact with binder are lost to the system without damaging vehicles or windscreens, resulting in increasing surface texture despite some coincident embedment of the larger chipping.

On a racked-in dressing, vehicle tyre contact is principally on the large aggregate with little contact on the small aggregate. Thus, the microtexture of the small aggregate is less critical and consideration should be given to the use of cheaper lower PSV aggregates, such as limestone, for the small chippings.

The advantages of the racked-in system are:

- virtual elimination of 'flying' 8/14 mm or 6/10 mm chippings, thereby reducing the incidence of broken car windscreens
- early stability of the dressing through good mechanical interlock
- good adhesion of larger size chippings

395

- a rugous surface texture with an initial texture depth exceeding 3·0 mm.

Double surface dressing

In this technique, two surface dressings are laid consecutively, the first consisting of 6/10 mm or 8/14 mm chippings and the second of 2·8/6 mm chippings. It is used on main roads as an alternative to the racked-in system to provide a mechanically strong dressing with a texture that is marginally less than a racked-in dressing, and therefore quieter. The method is also suitable for minor roads that have become very dry and lean in binder.

Inverted double surface dressing

In this system, two surface dressings are carried out consecutively. The first uses 2·8/6 mm chippings followed by a second application of 6/10 mm or 8/14 mm chippings. This technique is appropriate for use on minor concrete roads where chipping embedment does not occur and surface texture is not an important issue or on sections of road that have been widely patched.

Sandwich surface dressing

Sandwich surface dressing is used in situations where the road surface condition is very rich in binder. The first layer of 8/14 mm or 6/10 mm chippings is spread onto the existing road surface before any binder is applied. The binder is then sprayed over these chippings followed by an application of 2·8/6 mm chippings.

Sandwich dressings can be considered as double dressings in which the first binder film has already been applied. The degree of binder richness at the surface has to be sufficient to hold the first layer of chippings in place until the remainder of the operation has been completed. The sizes of the chippings must be chosen such that they are appropriate for the quantity of excess binder on the surface and the rate of application of the second coat of binder.

19.1.5 Surface dressing equipment

Improvements in the design of surface dressings and advances in binder technology and aggregate quality have been major contributory factors in improving the performance of surface dressings. Developments in application machinery in conjunction with the above system developments have enabled the whole process to be improved in terms of the speed of operation, compliance with specification and continued improvements in standards of safety.

Fig. 19.7 A modern surface dressing sprayer
Courtesy of Colas Ltd

Sprayers

In the UK, surface dressing sprayers must satisfy the requirements of BS 1707: 1989[513]. This sets requirements for evenness of distribution of the spray bar, heat retention, capacity of heaters and other important issues.

In the UK, it has been traditional for spray bars to have swirl jets. However, the increased use of polymer modified binders has resulted in some sprayers being equipped with bars having slot jets. The need to undertake work quickly on main roads with the minimum number of longitudinal joints has led to the use of expanding spray bars of length up to 4·2 m. In the UK, the usual way of checking that the rate of spread across the bar is uniform within 15% or better has been by the depot tray test that is detailed in BS 1707. This is normally checked before the start of the surface dressing season.

Figure 19.7 shows a modern surface dressing sprayer. Programmable bitumen distributors have been developed. These are extremely sophisticated spray machines using two bars to achieve differential rates of spread across the width of the machine. Application rates can be changed for areas of patching, wheel tracks, binder-rich fatty areas and binder-lean areas. The rate of spread is automatically reduced by up to 30% in the wheel tracks and can be increased between wheel tracks or in shaded

513 BRITISH STANDARDS INSTITUTION. *Specification for hot binder distributors for road surface dressing, BS 1707: 1989.* BSI, London.

areas. Both longitudinal and transverse binder distributions can be pre-
programmed. Vehicle speed, binder temperature, spray bar width and
application rates are controlled by an on-board computer.

A third type of spray bar is now available. Being computer controlled,
it is very accurate. The computer controls the management of the bar
pump and the forward speed of the sprayer. If sensors indicate that
conditions have changed outwith certain parameters, e.g. up or down
steep hills, then spraying stops. Accuracy of this equipment has been
shown to be ±5%.

Chipping spreaders

Figure 19.8 shows an example of a self-propelled forward-driven
chipping spreader. These machines not only allow work to proceed
rapidly but also spread the chippings more evenly and accurately at the
required rate of spread. The chippings are released much closer to the
road surface reducing the likelihood of chippings bouncing either elsewhere on the road surface or off the road surface altogether and exposing
binder.

Fig. 19.8 A modern chipping spreader
Courtesy of Colas Ltd

The latest developments in self-propelled chipping spreaders include:

- a four-wheel drive for pulling heavy tippers up steep inclines without juddering;
- a mechanism for breaking lumps of lightly coated chippings within the hopper;
- methods for spraying additive onto the binder film;
- incorporation of a pneumatic-tyred roller to ensure that the chippings are pressed into the binder film at the earliest possible moment while it is still hot.

Rollers

Rubber-covered steel-drummed vibratory rollers (see Fig. 19.9) are now regarded as the best method of establishing a close mosaic and ensuring initial bond of the chipping to the binder film without crushing the carefully selected aggregate. Pneumatic-tyred rollers are also used and some rolling is still carried out using steel-wheeled rollers. One of the risks of using this type of roller is that chippings can be crushed as a result of point loading. It should be noted that rolling is largely a preliminary process undertaken before the main stabilisation of the dressing by subsequent trafficking at slow speed.

Sweepers

It is necessary to remove surplus chippings before a surface dressing is opened to traffic travelling at uncontrolled speed. This is normally

Fig. 19.9 Rubber-covered steel-drummed vibratory roller
Courtesy of Colas Ltd

achieved with full-width sweepers that often have a full-width sucking capability. Where sections of motorways or other major roads have been closed to allow surface dressing to take place, sweepers are often used in echelon in order to ensure the rapid removal of surplus chippings. Sweeping that is properly organised and implemented will do much to alleviate the damage caused to surface dressings in their early life through chipping loss.

Traffic control and aftercare

Traffic control and aftercare are vital if the dressing is to be properly embedded. Ideally, when traffic is first allowed onto the new dressing, it should be behind a control vehicle that travels at a speed of about 15 km/h. If there are gaps in the traffic, it may be necessary to introduce additional control vehicles. The objectives should be to ensure that vehicles do not travel at more than 15 km/h on the new dressing and to prevent sharp braking and acceleration of these vehicles.

The length of time for which control is necessary will depend largely on the type of binder used and the prevailing weather conditions. It is likely to be longest where emulsions are being used and the weather conditions are humid. The use of wet or dusty chippings or the early onset of rain delays adhesion and necessitates a longer period of traffic control. Thus, ensuring that the chippings are clean, by washing if necessary, and that they comply with grading requirements will minimise the period necessary for traffic control and reduce disruption. Traffic passing slowly over a dressing immediately after completion creates a strong interlocking mosaic of chippings, a result only otherwise obtained by long periods of rolling.

19.1.6 Types of surface dressing failures

The majority of surface dressing failures fall into one of the following five categories:

- loss of chippings immediately or soon after laying
- loss of chippings during the first winter
- loss of chippings in later years
- bleeding during the first hot summer
- fatting-up in subsequent years.

The type and rate of application of the binder and size of chippings has a large influence on performance. Incorrect application rates are a frequent cause of premature failure. Every care should be taken to ensure that the selected application rates are maintained throughout the surface dressing operation.

Very early loss of chippings may be due to slow breaking in the case of emulsions or poor wetting in the case of cutback bitumens. The use of cationic rapid setting emulsions with high binder contents (more than 70% binder) largely overcomes the former problem and the use of adhesion agents and/or pre-coated chippings should negate the latter difficulty.

Many surface dressing failures only become apparent at the onset of the first prolonged frosts when the binder is very stiff and brittle. These can normally be attributed to a combination of inadequate binder application rate, the use of too large a chipping or inadequate embedment of the chippings into the old road due to the surface being too hard and/or there being insufficient time between applying the dressing and the onset of cold weather. Inadequate binder application and/or the use of excessively large chippings will exacerbate this problem. Loss of chippings in the long term usually results from a combination of low-durability binders and, again, poor embedment of chippings.

Bleeding may occur within a few weeks in dressings laid early in the season or during the following summer in dressings that were laid late the previous year. It results from the use of binders that have a viscosity that is too low for the high ambient temperatures or from the use of binders that have high temperature susceptibility.

Fatting-up is complex in nature and may result from one or more of the following factors.

- *Application of excessive binder*.
- *Embedment of chippings*.   Gradual embedment of chippings into the road surface causing the relative rise of binder between the chippings to a point where the chippings disappear beneath the surface. This is largely dependent on the intensity of traffic particularly the proportion of heavy commercial vehicles in the total traffic. The composition of the binder also affects the process. Cutback bitumens that have a high solvent content can soften the old road surface and accelerate embedment.
- *Crushing of chippings*.   Most chippings will eventually split or crush under heavy traffic with the loss from the dressing of many small fragments. The binder-to-chipping ratio therefore tends to increase adding to the process of fatting-up. A dilemma here is that many of the best aggregates for skid resistance often have a poor resistance to crushing.
- *Absorption of dust by the binder*.   Binders of high durability tend to absorb dust that falls upon them. Soft binders can absorb large quantities of dust thereby increasing the effective volume of the binder. This effect coupled with chipping embedment, leads to the eventual loss of surface texture.

401

19.1.7 Surface dressing season

It is clear from the above that the success of a surface dressing operation may well depend on the weather conditions during and following the application of the dressing. Generally speaking, surface dressing is undertaken from May to September in the UK. This varies throughout Europe and there are areas where the period is much shorter. In the Western Isles of Scotland, surface dressing using both conventional and polymer modified emulsions can only be carried out from mid May until the end of June.

At the start of the surface dressing season, environmental factors are critical. The weather conditions during the early life of the dressing have a significant influence on performance. For example, high humidity can significantly delay the break of a surface dressing emulsion that, in turn, delays opening the road to traffic. Heavy rainfall immediately after a road has been surface dressed can have a detrimental effect on the bond between the binder and the aggregate, which can result in stripping of the binder from the aggregate.

In the first summer, it is important that some chipping embedment into the existing road surface occurs as this will aid chipping retention during the first winter when the binder is at its stiffest and most vulnerable to brittle failure. Therefore, the later in the season the dressing is applied the more likely it is that chipping loss will occur during the first winter, particularly where the traffic volumes are insufficient to promote initial embedment. This explains why, in areas such as the Western Isles of Scotland, the season finishes at the end of June rather than September.

19.1.8 Safety

The following briefly highlights some of the potential hazards associated with the handling of surface dressing binders. However, this information is no substitute for the health and safety information available from individual suppliers or the advice contained in the relevant codes of practice[514,515].

Handling temperatures

Cutback bitumens are commonly handled at temperatures that are well in excess of their flash points, i.e. the temperature at which the vapour given off will burn in the presence of air and an ignition source. Accordingly, it is essential to exclude sources of ignition in the proximity of cutback handling operations by displaying suitable safety notices.

514 ROAD SURFACE DRESSING ASSOCIATION. *RSDA code of practice for surface dressing, 3rd ed, Oct 2002.* RSDA, Colchester.

515 INSTITUTE OF PETROLEUM. *The Institute of Petroleum bitumen safety code, Model code of safe practice, part 11, 3rd ed, 1991.* IP, London.

Table 19.6 Recommended handling temperatures for cutback binders[516]

Grade of cutback binder: sec	Temperature: °C		
	Minimum pumping[a]	Spraying[b]	Maximum safe handling[c]
50	65	150	160
100	70	160	170
200	80	170	180

[a] Based on a viscosity of 2000 cSt (2 Pa s).
[b] Based on a viscosity of 30 cSt (0·03 Pa s). Conforms with the maximum spraying temperatures recommended in Road Note 39.
[c] Based on generally satisfactory experience of the storage and handling of cutback grades in contact with air. Subject to the avoidance of sources of ignition in the vicinity of tank vents and open air operations.

Every operation should be carried out at a temperature that is as low as possible to minimise the risks from burns, fumes, flammable atmosphere and fire. Such temperatures must always be less than the maxima given in Table 19.6. Some bitumen emulsions are applied at ambient temperatures but the majority of high bitumen content cationic emulsions are applied at temperatures up to 85°C.

Precautions, personal protective clothing and hygiene for personnel
All operatives should wear protective outer clothing when spraying emulsions or cutback bitumen binders. This includes eye protection, clean overalls and impervious shoes and gloves to protect against splashes and avoid skin contact. In addition, operatives working near the spraybar should wear orinasal fume masks.

Although bitumen emulsions are applied at relatively low temperatures, the protection prescribed above is essential because emulsifiers are complex chemical compounds and prolonged contact may result in allergic reactions or other skin conditions. In addition, most surface dressing emulsions are highly acidic and, therefore, require the use of appropriate personal protective equipment (PPE). Barrier creams, applied to the skin prior to spraying, assist in subsequent cleaning should accidental contact occur but these are no substitute for gloves and other PPE. If any bitumen comes into contact with the skin, operatives should wash thoroughly as soon as practicable and always before going to the toilet, eating or drinking.

19.2 Slurry surfacings/microsurfacings
Slurry surfacing was first introduced into the UK in the late 1950s and was primarily used as a preservative treatment for airfield runways.

516 INSTITUTE OF PETROLEUM. *The Institute of Petroleum bitumen safety code, Model code of safe practice, part 11, 3rd ed, 1991.* IP, London.

The introduction of quick-setting cationic emulsions and the development of polymer modified emulsions for thick film slurry applications, now termed microsurfacings, has broadened their use in highway maintenance. Development of slurry surfacings has largely been on a proprietary basis with individual companies producing their own formulations, specifications, manufacturing and laying techniques. The basic specifications are included in both BS 434-1: 1984[517] and the SHW[518].

The purpose of slurry surfacing is to:

- seal surface cracks and voids from the ingress of air and water
- arrest disintegration and fretting of an existing surface
- fill minor surface depressions to provide a more regular surface
- provide a uniform surface appearance to a paved area.

19.2.1 Applications

- *Footway slurry*. Usually this is mechanically mixed and hand-applied using a brush to provide surface texture. This protects the footway and is aesthetically pleasing. Modifiers can be used to give the material greater cohesive strength and coloured slurries can be used to denote cycle tracks.
- *Thin carriageway slurries*. These are usually laid 3 mm thick and are applied by a continuous-flow machine. This type of slurry can be laid rapidly with minimal disruption to traffic. The texture depth that is achieved with this material is relatively low so it is only suitable for sites carrying relatively slow-moving traffic.
- *Thick carriageway slurries*. These are normally single treatments using 2·8/6 mm aggregates blended with fast-breaking polymer modified emulsions. They are more durable than thin slurries and are, therefore, suitable for more heavily trafficked locations.
- *Micro-asphalts*. Micro-asphalts typically use aggregates up to 6/10 mm and are placed in two layers. The first layer regulates and reprofiles whilst the second layer provides a dense surface with reasonable texture.
- *Airfield slurries*. Slurry surfacing has been used for runways and taxiways for many years. It is an ideal maintenance technique for airfields as the application is relatively rapid and the aggregates used are too small to present significant foreign object damage hazard to jet engines.

517 BRITISH STANDARDS INSTITUTION. *Bitumen road emulsions (anionic and cationic), Specification for bitumen road emulsion, BS 434-1: 1984.* BSI, London.
518 HIGHWAYS AGENCY *et al. Manual of contract documents for highway works, Specification for highway works, vol 1, cl 918, 2002.* The Stationery Office, London.

19.2.2 Usage, materials and developments

The texture depth achieved with slurry surfacing is relatively low and therefore the material is not suitable for roads carrying high-speed traffic. Slurry surfacing only provides a thin veneer treatment to an existing paved area. Consequently, it does not add any significant strength to the road structure nor will it be durable if laid on an inadequate substrate. The process has been restricted to lightly trafficked situations such as housing estate roads, footways, light airfield runways, car parks, etc. although specially designed slurry surfacings providing a higher texture depth have been used on the hard shoulders of motorways.

The mixed slurry comprises aggregate (crushed igneous rock, limestone or slag), an additive and a bitumen emulsion. All the components have to be chosen very carefully as each affects the setting time and performance of the mixture. Accordingly, it is often the case that a dedicated emulsion is developed for each aggregate source and size. The additive is usually Portland cement or hydrated lime which is added to the mixture to control its consistency, setting rate and degree of segregation. It normally represents about 2% of the total aggregate. Where rapid setting is required to permit early trafficking or to guard against early damage from wet weather, a rapid-setting anionic emulsion (Class A1) or cationic emulsion (Class K3) is used. Where setting time is not critical a slow-setting anionic emulsion (Class A4) can be used. Depending on the characteristics of the aggregate, between 180 and 250 litres of emulsion are required per tonne of aggregate. The composition of the cured material is about 80% aggregate and 20% bitumen.

Specialist continuous-flow machines that meter and deliver aggregate and emulsion and lay the mixture are used. In these machines, the ingredients are transported into a mixer that feeds a spreader box towed behind the machine. They are capable of laying up to $8000\,m^2$/day. Light compaction with a pneumatic-tyred roller may be required depending on the composition of the slurry. Rolling is applied as soon as the material has set sufficiently to support the compaction plant.

Recent developments in binder technology utilising polymer modification are now rapidly changing the scope and use of slurry sealing and further development work continues to enable the use of larger aggregate and thicker layers. Aggregate grading up to 6/10 mm has been tried and thicknesses up to 20 mm have been laid successfully. The use of fibre reinforcement is being developed and it seems likely that increasing the use of this technique will help to overcome some of the problems of heavily cracked surfaces.

19.3 High-friction surfaces

A study carried out by the Greater London Council (GLC) in 1965 showed that 70% of all road accidents occurred at or within 15 m of conflict locations such as road junctions, pedestrian crossings[519]. Les Hatherly, then the Chief Engineer of the GLC, approached Shell to produce a surfacing suitable for this type of site. Shell developed a bitumen-extended epoxy resin system and, following various successful road trials, the result was Shellgrip®.

In recent years, alternative systems have been developed including polyurethane resin, acrylic resin and thermoplastic rosin ester materials. A resin is generally a manufactured material or a natural secretion from certain plants whereas rosin is a hard residue from the distillation of turpentine.

- *Bitumen-extended epoxy resins.* These are produced by blending two components, one containing a resin and the other a hardener. When combined, they react chemically to form a very strong three-dimensional structure. A specially designed machine applies the material onto the road followed by the application of calcined bauxite. The curing time is usually between 2 and 4 hours. The excess aggregate is swept off the surface and the road can be opened to traffic.
- *Epoxy resin.* Such systems are a comparatively recent development. They have the advantages that they set very rapidly and adhere well to most surfaces including concrete.
- *Polyurethane resin.* These resins are normally two- or three-part binder systems with good adhesion to most surfaces. Some systems require a primer to be applied prior to application of the resin. They are normally applied by hand.
- *Acrylic resins.* These are fast-setting two-component systems that adhere well to most surfaces. They are transparent, which makes them ideal for pigmenting.
- *Thermoplastic rosin esters.* These are normally blended with calcined bauxite and heated and hand-screeded onto the road surface. The material cures rapidly and may be pigmented.

The UK's national road construction and maintenance specification recognises the benefits of these materials and permits the use of systems covered by HAPAS certification[520].

519 HATHERLY L W and D R LAMB. *Accident prevention in London by road surface improvements, Reprint of a paper given to the International Road Federation 6th World Highway Conf, Montreal, Oct 1970.* Shell International Petroleum Company, London.

520 HIGHWAYS AGENCY *et al. Manual of contract documents for highway works, Specification for highway works, vol 1, series 900, cl 924, 2002.* The Stationery Office, London.

19.4 Foamed bitumen

In order to mix bitumen with aggregates, it is necessary for the bitumen to have a viscosity around 0·2 Pa s. Traditionally, this is done by heating the bitumen and mixing it with heated aggregates to produce hot mix asphalt. To reduce the binder viscosity a number of methods can be used:

- addition of solvents, e.g. cutbacks (see Section 19.1.2), but this may have disadvantages in relation to health and safety considerations;
- emulsification of bitumen, a dispersion of bitumen, in very small droplets, in a bitumen-in-water emulsion as discussed in Chapter 6;
- foaming of bitumen.

Foaming of bitumen is a means of temporarily reducing the binder viscosity and increasing the binder volume.

19.4.1 Foam production

Foaming of bitumen and its application was first introduced at the end of the 1950s. Professor Csanyi at the Engineering Experiment Station of Iowa State University studied the potential of foamed bitumen in cold asphalt applications[521]. At that time, steam was injected under pressure into hot bitumen through specially designed nozzles yielding bitumen in the form of foam. This process was found to be impractical due to the complexity of the equipment and the difficulties in accurately metering the steam. Mobil Oil refined the foaming technology in the 1960s and developed an expansion chamber in which cold water at a level between 1 and 5% by weight of bitumen is injected under pressure into hot bitumen to produce foam[522–524].

19.4.2 Foam characterisation

The rapid evaporation of water produces a very large volume of foam that slowly collapses with time. The foam is usually characterised in

521 CSANYI L H. *Foamed asphalt in bituminous paving mixtures, Bulletin 160, pp 108-122, 1957*. Highway Research Board, Washington DC.

522 BOWERING R H and C L MARTIN. *Foamed bitumen production and application of mixtures evaluation and performance of pavements, Proc Assoc Asph Pav Tech, vol 45, pp 453–477, 1976*. Association of Asphalt Paving Technologists, Seattle.

523 MACCARRONE S, G HOLLERAN and A KEY. *Cold asphalt systems as an alternative to hot mix, 9th AAPA Int Asphalt Conf, 1995*.

524 MACCARONE S, G HOLLERAN, D J LEONARD and S HEY. *Pavement recycling using foamed bitumen, Proc 17th ARRB Conf, Australia, vol 17, part 3, pp 349–365, 1994*.

terms of its expansion ratio and its lifetime (or half-life)[525,526]. The expansion ratio is defined as the ratio of the volume of foam produced to the volume of liquid bitumen injected. The lifetime is expressed as the 'half-life' of the foam, defined as the time after which the maximum volume of foam is reduced by a factor of two. It is measured in seconds. More recently, a 'foamindex' has been proposed. This is the area under the curve of a plot of expansion ratio v. the half-life of the foam. A minimum expansion ratio for determining the foamindex has been proposed at a value of 4[525]. Laboratory foam equipment can be used to determine the 'foamability' of the bitumen.

An important factor in foaming is the design of the nozzle and the injection pressure necessary to obtain a good water droplet spray in contact with the hot bitumen. The foaming characteristics of a specific bitumen are further influenced by the following.

- *The temperature of the bitumen.* For most bitumens, the foaming characteristics improve with higher temperatures.
- *The expansion ratio.* As more water is added the expansion ratio increases whilst the half-life decreases. The water helps in creating the foam but the foam can collapse quickly due to rapidly escaping steam.
- *Addition of compounds.* Some compounds can be effective anti-foaming agents, e.g. silicone compounds. In contrast, some compounds can increase the half-life of the foam from seconds to minutes.

An example of the relationship between expansion ratio and water content and also half-life and water content is shown in Fig. 19.10.

19.4.3 *Mixing foamed bitumen and aggregate*
The foamed bitumen exits through nozzles and can be mixed directly with the mineral aggregate. It is widely used in stabilising granular materials and for cold-in-place recycling[526–529]. In these applications,

525 JENKINS K J, J L A DE GROOT, M F C VAN DE VEN. *Characterization of foamed bitumen, 7th Conf on asphalt pavements for Southern Africa, CAPSA '99, Victoria Falls, Aug–Sep 1999.*

526 WIRTGEN GROUP. *Foamed bitumen – the innovative technology for road construction, Wirtgen cold recycling manual, Windhagen, Germany, Nov 1998.*

527 BROSSEAUD Y, J-C GRAMSAMMER, J-P KERZREHO, H GOACOLOU and F LE BOURLOT. *Experimentation (premiere partie) de la Grave-Mousse sur le manege de fatigue, RGRA no 752, Jun 1997, no 754.*

528 BONVALLET J. *La mousse de bitume, une technique émergente, probablement incontourn-able, Rev Gen des Routes et Aer, no 789, Nov 2000.*

529 KHWEIR K, D FORDYCE and G MCCABE. *Aspects influencing the performance of foamed bitumen, Asphalt Yearbook 2001, pp 27–34.* Institute of Asphalt Technology, Dorking.

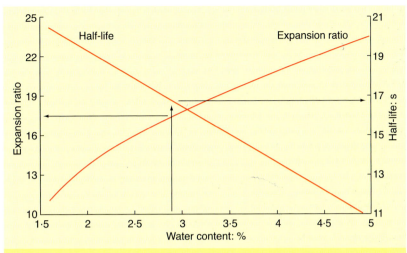

Fig. 19.10 Example of the relationship between expansion ratio/water content and half-life/water content for a bitumen

the foamed bitumen is applied to cold, moist aggregate. The bitumen 'bubbles' burst into tiny 'patches' that adhere, mainly, to the fine aggregate fraction. The resulting mortar of fine aggregate and bitumen binds the coarse aggregate particles after compaction. The foaming characteristics, the moisture content and aggregate grading curve together with adequate compaction are key factors in maximising the performance of aggregate mixtures stabilised with foamed bitumen.

Mixtures with low fines content will not mix well with foamed bitumen because the bitumen will not disperse properly. The expansion ratio of the foam should be high enough to coat as much of the aggregate as possible after the introduction of the foam into the mixer. The type and size of the mixer is important in order to obtain a homogeneous asphalt. The half-life of the foam is an important parameter for the distribution of the foam into the aggregate mixture. If the half-life is too short, the bitumen reverts to a high-viscosity liquid, with the risk of lumps of bitumen or pieces of mortar/mastic being formed. A high value of half-life may produce a mixture that has good workability. The lifetime of the bitumen foam can range from about 20 s up to a few minutes. Long lifetimes are obtained through additives. Optimum values of expansion ratio and foam half-life will depend on the application. A key determinant is the residence time in the pugmill[530].

---

530 BONVALLET J. *Fabrication industrielle et contrôlée d'enrobes à froid avec des mousses de bitume, IRF congress, paper O027, Paris, Jun 2001.*

Traditional mixing procedures that are adequate for the production of hot mixed (and cold mixed) materials have to be adapted in order to accommodate the different characteristics of foamed bitumen.

- In mixing with hot bitumen or bitumen emulsions, the actual mixing has to ensure good coverage. This consumes mixing energy and time.
- Foam properties and mixing times must be set to ensure a homogeneous mixture. Good distribution of the bitumen in the mixture is obtained very quickly as a result of high expansion of the bitumen foam.
- The volume expansion must be such that the bitumen is distributed evenly in the asphalt mixture.

19.4.4 Uses of foamed bitumen

The use of foamed bitumen in base stabilisation and in situ recycling in base layers is widely accepted as an effective technique[531–535]. The process is adaptable to treat virgin materials and recycled aggregates without heating and drying, thus saving energy compared to traditional hot mix production.

Mixed material can be stockpiled for a period before being used, reducing the amount of waste. Although foamed bitumen is suitable for stabilisation and recycling work, it cannot be objectively regarded as being equivalent to traditional hot mixed asphalts. In general terms, most cold products are currently inferior to hot mixed asphalts. Three factors will influence the possibility of a move from traditional asphalts to cold mixed mixtures – cost, market demand and the performance of the cold mixed materials.

19.4.5 Warm asphalt mixtures

Recent developments have resulted in improvements to the efficiency of the foam mixing process. This has been achieved by warming the mixture

531 WIRTGEN GMBH. *Foamed bitumen – the innovative technology for road construction, Wirtgen cold recycling manual, Windhagen, Germany, Nov 1998.*

532 BROSSEAUD Y, J-C GRAMSAMMER, J-P KERZREHO, H GOACOLOU and F LE BOURLOT. *Expérimentation (première partie) de la Grave-Mousse sur le manege de fatigue, Rev Gen des Routes et Aer, no 752, Jun 1997, no 754.*

533 BONVALLET J. *La mousse de bitume, une technique émergente, probablement incontournable, Rev Gen des Routes et Aer, no 789, Nov 2000.*

534 KHWEIR K, D FORDYCE and G MCCABE. *Aspects influencing the performance of foamed bitumen, Asphalt Yearbook 2001, pp 27–34.* Institute of Asphalt Technology, Dorking.

535 BONVALLET J. *Fabrication industrielle et contrôlée d'enrobes à froid avec des mousses de bitume, IRF congress, paper O027, Paris, Jun 2001.*

during manufacture to temperatures of between 80 and 100°C and by using two bitumen components to achieve full coating while maintaining workability during paving operations[536–540].

Increasing the temperature to an intermediate level improves the degree of coating. This is particularly true in regard to the larger aggregate fraction. Further handling and quality improvements are under investigation using two bitumen components. Laboratory studies in conjunction with small-scale and large-scale site trials show promising results in achieving a high degree of coating and a performance matching that of hot mixtures.

Reduction of operating temperatures results in lower fuel consumption and reduced emission of carbon dioxide. Energy savings obtained at 100°C are in the range of some 25 to 30% but are dependent on the moisture content of the aggregates. Significant further reductions can be obtained at temperatures below 100°C. This offers opportunities in the design of asphalt production plants in optimising intermediate-temperature processes. A secondary benefit of lower asphalt operating temperatures is the reduction of emissions from the plant.

19.5 Application of a coloured surface treatment

There are four surface treatments that can be applied to an asphalt surfacing to provide a decorative finish:

- pigmented slurry surfacings
- pigmented thermosetting or thermoplastic systems
- surface dressing
- coloured paints.

536 KOENDERS B G, D A STOKER, C BOWEN, P DE GROOT, O LARSEN, D HARDY and K P WILMS. *Innovative process in asphalt production and application to obtain lower operating temperatures, 2nd Eurasphalt & Eurobitume Congress, Barcelona, Book 2, Session 3, pp 831–840, 20–22 Sep 2000.*

537 DE GROOT P C, C BOWEN, B G KOENDERS, D A STOKER, O LARSEN and J JOHANSEN. *A comparison of emissions from hot mixture and warm asphalt mixture production, Proc IRF congress, paper O022, Paris, Jun 2001.*

538 JENKINS K J, A A A MOLENAAR, J L A DE GROOT and M F C VAN DE VEN. *Foamed bitumen treatment of warmed aggregates, 2nd Eurasphalt & Eurobitume Congress, Barcelona, Book 2, Session 2, pp 280–288, 20–22 Sep 2000.*

539 JENKINS K J, J L A DE GROOT, M F C VAN DE VEN and A A A MOLENAAR. *Half-warm foamed bitumen treatment, a new process, 7th Conf on asphalt pavements for Southern Africa, CAPSA '99, Victoria Falls, Aug–Sep 1999.*

540 JENKINS K J, M F C VAN DE VAN. *Mix design considerations for foamed bitumen mixes, 7th Conf on asphalt pavements for Southern Africa, CAPSA '99, Victoria Falls, Aug–Sep 1999.*

Table 19.7 The uses, benefits and disadvantages of thermosetting and thermoplastic coloured surfacings[541]

	Thermosetting systems	Thermoplastic systems
Uses	Suitable for coloured anti-skid surfacing and all traffic calming and delineation schemes	Most traffic calming and traffic delineation schemes
Benefits	• Suitable for high-stress sites • Anti-skid coloured surfacing to Cl 924 specification • Machine or hand application • Provides continuous seam-free textured surface	• Versatile material extensively used in traffic calming and low-cost accident remedial measures • Can be used all year round and can be applied at temperatures as low as 0°C • Fast setting – can normally be trafficked within 15 to 45 minutes • Good retention of colour as these systems encapsulate the pigment-enhancing colour durability • Lower initial cost than thermosetting systems
Disadvantages	• Extended curing times at low ambient temperature • Loss of colour under heavy trafficking • Relatively high initial installation cost	• Not as durable as thermosetting systems and tend to wear faster at locations subjected to high densities of commercial vehicles • As it is hand-screeded butt joints and screed lines are regarded by some as aesthetically undesirable • Careful control of heating required to avoid overheating of the material resulting in degradation of the pigment and possible colour changes

Pigmented slurry surfacings are available as proprietary products in a range of colours. They are applied in a very thin layer, about 3 mm, and are only suitable for pedestrian and lightly trafficked areas.

Thermosetting and thermoplastic materials are probably the most common method used to produce a coloured contrast for traffic calming, bus and cycle delineation. Thermosetting systems are two-component epoxy or polyurethane resin adhesives with a naturally coloured or dyed bauxite applied to the surface. Thermoplastic systems are hot applied with pigment and aggregate dispersal within the system that is screeded by hand onto the surface. A summary of the uses, benefits and disadvantages of these two systems is given in Table 19.7.

Surface dressings are suitable for most categories of road application but are less suitable for pedestrian situations. The final colour of the surface dressing will be that of the aggregate used. Proprietary coloured paints are available for overpainting conventional black surface courses. These are only suitable for pedestrian and games areas, e.g. tennis courts.

---

541 LILES P A. *Delivering value – coloured surfacings, CSS Conf Value Engineering in Maintenance, Leamington Spa, Nov 1999.*

19.6 Recycling asphalts

A limited amount of re-use or recycling of old asphalt mixtures has been carried out in the UK for many years. Material removed from roads prior to resurfacing has been used as a fill material and for regulating layers for paved areas carrying light traffic. A form of in situ recycling involving the scarification of the existing road surface, mixing this material with fresh bitumen, followed by profiling and compaction, was carried out in the UK as early as 1937.

Recycling of asphalts can be broadly divided into two categories – in situ (or surface) recycling and off-site (or hot mix) recycling.

19.6.1 In situ recycling

In the UK, in situ recycling can be further subdivided into four processes known as 'repaving', 'remixing', 'retread' and 'deep recycling'. Repaving and remixing are hot in situ recycling processes whereas retread and deep recycling are cold processes.

The repave process, which was introduced into the UK in 1975, involves heating and scarifying the existing surface to a depth of about 20 mm. Approximately 20 mm of new asphalt is laid directly onto the hot scarified material and then the material is compacted. These operations can be completed by a single pass of a purpose-built machine. As both layers are hot when compacted, a good bond is achieved between the recycled material and the new material. In the remixing process, a hot scarified material is mixed with new material in the pugmill mixer of a machine and the blended mixture is paved and compacted on the scarified surface. A report on the surface characteristics of roads using the repave process has been published by the TRL[542]. The current UK national design and specification standards contain provisions relating to the use of the repave and remix processes[543,544].

Retread, a cold in situ procedure, has been established UK practice since the middle of the twentieth century. This process consists of scarifying the existing road to a depth of approximately 75 mm, breaking down the scarified material to the required size and reshaping of the road profile. The material is then oversprayed two or three times with a bitumen emulsion. After each application of emulsion the material is

542 COOPER D R C and J C YOUNG. *Surface characteristics of roads resurfaced using the repave process, Supplementary Report 744, 1982.* Transport and Road Research Laboratory, Crowthorne.

543 HIGHWAYS AGENCY *et al. Design manual for roads and bridges, Pavement design and maintenance, Traffic assessment, vol 7, HD 31/94, ch 6, 1996.* The Stationery Office, London.

544 HIGHWAYS AGENCY *et al. Manual of contract documents for highway works, Specification for highway works, vol 1, cl 926, 2002.* The Stationery Office, London.

413

harrowed to distribute the emulsion. When harrowing has been completed, the material is compacted with an 8 to 10 tonne dead weight roller and finally the surface is sealed with a surface dressing. This process has been successfully carried out on minor roads, housing estate roads, etc. for many years. Currently, approximately 1 million m^2 of this type of carriageway are annually recycled using the retread process. Further information on the process is available in the literature[545,546]. Deep recycling is a cold in situ process for full-depth reconstruction of pavements. The road is pulverised, graded if necessary and mixed with cement as stabiliser and bitumen, usually with compaction between each stage of the process. The binder can be foamed bitumen, produced in situ by injecting a very small proportion of water and additives into the bitumen causing it to expand to 10 to 15 times its original volume. The resulting low viscosity makes it easy for the binder to mix with cold or damp aggregates. Alternatively, a bitumen emulsion can be used. Deep recycling is suitable for treating material depths between 125 and 300 mm.

19.6.2 *Off-site plant recycling*

In this process, material removed from the surface of an existing road is transported to a hot mix plant where it may be stockpiled for future use or processed immediately. Both batch and continuous plants have been successfully converted to produce recycled mixtures[547] utilising a range of methods to heat the reclaimed material prior to mixing (see Section 14.2). This success manifests itself in the fact that recycling can now be undertaken without excessive fuming and blue smoke emissions during manufacture and laboratory performance tests on asphalt containing a proportion of recycled material were found to be identical to mixtures using virgin components[548].

Recycling in batch plants is achieved by superheating the virgin aggregate and then adding the material to be recycled either immediately after the drier or directly into the pugmill. Heat transfer takes place during the mixing cycle. Although this method successfully overcomes the problem of blue smoke, it entails keeping aggregate in the heating drum for longer, which means that the output from the plant is reduced.

545 ROAD EMULSION ASSOCIATION LTD. *The retread process, Technical Data Sheet No 10, Mar 1998.* REAL, Clacton-On-Sea.

546 DINNEN A. *Recycling – the retread process, Shell Bitumen Review 60, p 12, Mar 1983.* Shell International Petroleum Company Ltd, London.

547 SERVAS V. *Hot mix recycling of bituminous pavement materials, The Highway Engineer, vol 27, no 12, pp 2–8, Dec 1980.*

548 STOCK A F. *Structural properties of recycled mixes, Highways and Transportation, vol 32, pp 8–12, Mar 1985.* Institution of Highways & Transportation, London.

As it is not possible to obtain adequate heat transfer with high percentages of recycled material, it is widely accepted that the maximum quantity of recycled material that can be economically added is between 25 and 40%. Other disadvantages using batch plants for recycling are high heating costs and accelerated wear and tear on the drum and dust collectors due to the higher manufacturing temperatures.

Recycling using continuous mixers involves introduction of the reclaimed material into the drum itself. During early trials using this type of plant, environmentally unacceptable blue smoke was produced. Blue smoke is produced when vapourised bitumen condenses. The condensate takes the form of particles that are too small to be removed by conventional emission control equipment and so escape through the stack. Bitumen vaporises at about 450°C and as gas temperatures in the drum of a continuous mixer can reach 2000°C, the introduction of reclaimed material has to be carefully controlled. By various modifications, continuous mixers that can take up to 60% recycled material without exceeding statutory pollution standards have been designed. The entry point for the recycled material is approximately half way down the drum. The flights in the drum have been modified to produce a homogeneous mixture of virgin and recycled material prior to the addition of the bitumen and filler. In addition, the redesigned flights shield the recycled material from the intense radiant heat originating from the heating unit.

Environmental and governmental pressures are certain to continue to encourage industries to increase the amount of material that it regularly recycles. Consequently, technology associated with recycling will continue to develop.

19.7 Grouted macadams

Grouted macadams are proprietary products, the most well known being 'Hardicrete'. They are not controlled by a British Standard but are the subject of an Agrément Certificate. Grouted macadams are available in two variants, cementitious grouted macadams and asphaltic grouted macadams.

19.7.1 Cementitious grouted macadams

Grouted macadams are used in areas where loading is particularly heavy or concentrated, where there is likely to be spillages of aggressive materials or in areas that require high surface rigidity.

Typical uses include:

- container handling and storage areas and docks
- vehicle maintenance and refuelling areas

415

- runway thresholds and areas subjected to jet blast
- areas used by tracked vehicles
- bus lanes
- industrial areas and flooring
- roundabouts
- car parks.

19.7.2 Asphaltic grouted macadams

Asphaltic grouted macadams evolved from the long-term use of cementitious grouted macadams. The essential difference between asphaltic and cementitious grouted macadams is the nature of the grout which, in the former, is based on bitumen rather than cement. Manufacturers claim that the materials, when laid, improve flexibility and resistance to permanent deformation whilst, via increased bitumen content, possessing increased resistance to oxidisation leading to longer life. The material is marketed as a carriageway surface course and is particularly suited to use on airfields.

19.7.3 Development

Grouted macadams originated in France where development of the first 'Salviacim' took place in the early 1950s. 'Gercim' appeared in the early 1960s and others were developed and appeared over time. The use of these French materials has now spread to other countries through licensing agreements. Their use in the UK started in the mid 1960s, mostly for industrial flooring and areas subject to oil spillage. UK use remained relatively constant until the late 1990s when a need for materials capable of withstanding more severe working conditions prompted a rise in their usage.

19.7.4 Laying

There are two stages in laying these materials. The first stage is to lay a single layer of open textured coated macadam designed with a controlled void content. A resin/cementitious grout or asphaltic grout is then vibrated into this 'receiving coat', filling the voids and sealing the surface. The resulting product, in the case of the cementitious grouted macadam, exhibits properties that lie between those of a flexible asphaltic layer and one constructed of rigid concrete.

Grouted macadams may be laid onto any existing clean, sound and level surface. A layer of regulating course is normally required over an irregular surface.

When laying over existing concrete, all joints should be inspected and any loose joint compound removed. Where edges have spalled, these should first be treated with a suitable material such as fine graded asphalt or bituminous sealant before the application of the receiving coat. For new construction, grouted macadams should be considered as an alternative surface course.

Grouted macadams are generally laid at a nominal thickness of 40 mm but this may be varied to suit the particular application. The thickness of this layer and the necessary void content controls the aggregate size used in the receiving course. The design of the traditional matrix can be altered to produce a range of material strengths suitable for use from general carriageways to heavy industrial areas.

The type of coarse aggregate used is governed by the application. However, it is essential to use an aggregate that is hard and durable. Thus, if polishing of the aggregate is not a consideration then macadams containing carboniferous limestone can be successfully used. However, in locations where severe abrasion is likely, igneous rock sources would produce a superior material.

Normally 100/150 penetration grade bitumen is used as the binder, although 160/220 penetration grade bitumen may be used where ambient temperatures so warrant. Use of a softer bitumen will assist where the material is laid by hand. Conversely, the use of a harder grade of bitumen such as 40/60 pen may be adopted in areas where high ambient temperatures occur.

Chapter 20

Other important uses of bitumens and asphalts

The UK produces some 26 million tonnes of asphalt each year and the bulk of this is used in the construction of roads and airfield runways and taxiways. However, in addition to these well-known applications, asphalts are also employed for a wide range of specialist uses. This chapter considers the use of asphalts in airfields and other less orthodox uses. Some of the text in this chapter [549–552] was taken from a series of publications produced by the Quarry Products Association. These can be downloaded from the internet[553].

20.1 Airfield pavements

In Europe, there are some 635 airports with roughly 9 million take-offs and landings each year. Uninterrupted flow of air traffic and safe traffic conditions are essential for all airport owners and their customers. The combination of the large number of airports and the increase in air travel encouraged by low-cost flights means that construction works are ongoing at many of those airports. This is necessary for reasons of expansion and maintenance to keep the taxiways and runways in a serviceable condition.

In airports, both asphalt pavements and concrete pavements are used. Consideration of a number of factors related to cost, performance and environment will assist in explaining the increasing use of asphalt pavements on airfields.

549 QUARRY PRODUCTS ASSOCIATION. *Decorative and coloured finishes for asphalt surfacings, Asphalt Applications 4, 2001*. QPA, London.

550 QUARRY PRODUCTS ASSOCIATION. *Use of asphalt in the construction of games and sport areas, Asphalt Applications 7, 2001*. QPA, London.

551 QUARRY PRODUCTS ASSOCIATION. *Miscellaneous uses of asphalts, Asphalt Applications 9, 2001*. QPA, London.

552 QUARRY PRODUCTS ASSOCIATION. *Airfield uses of asphalts, Asphalt Applications 10, 2001*. QPA, London.

553 *www.qpa.org*

In matters of cost, flexible pavements have the following characteristics.

- Initial construction costs for flexible pavements are lower than those for concrete pavements.
- Less time is required for curing before the facility can be opened to traffic.
- Although, in general, maintenance costs for asphalt pavements are slightly higher than those for concrete pavements, repair of flexible pavements is far easier and quicker than repair of concrete pavements meaning less downtime for airport operation. This latter consideration may completely reverse the maintenance cost argument. Maintenance of concrete slabs is not cheaper than asphalt.
- Whole-life costs of an asphalt pavement compared to those of a concrete pavement show that flexible pavements are the most cost effective solution.

In relation to performance and environmental issues, asphalt pavements exhibit the following characteristics in comparison to concrete pavements:

- lower noise levels;
- less cracking and thus less ravelling or loss of aggregate, an important safety aspect for aircraft;
- lower spray and aquaplaning risks particularly with negatively textured asphalts;
- higher friction when wet;
- vastly superior surface regularity;
- recycling of existing asphalt in new construction is easy and already common practice whereas concrete recycling is uncommon;
- less energy is required to build asphalt-based works; and
- life cycle inventory studies have shown that CO_2 emissions when constructing asphalt pavements are roughly 50% of those emanating from a similar concrete construction.

The use of flexible pavements manufactured with polymer modified bitumen has become much more common in airfield works. Traditional mixture design and flexible pavement design procedures do not take account of the improved characteristics of asphalts that are modified in this manner.

Many UK airports now use an open graded macadam[554] for the surface course. This material, known as the 'friction course', allows

---

554 BRITISH STANDARDS INSTITUTION. *Coated macadam for roads and other paved areas, Specification for constituent materials and for mixtures, BS 4987-1: 2003.* BSI, London.

water to permeate through the layer, thus maximising the frictional characteristics of the running surface. An impervious binder course is used thereunder and laid to falls to take the water away from the pavement. On airfields carrying aircraft that are relatively light the use of the standard road mixtures may well be quite adequate.

20.1.1 Design of flexible airfield pavements

The design of airfield pavements differs significantly from the design of roads due to the fact that the traffic and loading cycles are entirely different from those that are typically encountered on highways. The number of loads or passes on airfields is, in general, much lower than that found on roads carrying high volumes of traffic but the loads and tyre pressures on aircraft can be significantly higher. This has important consequences for the pavement design and choice of materials. Figure 20.1 illustrates the significant difference in loading and tyre pressures on roads and airfields.

The high loads on airfields will produce critical stresses and strains in the subgrade and lower layers of the pavements if not properly designed. The extent to which critical stresses and strains are produced in the upper layers of the pavement is dependent on the wheel configuration of the aircraft.

For design purposes, aircraft loads are transformed into 'equivalent single wheel loads' (ESWLs). An ESWL is defined as that single wheel load that will produce the same stresses and strains, in a given pavement

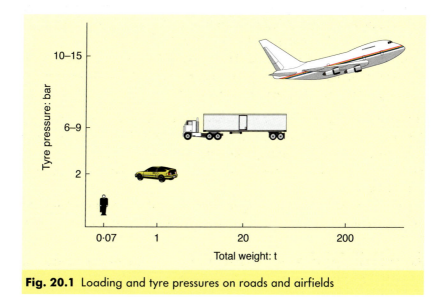

Fig. 20.1 Loading and tyre pressures on roads and airfields

Table 20.1 Examples of aircraft loads and wheel configurations

	Boeing 747-200C		Boeing 737		Illyushin 62	
Gross weight: t	373		65		161	
Tyre pressure: bar	13·2		12·5		16·8	
Wheel configuration	Double dual tandem/ complex		Dual		Dual tandem	
	Main wheel	Nose wheel	Main wheel	Nose wheel	Main wheel	Nose wheel

structure, as would be produced by the wheel groups and wheel config-
uration of real aircraft given realistic wheel loads and tyre pressures.
Examples of aircraft loads and wheel configurations are shown in
Table 20.1.

The traffic situation and loading severity varies greatly at different
locations in an airfield with the ends of runways and taxiways being
subject to the greatest concentrations of traffic. In particular, design of
taxiways that are used for departing traffic, is critical. In taxiways,
channelisation of fully loaded aircraft occurs. Runways are more critical
with respect to cracking and adhesion but have a lesser priority from a
design perspective. Detached pieces of aggregate may lead to severe
damage to aircraft engines and, consequently, present a significant
safety risk.

Design methods

The International Civil Aviation Organization (ICAO) classifies the
bearing capacity of airfield pavements in terms of a 'pavement classification number' (PCN). Similarly, the severity of loading from aircraft is
expressed in terms of an 'aircraft classification number' (ACN). Several
design procedures can be used to select materials, layer thickness, mixture
type, etc. that will provide a pavement structure with PCN values that
exceed the ACN values required by the ICAO.

Some of the most commonly used design procedures for flexible pavements on military and civilian airfields are:

- FAA method (Federal Aviation Administration, USA)
- LCN method (UK load classification number system)
- The French method (ICAO French practice)
- Canadian Ministry of Transportation method
- Corps of Engineers method (CBR method)
- Asphalt Institute method
- Shell Pavement Design Method (SPDM).

Most of these systems are empirical with strong links to the Californian bearing ratio (CBR) method. (The CBR test has been discussed previously, in particular in Section 18.4.3.) Exceptions are the Asphalt Institute method and SPDM, which are analytical pavement design methods based on calculations and criteria for stresses and strains in the various pavement layers (see Chapter 18).

These design systems recommend a minimum thickness of asphalt pavement. The methods listed above have few or no specific requirements for the asphalt layers. For this reason, theoretical models for pavement responses are often utilised as a supplementary tool to optimise the pavement construction and selection of materials. In relation thereto, analytical software such as BISAR is often used (BISAR is discussed in Section 18.5).

20.1.2 Requirements for airfield pavements

Airfield pavements must perform certain functions in line with agreed international standards. The consequences of a pavement that performs poorly may be dangerous and very costly. Detached pieces of aggregate may be ingested by jet engines or strike aircraft moving at speed causing serious and possibly fatal foreign object damage (FOD). As a consequence, maintenance on airfield pavements is often focused on minimising the risk of FOD.

The most important requirements for asphalt pavements on airfields are:

- adequate wet friction for landing aircraft
- resistance to ravelling and cracking to minimise the risk of FOD
- surface regularity
- surface water drainage
- resistance to rutting and fatigue
- pavement bearing capacity
- resistance to deterioration due to spilled fuel (especially on military airfields), de-icing chemicals and other chemicals
- sufficient bonding strength between asphalt layers to avoid slippage cracking.

The asphalt/bitumen solutions used to meet specific airport requirements will vary substantially from location to location depending on several factors such as climate, traffic intensity, types of aircraft, etc. Asphalts offer great flexibility in their composition and physical characteristics and may be specially designed to meet the needs at individual airports. Bitumen, obviously a very important component in relation to the properties of the asphalt, can be selected in different stiffness grades, specially designed to meet climatic challenges, improved resistance to damage

423

resulting from fuel spillage and improved rutting, low-temperature cracking and fatigue properties. Conventional bitumens cannot meet these demands. Thus, the use of polymer modified bitumens is, in most cases, necessary to meet the specific needs of an airport pavement.

Asphalt concrete with specification requirements for Marshall stability, flow and void content has normally been used for airfields in the past. The surfaces of these pavements were usually grooved to provide sufficient macrotexture and to create drainage channels leading to improved frictional characteristics. However, the introduction of stone mastic asphalt and porous asphalt on airfields has resulted in improved wet friction properties. Grooves, which often lead to premature ravelling, are unnecessary when stone mastic asphalt or porous asphalt is used as the surface course. The use of an optimised and adequate asphalt composition may offer improved pavement service life. Polymer modified binders based on SBS (styrene–butadiene–styrene) are generally used in stone mastic asphalt and porous asphalts, e.g. Shell's Cariphalte TS and Cariphalte DA. These bitumens provide the best possible aggregate adhesion to prevent cracking and ravelling while at the same time providing the pavement with an improved anti-rutting performance.

The development of new asphalt mixtures has been followed by improvements in mechanical tests for bitumen and asphalts as well as in specifications and mixture design procedures. Specifications for some recent airport works have been based on various sets of special binder specifications (Kuala Lumpur in Malaysia, Gardermoen Airport in Norway, Schiphol Airport in the Netherlands, Arianda Airport in Sweden and various airports in Greenland). At Schiphol Airport, specifications for rutting and crack resistance were used based on the dynamic creep test and indirect tensile test while the binder for Gardermoen Airport was specified using SHRP specifications (see Section 4.7). These developments in tests and test criteria for airports have led to new asphalt and bitumen solutions that will further encourage clients to choose asphalts for airfield applications.

The nature of the loading on airfield pavements may produce critical shear stresses in the interface between asphalt layers and result in slippage cracking if the bonding strength is insufficient. For main airfields with severe traffic, it is advisable to use polymer modified tack coat with an application rate resulting in $300 \, \text{g/m}^2$ residual binder.

20.2 Railway applications

In the USA, Japan and many European countries, asphalts have been tested and used in the construction of railway track beds for many years. One of the earliest known applications dates back to 1894 when an asphalt base, approximately 150 mm thick surfaced with a 30 mm surface course, was laid as a tramway bed in Visalia, California. In Japan, asphalt has been used in railway construction since the late 1950s and has been the subject of numerous tests and commercial applications in Europe.

In railway construction, asphalts are used in one of four ways[555]:

- to stabilise the aggregate ballast
- as an asphalt layer below the aggregate ballast
- as a direct base for the track sleepers
- as a direct base for the rails.

The reason for using asphalts is to improve the stability of the track, providing an improved running surface for the carriages which requires less maintenance.

In 1974, an experimental length of track incorporating an asphalt layer was laid in the UK at Radcliffe-on-Trent[556]. The experimental track consisted of a 230 mm thick rolled asphalt base onto which various types of concrete superstructures were laid to support the rails. Since installation, the performance of the asphalt track bed has been very good. Despite this, widespread use of asphalt-based railway tracks has been limited by the higher initial costs compared with those of a conventional ballasted track.

Germany now has a number of railway sections where asphalt has been used as a direct base for the rails. One of the main disadvantages of ballast, traditionally used to fix rails, is that it is damaged by erosion and this affects the position of sleepers and rails. Accordingly, building solid rail beds and replacing ballast with asphalt leads to a considerable reduction in maintenance costs. These savings mean that the higher initial costs of asphalt rail beds are recovered in the short to medium term.

Constructing solid rail beds requires the use of the usual items of paving equipment but particular skills are necessary for success. There are numerous systems for fixing sleepers on an asphalt base layer. Modern practice is to place the sleepers directly on the asphalt layer. Such layers require very close control in terms of height, evenness and bearing. For example, the tolerance in Germany is 0·1 mm for the height and 2 mm for the bearing. The bearing and the height have to be controlled constantly by a surveyor. The lifespan of an asphalt solid rail bed is expected to be as long as 60 years. In order to meet

555 BEECKEN G. *Asphaltic base course for railway track construction, Shell Bitumen Review 61, pp 10–21, Jun 1986.* Shell International Petroleum Company Ltd, London.

556 COPE D L. *Construction and experience of asphalt based track on British railways, Proc 2nd Eurobitume Symp, Cannes, pp 241–244, Oct 1981, Reprinted in a special publication of Asphalt Technology, Jan 1982.* Institute of Asphalt Technology, Dorking.

this requirement, adequate stability and durability of the asphalt are vital. For durability, it is important to have thick binder films and a low void content which would normally compromise the stability of the mixture. However, optimisation is achieved by undertaking a special mixture design that balances the long-term durability and stability of the asphalt. These asphalts use polymer modified bitumens.

In Germany, several stretches of rail track have been constructed with a solid asphalt track bed and the use of solid asphalt rail beds has expanded over recent years. Such growth will continue because it is innovative and, in whole-life terms, it is economic. It also makes travelling by train safer and more comfortable.

20.3 Bridges

Bridges can be divided into two classes, those that have steel decks and those that have concrete decks. Apart from the standard requirements for asphalt layers, bridge deck surfacings have to meet additional specifications. Some asphalts on bridge decks undergo large deflections due to dynamic traffic loading and this is particularly true of asphalt on steel bridge decks. At low ambient temperatures, high tensile stresses occur, resulting in the formation of cracks. In bridges over sea water or those whose decks are salted to treat frost and snow, the water containing salt can make its way to the steel deck or the reinforcement. This cycle will threaten not only the asphalt but also the deck or the reinforcement, thus requiring an expensive repair. Accordingly, the asphalt layer has to have adequate rut resistance and also be sufficiently flexible to avoid cracking. An excellent material to meet these criteria is güssasphalt (mastic asphalt) in combination with polymer modified bitumen. The advantages of such a combination are as follows.

- Güssasphalt is free of voids so there is no penetration by water. It can, therefore, be used not only for the surface course but also for the waterproofing.
- The use of a hard polymer modified bitumen maximises the stability of the güssasphalt. At the same time, good low-temperature behaviour (flexibility) is achieved which means that cracking should not occur.
- From a physical perspective, güssasphalt can be regarded as a fluid. Thus, it is unnecessary to compact the layer with rollers, which removes the possibility of the bridge deck waterproofing being damaged by aggregate indentation during compaction.

The system cited above is a means of extending maintenance intervals. Given the great expense associated with the replacement or repair of waterproofing, this is a very attractive feature. However, güssasphalt is not the only possible solution. Other asphalts, particularly stone mastic asphalt, are also suitable for bridge deck surfacings.

Regardless of the type of asphalt layer, protection of the bridge deck waterproofing and the deck itself is of paramount importance.

20.4 Recreational areas

Asphalts are used extensively for recreational areas, e.g. tennis, netball and volleyball courts, five-a-side football pitches, etc. They are used in two ways:

- as the final surfacing over an appropriate base which may itself be asphalt
- as a firm base beneath a synthetic surfacing.

By choosing an appropriate mixture, a pleasing and durable finish requiring little maintenance over long periods can be obtained. Asphalts are generally black or dark grey, providing a good contrast for line paints. If a coloured surfacing is required this can be achieved by using a special pigmentable binder (discussed in Section 5.4) or by painting the surface with a special proprietary paint.

Synthetic sports surfaces are very popular for athletic tracks, football pitches and other specialised applications. Asphalts have been found to be ideal for forming the base for these applications.

Impact absorbing asphalt (IAA) is a patented technology[557] consisting of a relatively conventional asphalt macadam with a large proportion of the aggregate (typically up to 30%) replaced by dry crumb rubber obtained from the recycling of truck tyres. It has many potential applications. One example is as a surface layer for play areas, which has resulted in significant reductions in injurious incidents. Another is its use as a shock pad underneath artificial sports surfaces where it provides the required impact absorption, longitudinal and transverse regularity.

The technology is based on the dry process developed in the late 1960s in Sweden and traded in Europe under the name Rubit[558–560].

557 *European patent application no. 828 893 (claiming a priority date of 26.5.1995; title: Impact Absorbing Macadam)*

558 HEITZMAN M. *Design and construction of asphalt paving materials with crumb rubber modifier, Transportation Research Record 1339, 1992.* Transportation Research Board, Washington.

559 GREEN E L and W J TOLONEN. *The chemical and physical properties of asphalt–rubber mixtures, Arizona Department of Transport, Report ADOT-RS-14 (162), Jul 1977.*

560 TAKALLOU H B. *Development of improved mix and construction guidelines for rubber modified asphalt pavements, Transport Research Record 1171, Jan 1988.* Transportation Research Board, Washington.

Production of the material involves addition of crumb rubber to hot aggregate prior to mixing with bitumen. The material is then mixed and maintained at a high temperature. It is laid and compacted using conventional paving techniques.

20.5 Motor racing tracks

International motor racing circuits are generally surfaced with asphalt. The main requirements are very good surface regularity, a high value of skid resistance and sufficient cohesion to resist the high tangential forces exerted by wide racing tyres. This last property is normally achieved with dense asphalt mixtures such as asphalt concrete. However, stone mastic asphalt has also found its way onto racing track surface courses. Cohesion of the asphalt is greatly improved by using polymer modified binders. Use of these binders may permit the use of porous asphalt on racing tracks which would result in a significant improvement in safety during races held during inclement weather.

20.6 Vehicle testing circuits

Many vehicle testing circuits have steep banked curves and special techniques have been developed, by experienced contractors, to ensure that the finished pavement is fully compacted, meets the surface tolerances and has a high degree of surface regularity. Figure 20.2 shows a surfacing

Fig. 20.2 Surfacing the Rockingham speedway track
Courtesy of Colas Ltd

Fig. 20.3 Bituminous grouting on the face of Megget dam in Scotland

being laid on the Rockingham Speedway track where considerable skill and expertise was required to satisfactorily lay and compact materials under the prevailing conditions.

20.7 Hydraulic applications

The waterproofing properties of bitumen and asphalts have been recognised from ancient times. The ancient civilisations of Ur, Egypt, Babylon, Assyria and the Indus all used the naturally occurring surface seepages of bitumen for waterproofing and building. The reservoir dam at Mohenjo Daro in the Indus basin is particularly well preserved and demonstrates that, in this field, asphalt can claim a life of 5000 years[561]!

Bitumens and asphalts have been used effectively for a range of hydraulic applications such as canal lining, reservoir lining, sea wall construction, coastal groynes, dam construction and the lining of leisure lakes. In the UK, dams at Dungonnell, Colliford, Winscar, Marchlyn and Sulby incorporate asphalt mixtures. So too, does the Megget dam, shown in Fig. 20.3. Reservoirs at Shotton, Leamington and Towey also contain asphalts.

The two principal properties that make bitumen ideal for hydraulic applications are its impermeability and the fact that it is chemically

561 MENZIES I. *Waterproofing with asphalt, Water and Waste treatment, vol 31, pp 20–21, 44, Nov 1988.*

inert. In combination with suitable aggregates in mixtures that are specifically designed for particular applications, bitumen can impart these properties to the structures in which the asphalt is used. In many of these structures, it is essential to have impermeability under pressure and asphalt mixtures with void contents below 3% can successfully contain water column pressures up to 200 m.

Reservoir and canal embankments normally have a slope of 1:1.75 but can be as steep as 1:1.25. Thus, asphalts used in such applications must have sufficient stability to be laid and compacted on such slopes without cracking during application and in service. The material must be flexible enough to accommodate differential settlement of the substrate.

One great advantage of asphalt is that it can be laid on the dam or reservoir face in a continuous manner thus eliminating joints. Any joint constitutes a discontinuity and, at the high pressures that exist at the base of dams and reservoirs, a joint provides the opportunity for the water to find an outlet. A book published by Shell provides substantial detail on hydraulic applications[562].

20.8 Coloured surfacings

The appearance of traditional asphalts, especially when finished with the normal contrasting white lines, is generally very pleasing to the eye. However, there are some locations where a specific colour is desired and there are a number of ways in which this can be achieved:

- incorporation of coloured pigments into the mixture during manufacture
- application of suitably coloured chippings to the surfacing during laying
- application of a coloured surface treatment after laying
- use of a conventional bitumen with a coloured aggregate
- use of a suitably coloured aggregate with a translucent binder.

20.8.1 Incorporating coloured pigments

These are discussed in Chapter 5.

20.8.2 Application of coloured chippings

Hot rolled asphalt and fine graded macadam (previously termed fine cold asphalt) can have decorative coloured chippings rolled into the surface during compaction. Hot rolled asphalt is suitable for most traffic conditions but fine graded macadam is only appropriate for lightly trafficked areas, private drives and pedestrian areas.

562 SCHÖNIAN E. *The Shell Bitumen Hydraulic Engineering Handbook, 1999*. Shell International Petroleum Company Ltd, London.

To provide a decorative finish to both of these types of surfacing, pigmented bitumen coated or clean resin coated chippings can be applied during laying. The bitumen or resin coating promotes adhesion to the surfacing and it is not recommended that uncoated chippings are used. However, in the case of fine graded macadam that is laid on areas subject to little traffic, a light application of uncoated chippings, e.g. white spar, produces a very attractive finish if they are uniformly applied and properly embedded. Many footways in Scotland are finished with a 20 mm thick layer of hot rolled asphalt surface course into which white limestone chippings are added. Decorative chippings cannot be successfully rolled into the surface of high stone content mixtures such as dense macadams or high stone content rolled asphalts.

20.8.3 Application of a coloured surface treatment

There are three surface treatments that can be applied to an asphalt surface course to provide a decorative finish:

- pigmented slurry seals
- surface dressing
- application of a coloured paint.

Pigmented slurry seals are available as proprietary products in a range of colours. However, as they are applied in a very thin coat, approximately 3 mm thick, they are only suitable for pedestrian and lightly trafficked areas.

Surface dressings are suitable for most categories of road application[563] but are less suitable for pedestrian situations. The final colour of the surface dressing will be that of the aggregate used.

Proprietary coloured paints are available for overpainting conventional black surface courses. These are only suitable for pedestrian and games areas, e.g. tennis courts.

20.8.4 Coloured aggregate bound by a conventional bitumen

When a conventional bitumen is used, the depth of colour achieved is dependent upon:

- the colour of the aggregate itself
- the thickness of the binder film on the aggregate
- the rate at which the binder exposed on the road surface is eroded by traffic.

---

563 NICHOLLS J C *et al. Design guide for road surface dressing, Road Note 39, 5th ed, 2002.* TRL Ltd, Crowthorne.

In medium and heavily trafficked situations, the natural aggregate colour will start showing through fairly quickly but in lightly trafficked situations, where coloured surfacings are often specified, it may take a significant time for the aggregate colour to become apparent.

20.8.5 Coloured aggregate bound by a translucent binder

Several proprietary macadams are available in which the binder is a clear resin rather than a bitumen. A range of coloured surfacings can be manufactured by selecting appropriately coloured aggregates. The major advantage of such a system is that the colour is obtained immediately.

20.9 Kerbs

Extruded asphalt kerbs are very simple to construct. They are produced by extruding a rolled asphalt mixture through specialised equipment directly onto a binder course or surface course which has been treated with tack coat. They are unsuitable for the edge of carriageways carrying heavy traffic as they may be damaged if hit by vehicles, particularly commercial vehicles. However, they do provide a convenient means of producing a raised edge on other paved areas, e.g. perimeter kerbing to drives or car parks. In the UK, BS 5931: 1980[564] gives guidance on the use of asphalt for producing kerbs in situ.

---

564 BRITISH STANDARDS INSTITUTION. *Code of practice for machine laid in situ edge details for paved areas, BS 5931: 1980.* BSI, London.

Appendices

Appendix 1
Physical constants of bitumens[565]

A1.1 Specific gravity

The specific gravity of a bitumen is primarily dependent on the grade of the bitumen and temperature. Typical values of specific gravity for a range of grades of bitumen are given in Table A1.1 and the effect of temperature on specific gravity is detailed in Table A1.2. The conversion from volume of bitumen in litres to weight in tonnes at various temperatures and values of specific gravity is detailed in Table A1.3.

A1.2 Coefficient of cubical expansion

The coefficient of cubical expansion of bitumen is effectively independent of the grade and is virtually constant at $0.00061/°C$ over the temperature range 15 to 200°C.

A1.3 Electrical properties
Electrical conductivity

Bitumen has a low electrical conductivity and is therefore an ideal insulating material. Hard grades of bitumen have slightly lower values of electrical conductivity than soft grades. However, these differences are insignificant. The electrical conductivity of all grades increases with increasing temperature, as shown in Table A1.4.

The influence of fillers on electrical conductivity is negligible unless conductive fillers such as graphite, coke or metal powders are used in significant quantities.

Dielectric strength

The dielectric strength is measured in kilovolts per millimetre and depends upon the conditions of measurement and the shape of the

---

565 PFEIFFER J P H (Ed). *The properties of asphaltic bitumen, vol 4, Elsevier's Polymer Series, 1950*. Elsevier Science, Amsterdam.

Table A1.1 Typical specific gravities of bitumens at 25°C

Grade of bitumen	Typical specific gravity at 25°C
Penetration grades	
250/330	1·010–1·020
160/220	1·015–1·025
100/150	1·020–1·030
50/70	1·020–1·030
40/60	1·025–1·035
30/45	1·025–1·035
20/30	1·030–1·040
Oxidised grades	
75/30	1·015–1·030
85/25	1·020–1·035
85/40	1·010–1·025
95/25	1·015–1·030
105/35	1·000–1·015
115/15	1·020–1·035
Hard grades	
H 80/90	1·045–1·055
H 110/120	1·055–1·065
Cutback grades	
50 sec	0·992–1·002
100 sec	0·995–1·005
200 sec	0·997–1·007

Table A1.2 Typical specific gravities of bitumen at various temperatures

Temperature, °C	Specific gravity at 25°C					
	1·00	1·01	1·02	1·03	1·04	1·05
15·5	1·006	1·016	1·026	1·036	1·046	1·056
25	1·000	1·010	1·020	1·030	1·040	1·050
45	0·988	0·998	1·008	1·018	1·028	1·038
60	0·979	0·989	0·999	1·009	1·019	1·029
90	0·961	0·971	0·981	0·991	1·001	1·011
100	0·955	0·965	0·975	0·985	0·995	1·005
110	0·949	0·959	0·969	0·979	0·989	0·999
120	0·943	0·953	0·963	0·973	0·983	0·993
130	0·937	0·947	0·957	0·967	0·977	0·987
140	0·931	0·941	0·951	0·961	0·971	0·981
150	0·925	0·935	0·945	0·955	0·965	0·975
160	0·919	0·929	0·939	0·949	0·959	0·969
170	0·913	0·923	0·933	0·943	0·953	0·963
180	0·907	0·917	0·927	0·937	0·947	0·957
190	0·901	0·911	0·921	0·931	0·941	0·951
200	0·895	0·905	0·915	0·925	0·935	0·945

Table A1.3 Conversion factors linking volume to weight for bitumen at various temperatures and specific gravities

Temperature: °C	Specific gravity at 25°C					
	1·00 litres/tonne	1·01 litres/tonne	1·02 litres/tonne	1·03 litres/tonne	1·04 litres/tonne	1·05 litres/tonne
25	995	984	973	963	953	943
45	1010	999	988	978	968	958
60	1020	1009	998	988	978	968
90	1041	1030	1019	1009	999	989
100	1047	1036	1026	1015	1005	995
110	1054	1043	1032	1022	1011	1001
120	1060	1049	1038	1028	1017	1007
130	1067	1056	1045	1034	1024	1013
140	1074	1063	1052	1041	1030	1019
150	1081	1070	1058	1047	1036	1026
160	1088	1076	1065	1054	1043	1032
170	1095	1083	1072	1060	1049	1038
180	1103	1091	1079	1067	1056	1045
190	1110	1098	1086	1074	1063	1052
200	1117	1105	1093	1082	1070	1058

Table A1.4 Relationship between temperature and electrical conductivity

Temperature, °C	Conductivity, S/cm
30	10^{-14}
50	10^{-13}
80	10^{-12}

Table A1.5 Relationship between temperature and dielectric strength

Temperature: °C	Dielectric strength: kV/mm (flat electrodes)
20	20–30
50	10
60	5

electrodes. Hard grades of bitumen tend to have a higher dielectric strength than soft grades. The dielectric strength of all grades decreases with increasing temperature, as shown in Table A1.5.

Dielectric constant (permittivity)

The dielectric constant (or permittivity) of bitumen is about 2·7 at 25°C rising to about 3·0 at 100°C. The dielectric losses in bitumen rise with increasing temperature but fall with increasing frequency. The Transport and Road Research Laboratory showed that the rate at which bitumen weathers under the combined effects of oxygen, rain and oil deposition from traffic and ultraviolet light in sunlight is related to the dielectric constant of the bitumen[566].

A1.4 Thermal properties
Specific heat
The specific heat of bitumen is dependent on both the grade of the bitumen and the temperature. Values of specific heat vary from 1675 to 1800 J/kg/°K at 0°C. The specific heat is increased by 1·67 to 2·51 J/kg for every 1°C increase in temperature.

Thermal conductivity
Bitumen is a good thermal insulating material and typically has a thermal conductivity of 0·15 to 0·17 W/m/°K.

566 GREEN E H. *An acceptance test for bitumen for rolled asphalt wearing courses, Laboratory Report 777, 1977.* Transport and Road Research Laboratory, Crowthorne.

Appendix 2
Conversion factors for viscosities

The viscosity of bitumen is usually measured by either capillary (kinematic viscosity) or cup viscometers (see Section 7.3). Kinematic viscosity is determined by timing the flow of bitumen through a glass capillary viscometer at a given temperature. The product of flow time and a calibration factor gives the kinematic viscosity in centistokes (cSt) or mm^2/s. In cup viscometers, the time is recorded in seconds for a standard volume of bitumen to flow out through the orifice in the bottom of the cup. The values given in Table A2.1 enable conversion of cup viscosities to kinematic viscosity for a number of different types of cup viscometer and vice versa.

Table A2.1 Conversion factors for viscosities

Known viscosity	To obtain unknown viscosity multiply by:						
	Kinematic: mm^2/s	Redwood I: s	Redwood II: s	Saybolt Universal: s	Saybolt Furol: s	Engler: °E	Standard tar viscometer: s
Kinematic: mm^2/s	–	4·05	0·405	4·58	0·458	0·132	0·0025
Redwood I: s	0·247	–	0·1	1·13	0·113	0·0326	–
Redwood II: s	2·47	10	–	11·3	1·13	0·326	0·0062
Saybolt Universal: s	0·218	0·885	0·0885	–	0·1	0·0287	–
Saybolt Furol: s	2·18	8·85	0·885	10	–	0·287	0·0054
Engler: °E	7·58	30·7	3·07	34·81	3·48	–	–
Standard tar viscometer (10 mm cup): s	400	–	162	–	183	528	–

Appendix 3
Blending charts and formulae

Bitumen may be blended with a wide variety of crude oil based fractions for different applications. Volatile light fractions, e.g. white spirit, are used where rapid curing is required. Less volatile fractions, such as cutback kerosene, are used for the manufacture of cutback bitumens. Figures A3.1 and A3.2 show the effect of adding cutback kerosene and

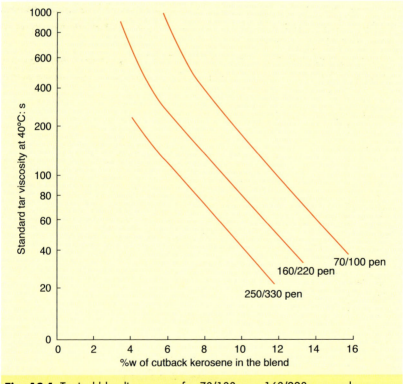

Fig. A3.1 Typical blending curves for 70/100 pen, 160/220 pen and 250/330 pen bitumen and cutback kerosene

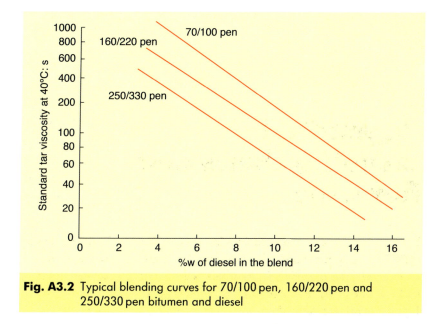

Fig. A3.2 Typical blending curves for 70/100 pen, 160/220 pen and 250/330 pen bitumen and diesel

diesel respectively to 70/100 pen, 160/220 pen and 250/330 pen bitumen. Similarly, Fig. A3.3 shows the effect on penetration when diesel is added to 40/60 pen bitumen.

Bitumens are miscible with each other in all proportions. The penetration and softening point of a blend of two bitumens can be estimated

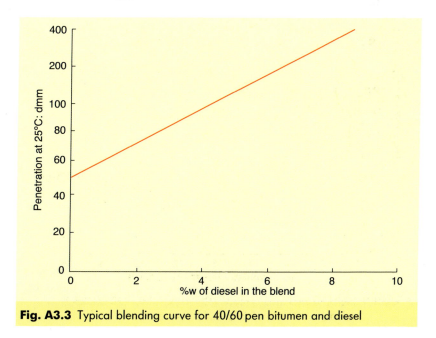

Fig. A3.3 Typical blending curve for 40/60 pen bitumen and diesel

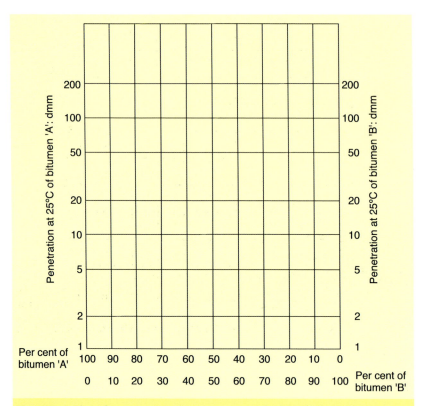

Fig. A3.4 Chart for estimating the penetration of a blend of two bitumens

using the blending charts shown in Figs. A3.4 and A3.5. These charts should only be used for blends of bitumen of the same type, i.e. those having the same penetration index. They should not be used for blends of oxidised and penetration bitumens. As an alternative to these charts the following formulae can be used to estimate the penetration and softening point of a blend.

$$\log P = \frac{A \log P_a + B \log P_b}{100}$$

where P = penetration of the final blend, P_a = penetration of component a, P_b = penetration of component b, A = percentage of component a in the blend and B = percentage of component b in the blend.

$$S = \frac{AS_a + BS_b}{100}$$

where S = softening point of the final blend, S_a = softening point of component a, S_b = softening point of component b, A = percentage of component a in the blend and B = percentage of component b in the blend.

440

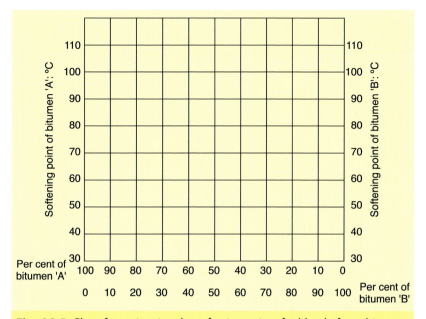

Fig. A3.5 Chart for estimating the softening point of a blend of two bitumens

Appendix 4
Calculation of bitumen film thickness in an asphalt

It is possible to estimate the surface area of a sample of aggregate by coating it with an oil and measuring the quantity of oil required for complete coating. Alternatively, the surface area of an aggregate can be calculated by assuming a specific aggregate particle shape. Hveem[567] calculated surface area factors, assuming a spherical particle shape and a specific gravity of 2·65. Typical surface area factors are shown in Table A4.1.

The surface area of the aggregate is calculated by multiplying the total mass expressed as a percentage passing each sieve size by the appropriate surface area factor and adding the resultant products together. All the factors must be used and if different sieves are used, different factors are necessary. The theoretical bitumen film thickness is then calculated as:

$$T = \frac{b}{100 - b} \times \frac{1}{\rho_b} \times \frac{1}{SAF}$$

where T = bitumen film thickness (in mm), ρ_b = density of bitumen (in kg/m^3), SAF = surface area factor (in m^2/kg) and b = bitumen content (in %).

Another method, developed in France[568], gives an approximation of the binder film thickness using the formula:

$$T = \frac{b}{a \times \sqrt[5]{\Sigma}}$$

where T = bitumen film thickness (in mm), b = bitumen content expressed as the percentage by total mass of the mixture, a is a correction

567 AMERICAN SOCIETY for TESTING and MATERIALS. *Standard test methods for resistance to deformation and cohesion of bituminous mixtures by means of Hveem apparatus, D1560-92, 1 Jan 1992*. ASTM, Philadelphia.
568 NORME FRANÇAISE. *Couches d'assis: Enrobé à élevé, NF P 98-140, Dec 1991*. NF, Paris.

Table A4.1 Typical surface area factors

Sieve size: mm	Surface area factor: m^2/kg
0·075	32·77
0·150	12·29
0·300	6·14
0·600	2·87
1·18	1·64
2·36	0·82
>4·75	0·41

coefficient taking into account the density of the aggregate and is given by

$$a = \frac{2650}{\text{SG}_a}$$

where SG_a = density of the aggregate (in kg/m^3), Σ is the specific surface area of the aggregate and is given by

$$\Sigma = 0·25G + 2·3S + 12s + 135f$$

where G = proportion by mass of aggregate over 6·3 mm, S = proportion by mass of aggregate between 6·3 and 3·15 mm, s = proportion by mass of aggregate between 3·15 and 0·80 mm and f = proportion by mass of aggregate smaller than 0·80 mm.

Index